Second Edition

GLOSSARY OF BIOTECHNOLOGY TERMS

Kimball R. Nill
Deputy Director of International Marketing
American Soybean Association
St. Louis, Missouri

TECHNOMIC PUBLISHING CO., INC.
LANCASTER · BASEL

Glossary of Biotechnology Terms

a **TECHNOMIC**® publication

Published in the Western Hemisphere by
Technomic Publishing Company, Inc.
851 New Holland Avenue, Box 3535
Lancaster, Pennsylvania 17604 U.S.A.

Distributed in the Rest of the World by
Technomic Publishing AG
Missionsstrasse 44
CH-4055 Basel, Switzerland

Printed in the United States of America

10 9 8 7 6 5 4 3 2 1

Main entry under title:
Glossary of Biotechnology Terms, Second Edition

A Technomic Publishing Company book
Bibliography: p.

Library of Congress Catalog Card No. 97-62206
ISBN No. 1-56676-580-3

To my wife, Janet J. Nill.

PREFACE

I began writing this book while employed by a major Midwestern company which was at that time venturing aggressively into the biotechnology arena. It grew out of an original work which was, in its infancy, a photocopy text containing quite a few handwritten notes, which attested to its evolving nature. Within the company, it was in demand.

As time went on, I noticed that a plethora of individuals, representing departments ranging from marketing and sales to engineering, legal, and human resources, were involved in various ways in the critically important process of gaining commercial success for our biotech products. It also became apparent that many of the players had little or no formal training in the bio and chemical sciences, yet their work required that they have an understanding of the biotech buzzwords and concepts in use. I felt that this pattern was bound to repeat itself in other companies, and that because of this there existed a large audience which could benefit from an understanding of the biovocabulary.

To this end I have attempted to write the text in terms which would enable the reader to conceptualize the idea embodied in the word, without the necessity of holding advanced biochemistry and molecular biology degrees. In order to accomplish this, however, I had to make certain compromises between scientific rigor and definitions based on analogy, with the inherent possibility of oversimplification. Nonetheless, throughout the text, emphasis has been placed on explanation by analogy whenever possible. And although I may be cited for oversimplification, I feel that this is compensated for in that it will be easier to grasp the underlying idea.

I offer this work in good faith and in the hope that it will assist those individuals who seek to gain an understanding of the terminology as it is currently used. However, the reader should be aware that the field of biotechnology is rapidly expanding and evolving, and that new terms are entering the mainstream nomenclature at a rapid pace. In fact, the exact meaning of some of these terms is still under dispute, while the meaning of others will undoubtedly be expanded or narrowed as the technology develops. Although I have endeavored to be as accurate as possible, this work is meant to provide a general introduction rather than to be absolute and legally definitive.

Since what is loosely known as biotechnology covers such a wide and diverse field, I have included terms from biology, biochemistry, and chemistry. In seeking terms and definitions I have borrowed freely from books, journals, industry periodicals and my friends.

I hope that this brief text will enable a better understanding of the complex ideas and techniques present in today's biotech work environment.

KIMBALL R. NILL
Quincy, Illinois

GLOSSARY OF BIOTECHNOLOGY TERMS

A

A. flavus See *ASPERGILLUS FLAVUS*.

A-DNA A particular right-handed helical form of DNA. A-form DNA is found in fibers at 75 percent relative humidity and requires the presence of sodium, potassium, or cesium as the counterion. Instead of lying flat, the bases are tilted with regard to the helical axis and there are more base pairs per turn. The A-form is biologically interesting because it is probably very close to the conformation adopted by DNA-RNA hybrids or by RNA-RNA double-stranded regions. The reason is that the presence of the $2'^{2}$ hydroxyl group prevents RNA from lying in the B-form. See also B-DNA, DNA-RNA HYBRID, and DEOXYRIBONUCLEIC ACID (DNA).

ABC See ASSOCIATION OF BIOTECHNOLOGY COMPANIES (ABC).

Abiogenesis Spontaneous generation. See also BIOGENESIS.

Abiotic Absence of living organisms.

AβPP See AMYLOID β PROTEIN PRECURSOR.

Abrin A toxin derived from the seed of the rosary pea. See also RICIN, GENISTEIN (GEN), and TOXIN.

Absolute Configuration The configuration of four different substituent groups around an asymmetric carbon atom, in relation to D- and L-glyceraldehyde. See also DEXTROROTARY (D) ISOMER and LEVOROTARY (L) ISOMER.

Absorbance (A) A measure of the amount of light absorbed by a substance suspended in a matrix. The matrix may be gaseous, liquid, or solid in nature. Most biologically active compounds (e.g., proteins) absorb light in the ultraviolet (UV) or visible light portion of the spectrum. Absorbance is used to quantitate (measure) the concentration of the substance in question (e.g., substance dissolved in a liquid). See also OPTICAL DENSITY (OD) and SPECTROPHOTOMETER.

Absorption Transport of the products of digestion from the intestinal tract across the cell membranes that comprise the gut, and into the blood. See also "ADME" TESTS.

Abzymes Catalytic antibodies that are synthetic constructs. They either stabilize the transition state of a chemical reaction or bind to a specific substrate, thereby increasing the reaction rate of that chemical reaction. See also CATALYTIC ANTIBODY, TRANSITION STATE (IN A CHEMICAL REACTION), and SUBSTRATE (CHEMICAL).

ACC Synthase One of the most critical enzymes in the metabolic pathway that creates the hormone ethylene inside fruit. Because ethylene causes certain fruit (e.g., tomatoes) to ripen (soften), it is possible to significantly delay the softening (i.e., spoilage) process by controlling creation of ACC synthase via manipulation of the ACC synthase gene. See also METABOLISM, ENZYME, METABOLITE, INTERMEDIARY METABOLISM, POLYGALACTURONASE (PG), and ENZYME.

Acceptor Control The regulation of the rate of respiration by the availability of ADP as phosphate acceptor. See also RESPIRATION and ADENOSINE DIPHOSPHATE (ADP).

Acceptor Junction Site The junction between the right 3′ end of an intron and the left 5′ end of an exon. See also INTRON and EXON.

Accession The addition of germ-plasm deposits to existing germ-plasm storage bands. See also AMERICAN TYPE CULTURE COLLECTION (ATCC).

Acclimatization The biological process whereby an organism adapts to a new environment. This adaptation actually occurs on a molecular level. One example is when natural microorganisms adapt so that they feed on and degrade toxic chemical wastes, or change from using one sugar as a fuel source to another. See also SUGAR MOLECULES and CATABOLISM.

Ac-CoA Acetyl–coenzyme A. See also COENZYME and COENZYME A.

Ac-P Acetylphosphate.

Acid A substance with a pH in the range from 0–6 which will react with a base to form a salt. Acids normally taste sour and feel slippery. For example, food product manufacturers often add citric acid, malic acid, fumaric acid, and itaconic acid in order to impart a "sharp" taste to food products. See also BASE, CITRIC ACID, and FUMARIC ACID.

Acidic Fibroblast Growth Factor (AFGF) See FIBROBLAST GROWTH FACTOR (FGF).

Acidosis A metabolic condition in which the capacity of the body to buffer changes in pH is diminished. Hence, acidosis is accompanied by decreased blood pH (i.e., the blood becomes more acidic than is normal).

ACP (acyl carrier protein) It binds acyl intermediates during the formation of long-chain fatty acids. ACP is important in that it is involved in every step of fatty acid synthesis. See also FATTY ACID and ACYL-COA.

Acquired Immune Deficiency Syndrome (AIDS) A disease in which a specific virus attacks and kills macrophages and helper T cells (thus causing collapse of the entire immune system). Once the immune system has been inactivated, other diseases, which under normal circumstances can be fought off, become fatal. See also HUMAN IMMUNODEFICIENCY VIRUS (TYPE 1 and TYPE 2), HELPER T CELLS (T4 CELLS), MACROPHAGE, and TUMOR NECROSIS FACTOR (TNF).

ACTH [adrenocorticotropic hormone (corticotropin)] A polypeptide secreted by the anterior lobe of the pituitary gland. This is an example of a protein hormone. See also POLYPEPTIDE (PROTEIN), ENDOCRINE GLANDS, and ENDOCRINE HORMONES.

Activation Energy The amount of energy (calories) required to bring all the molecules in one mole of a reacting substance to the transition state. More simply, it may also be viewed as the energy required to bring reacting molecules to a certain energy state from which point the reaction proceeds spontaneously. See also TRANSITION STATE (IN A CHEMICAL REACTION) and MOLE.

Activator A small molecule that stimulates (increases) an enzyme's catalytic activity when it binds to an allosteric site. See also ENZYME, EFFECTOR, and ALLOSTERIC SITE.

Active Site The region of an enzyme surface that binds the substrate molecule and transforms the substrate molecule into the new (chemical) product (entity). This site is usually located not on a protruding portion of the enzyme, but rather in a cleft or depression. This establishes a controlled environment in which the reaction may occur. See also CATALYTIC SITE, AGONISTS, PHARMACOPHORE, SUBSTRATE (CHEMICAL), ENZYME, and ANTAGONISTS.

Active Transport Cell-mediated, energy-requiring translocation of a molecule across a membrane in the direction of increasing concentration (i.e., opposite of natural tendency). See also OSMOTIC PRESSURE.

Activity Coefficient The factor by which the concentration of a solute must be multiplied to give its true thermodynamic activity.

Acute Transfection Short-term infection of cells with DNA.

Acyl-CoA Acyl derivatives of coenzyme A (acyl-S-CoA). See also COENZYME A.

Adaptive Enzymes See INDUCIBLE ENZYMES.

ADBF See AZUROPHIL-DERIVED BACTERICIDAL FACTOR (ADBF).

Additive Genes Genes that interact but do not show dominance (in the case of alleles) or epistasis (if they are not alleles). See also GENE, ALLELE, DOMINANT ALLELE, and EPISTASIS.

Adenine A purine base, 6-aminopurine, occurring in ribonucleic acid (RNA) as well as in deoxyribonucleic acid (DNA) and a component of adenosine diphosphate (ADP) and adenosine triphosphate (ATP). Adenine pairs with thymine in DNA and uracil in RNA. See also BASE PAIR (bp), RIBONUCLEIC ACID (RNA), and DEOXYRIBONUCLEIC ACID (DNA).

Adenosine Diphosphate (ADP) A ribonucleoside 5′-diphosphate serving as phosphate-group acceptor in the cell energy cycle. See also CATABOLISM, ADENOSINE TRIPHOSPHATE (ATP), and ADENOSINE MONOPHOSPHATE.

Adenosine Monophosphate (AMP) A ribonucleoside 5′-monophosphate that is formed by hydrolysis of ATP or ADP. See also HYDROLYSIS, ADENOSINE DIPHOSPHATE (ADP), and ADENOSINE TRIPHOSPHATE (ATP).

Adenosine Triphosphate (ATP) The major carrier of chemical energy in the cells of all living things on this planet. A ribonucleoside 5′-triphosphate functioning as a phosphate-group donor in the energy cycle of the cell. ATP contains three phosphate/oxygen molecules linked together. When a phosphate-phosphate bond in ATP is broken (hydrolyzed), energy that the cell can use to carry out its functions is produced. Thus, ATP serves as the universal medium of biological energy storage and exchange, in living cells. See also ATPase, ATP SYNTHETASE, HYDROLYSIS, BIOLUMINESCENCE, ATP SYNTHASE, and ADENOSINE MONOPHOSPHATE.

Adenovirus A type of virus that can infect humans. Like all viruses, it can reproduce only inside living cells (of other host, organisms). Adenovirus causes a protein (metabolite) to be made that disables the p53 gene. Because the p53 gene then cannot perform its usual function (i.e., prevention of uncontrolled cell growth caused by virus/DNA damage), the adenovirus thus "takes over" and causes the cell to make numerous copies of the virus until the cell dies (thus, releasing the virus copies into the body of the host organism to cause further infection). See also VIRUS, RETROVIRUSES, GENE DELIVERY (GENE THERAPY), CELL, PROTEIN, p53 GENE, and DEOXYRIBONUCLEIC ACID (DNA).

Adhesion Molecule A glycoprotein "chain" that protrudes from the surface membrane of certain cells, and causes cells (possessing "matching" adhesion molecules) to adhere to each other. For example, in 1952 Aaron Moscona observed that (harvesting enzyme-separated) chicken embryo cells did not remain separated, but instead coalesced again into an (embryo) aggregate. In 1955, Philip Townes and Johannes Holtfreter showed that "like" amphibian (e.g., frog) neuron cells will rejoin together after being physically separated (e.g., with a knife blade); but "unlike" cells remain segregated (apart).

Adhesion molecules also play a crucial role in guiding monocytes to sources of infection (e.g., pathogens) because adhesion molecules in the walls of blood vessels (after activation caused by pathogen invasion of adjacent tissue) adhere to like adhesion molecules in the membranes of monocytes in the blood. The monocytes pass through the blood vessel walls, become macrophages, and fight the pathogen infection. See also MONOCYTES, MACROPHAGE, POLYPEPTIDE (PROTEIN), CELL, PATHOGEN, CD4

PROTEIN, CD44 PROTEIN, GP120 PROTEIN, VAGINOSIS, HARVESTING ENZYMES, HARVESTING, SIGNAL TRANSDUCTION, SELECTINS, LECTINS, GLYCOPROTEINS, SUGAR MOLECULES, LEUKOCYTES, LYMPHOCYTES, NEUTROPHILS, ENDOTHELIUM, ENDOTHELIAL CELLS, P-SELECTIN, ELAM-1, INTEGRINS, and CYTOKINES.

Adhesion Protein See ADHESION MOLECULE and ENDOTHELIAL CELLS.

Adjuvant (to a herbicide) Any compound that enhances the effectiveness (i.e., weed-killing ability) of a given herbicide. For example, adjuvants such as surfactants can be mixed (prior to application to weeds) with herbicide (in water), in order to hasten transport of the herbicide's active ingredient into the weed plant. That is because the herbicide must move from an aqueous (water) environment into one (i.e., the weed plant's cuticle or "skin") comprised of lipids/lipophilic molecules, before it can accomplish its task. See SURFACTANT, LIPIDS, and LIPOPHILIC.

Adjuvant (to a pharmaceutical) Any compound that enhances the desired response by the body to that pharmaceutical. For example, adjuvants such as certain polysaccharides or surface-modified diamond nanoparticles, can be injected along with (vaccine) antigen in order to increase the immune response (e.g., production of antibodies) to a given antigen. Another example is that consumption of grapefruit juice by humans will increase the impact of certain pharmaceuticals. Those pharmaceuticals include some sedatives, antihypertensives, the antihistamine terfenadine, and the immunosuppressant cyclosporine. The adjuvant effect of grapefruit juice is thought to be caused via inhibition of the enzyme cytochrome P4503A4, which catalyzes reactions involved in the metabolism (breakdown) of those pharmaceuticals. See also CELLULAR IMMUNE RESPONSE, HUMORAL IMMUNITY, POLYSACCHARIDES, NANOTECHNOLOGY, ANTIGEN, ANTIBODY, ENZYME, METABOLISM, HISTAMINE, CYCLOSPORINE, and CYTOCHROME P4503A4.

"ADME" Tests Absorption, distribution, metabolism, and elimination tests required by America's Food and Drug Administration (FDA) for approval of new food ingredients. See also FOOD AND DRUG ADMINISTRATION (FDA), ABSORPTION, METABOLISM, INTERMEDIARY METABOLISM, PHARMACOKINETICS, and CODEX ALIMENTARIUS COMMISSION.

Adoptive Cellular Therapy The increase in immune response that is achieved by selectively removing certain immune system cells from (a patient's) body, multiplying them *in vitro* outside the body to greatly increase number, then re-inserting those (more numerous) immune system cells into the same body. See also CELLULAR IMMUNE RESPONSE, CELL CULTURE, *IN VITRO*, GENE DELIVERY (GENE THERAPY), and *EX VIVO* (THERAPY).

Adoptive Immunization The transfer of an immune state from one animal to another by means of lymphocyte transfusions. See also LYMPHOCYTE.

ADP See ADENOSINE DIPHOSPHATE (ADP).

Aerobe An organism that requires oxygen to live (respire).

Aerobic Exposed to air or oxygen. An oxygenated environment.

Affinity Chromatography A method of separating a mixture of proteins or nucleic acids (molecules) by specific interactions of those molecules with a component known as a ligand, which is immobilized on a support. If a solution of, say, a mixture of proteins is passed over (through) the column, one of the proteins binds to the ligand on the basis of specificity and high affinity (they fit together like a lock and key). The other proteins in the solution wash through the column because they were not able to bind to the ligand. Once the column is devoid of the other proteins, an appropriate wash solution is passed through the column, which causes the protein/ligand complex to dissociate. The protein is collected in a highly purified form. See also CHROMATOGRAPHY, ANTIBODY AFFINITY CHROMATOGRAPHY, and LIGAND (IN CHROMATOGRAPHY).

Aflatoxin The term that is used to refer to a group of related mycotoxins (i.e., metabolites produced by fungi that are toxic to animals and humans) produced by some strains of the fungi *Aspergillus flavus* and *Aspergillus parasiticus*. *Aspergillus flavus* and *Aspergillus parasiticus* are common fungi that typically live on decaying vegetation. Corn earworm (*Helicoverpa zea*) and European corn borer (*Ostrinia nubialis*) are vectors (carriers) of *Aspergillus flavus*. Aflatoxin B_1 is the most commonly occurring aflatoxin, and is the most potent carcinogen known to man.

When dairy cattle eat aflatoxin-contaminated feed, their metabolism process converts the aflatoxin (e.g., Aflatoxin B_1) into the mycotoxins known as Aflatoxin M_1 and Aflatoxin M_2. See also CARCINOGEN, TOXIN, FUNGUS, MYCOTOXINS, STRESS PROTEINS, LIPOXYGENASE (LOX), PEROXIDASE, *HELICOVERPA ZEA*, BETA CAROTENE, OH43, BRIGHT GREENISH YELLOW FLUORESCENCE (BGYF), CORN, and EUROPEAN CORN BORER (ECB).

Agar A complex mixture of polysaccharides obtained from marine red algae. It is also called agar-agar. Agar is used as an emulsion stabilizer in foods, as a sizing agent in fabrics, and as a solid substrate for the laboratory culture of microorganisms. Agar melts at 100°C (212°F) and when cooled below 44°C (123°F) forms a stiff and transparent gel. Microorganisms are seeded onto and grown (in the laboratory) on the surface of the gel. See also POLYSACCHARIDES and CULTURE MEDIUM.

Agarose A highly purified form of agar. Used as a stationary phase (substrate) in some chromatography and electrophoretic methods. See also CHROMATOGRAPHY, ELECTROPHORESIS, and AGAR.

Aging The process, affecting organisms and most cells, whereby each cell division (mitosis) brings that cell (or organism composed of such cells) closer to its *final* cell division (i.e., death). Notable exceptions to this aging process include cancerous cells (e.g., myelomas) and the single-celled organism; both of which are "immortal." See also TELOMERES, MITOSIS, HYBRIDOMA, MYELOMA, and CANCER.

Aglycon A nonsugar component of a glycoside. See also GLYCOSIDE.

Agonists Small protein or organic molecules that bind to certain proteins (i.e., receptors) at a site that is adjacent to the active site of the protein (i.e., receptor) to induce a conformational change in that protein, thus enhancing its activity. See also RECEPTORS, ACTIVE SITE, CONFORMATION, and ANTAGONISTS.

Agrobacterium tumefaciens A naturally occurring bacterium that is capable of inserting its DNA (genetic information) into plants, resulting in a type of injury to the plant known as crown gall. Among others, Monsanto Company has developed a way to stop *Agrobacterium tumefaciens* from causing crown gall, while maintaining its ability to insert DNA into plant cells, and now uses *Agrobacterium tumefaciens* as a vehicle to insert desired genes into plants (e.g., the gene causing overproduction of CP4 EPSP synthase, thus conferring resistance to glyphosate-containing herbicide). See also EPSP SYNTHASE, CP4 EPSPS, "SHOTGUN" METHOD [TO INTRODUCE FOREIGN (NEW) GENES INTO PLANT CELLS], BIOLISTIC® GENE GUN, "WHISKERS™," GENETIC ENGINEERING, GENE, BIOSEEDS, GLYPHOSATE, GLYPHOSATE-TRIMESIUM, and GLYPHOSATE ISOPROPYLAMINE SALT.

AHG Antihemophilic Globulin. Also known as FACTOR VIII or Antihemophilic Factor VIII. See also FACTOR VIII and GAMMA GLOBULIN.

AIDS See ACQUIRED IMMUNE DEFICIENCY SYNDROME (AIDS).

Alanine (ala) A nonessential amino acid of the pyruvic acid family. In its dry, bulk form it appears as a white crystalline solid. See also ESSENTIAL AMINO ACIDS.

Aldose A simple sugar in which the carbonyl carbon atom is at one end of the carbon chain. A class of monosaccharide sugars; the molecule contains an aldehyde group. See also MONOSACCHARIDES.

Algae A heterogeneous (i.e., widely varying) group of photosynthetic plants, ranging from microscopic single-cell forms to multicellular, very large forms such as seaweed. All of them contain chlorophyll and hence most are green, but some of them may be different colors due to the presence of other, overshadowing pigments.

Alkaline Hydrolysis A chemical method of liberating DNA from a DNA-RNA hybrid. See also HYDROLYSIS, RIBONUCLEIC ACID (RNA), DNA-RNA HYBRID, and DEOXYRIBONUCLEIC ACID (DNA).

Allele One of several alternate forms of a gene occupying a given locus on the chromosome, which controls expression (of product) in different ways. See also EXPRESS, GENE, CHROMOSOMES, and LOCUS.

Allelic Exclusion The expression in any particular manner of only one of the alleles, coding for the expressed immunoglobin. See also ALLELE, CODING SEQUENCE, and IMMUNOGLOBULIN (IgA, IgE, IgG, and IgM).

Allergies (airborne) See MAST CELLS.

Allergies (foodborne) An IgE-mediated (aggressive) immune system response to antigen(s) present on protein molecules in the particular food that (a given) person is allergic to. The antibodies (IgE) bind to those antigens and trigger a humoral immune response, which can cause vomiting, diarrhea, skin reactions, wheezing, and respiratory distress. In severe cases, the immune response can cause death. American's Food and Drug Administration (FDA) requires testing in advance to determine if a genetically engineered foodstuff has the potential to cause allergic reactions in humans, before that genetically engineered foodstuff (e.g., a modified crop plant) is approved by the FDA. See also FOOD AND DRUG ADMINISTRATION (FDA), GENETIC ENGINEERING, IMMUNOGLOBULIN (IgA, IgE, IgG, and IgM), and HUMORAL IMMUNITY.

Allogeneic With a different set of genes (but same species). For example, an organ transplant from one nonrelated human to another is allogeneic. An organ transplant from a baboon to a human would be xenogeneic. See also GENE, SPECIES, and XENOGENEIC ORGANS.

Allosteric Enzymes Regulatory enzymes whose catalytic activity is modulated by the noncovalent binding of a specific metabolite (effector) at a site (regulatory site) other than the catalytic site (on the enzyme). Effector binding causes a three-dimensional conformation change in the enzyme and is the root of the modulation. The term is used to differentiate this form of regulation from the type that may result from the competition between substrate and inhibitors at the catalytic site. See also STERIC HINDRANCE, EFFECTOR, CONFORMATION, and ACTIVE SITE.

Allosteric Site The "site" on an (allosteric) enzyme molecule where, via noncovalent binding to the site, a given effector can increase or decrease that enzyme's catalytic activity. Such an effector is called an allosteric effector because it binds at a site on the enzyme molecule that is other (allo) than the enzyme's catalytic site. See also ALLOSTERIC ENZYMES, ACTIVATOR, CATALYTIC SITE, EFFECTOR, CONFORMATION, ENZYME, METABOLITE, and CATALYST.

Allotypic Monoclonal Antibodies Monoclonal antibodies that are isoantigenic. See also MONOCLONAL ANTIBODIES (MAb) and ANTIGEN.

Alpha Helix (α-helix) A highly regular (i.e., repeating) structural feature that occurs in certain large molecules. First discovered in protein molecules by Linus Pauling in the late 1940's. See A-DNA, PROTEIN, PROTEIN FOLDING, and PROTEIN STRUCTURE.

Alpha Interferon Also written as α-interferon. One of the interferons, it has been shown to prolong life and reduce tumor size in patients suffering from Kaposi's sarcoma (a cancer that affects approximately 10 percent of people with acquired immune deficiency syndrome). It is also effective against hairy-cell leukemia and may work against other cancers. It has recently been approved by the FDA for use against certain types of sarcoma. Recent research indicates that injections of alpha interferon can limit the liver damage typically caused by hepatitis C, a viral disease. See also INTERFERONS.

ALS Gene See HTC and STS.

Alu Family A set of dispersed and related genetic sequences, each about 300 base pairs long, in the human genome. At both ends of these 300 bp segments there is an A-G-C-T sequence. Alu 1 is a restriction enzyme that recognizes this sequence and cleaves (cuts) it between the G (guanine) and the C (cytosine). See also GENOME and RESTRICTION ENDONUCLEASES.

Aluminum Resistance See CITRATE SYNTHASE (CSb) GENE, GENE, and CITRIC ACID.

Aluminum Tolerance See CITRATE SYNTHASE (CSb) GENE, GENE, and CITRIC ACID.

Aluminum Toxicity See CITRATE SYNTHASE (CSb) GENE, GENE, and CITRIC ACID.

Alzheimer's Disease Named after Alois Alzheimer who first described the Amyloid β Protein (AβP) plaques in the human brain that are caused by this disease, in 1906. Alzheimer's disease causes progressive memory loss and dementia in its victims as it kills brain cells (neurons). The drug Tacrine appears to slow the progression of Alzheimer's disease, but there is currently no way to stop the disease. See also AMYLOID β PROTEIN (AβP), AMYLOID β PROTEIN PRECURSOR (AβPP), and NEUROTRANSMITTER.

American Society for Biotechnology (ASB) A society founded for the purpose of "providing a multi- and inter-disciplinary forum for those persons from academia, industry, and government who are interested in any and all aspects of biotechnology, and will achieve its aims by cooperation with existing organizations active in the field." To join, write ASB, P.O. Box 2820, Sausalito, California, 94966-2820. See also BIOTECHNOLOGY, INTERNATIONAL SOCIETY FOR THE ADVANCEMENT OF BIOTECHNOLOGY (ISAB), and BIOTECHNOLOGY INDUSTRY ORGANIZATION (BIO).

American Type Culture Collection (ATCC) An independent, nonprofit organization that was established in 1925 for the preservation and distribution of reference cultures. See also CELL CULTURE, CULTURE, CULTURE MEDIUM, STAIN, TYPE SPECIMEN, and CONSULTATIVE GROUP ON INTERNATIONAL AGRICULTURAL RESEARCH (CGIAR).

Ames Test A simple bacterial-based test for carcinogens that was developed by Bruce Ames in 1961. Although this test evaluates mutagenesis (i.e., causation of mutations) in the DNA of bacteria, its results have been utilized to approve or not approve certain compounds for consumption by humans. See also BIOASSAY, BACTERIA, ASSAY, MUTUAL RECOGNITION AGREEMENTS (MRAs), GENOTOXIC CARCINOGENS, and CARCINOGEN.

Amino Acid There are 20 common amino acids, each specified by a different arrangement of three adjacent DNA nucleotides. These are the building blocks of proteins. Joined together in a strictly ordered chain, the sequence of amino acids determines the character of each protein (chain) molecule. The 20 common amino acids are: alanine, arginine, aspartic acid, glutamic acid, glutamine, glycine, histidine, isoleucine, leucine, phenylalanine, proline, serine, threonine, tryptophan, tyrosine, valine, cysteine, methionine, lysine, and asparagine. Note that virtually all of these amino acids (except glycine) possess an asymmetric carbon atom, and thus are potentially chiral in nature. See also PROTEIN, POLYPEPTIDE, STEREOISOMERS, CHIRAL COMPOUND, MESSENGER RNA (mRNA), ESSENTIAL AMINO ACIDS, DEOXYRIBONUCLEIC ACID (DNA), and ABSOLUTE CONFIGURATION.

Amino Acid Profile Also known as "protein quality," this refers to a quantitative delineation of how much of each amino acid is contained in a given source of (livestock feed or food) protein. For example, the amino acid profile of soybean meal is matched closest to the profile of amino acids needed for human nutrition, of all protein meals. See IDEAL PROTEIN CONCEPT, PROTEIN, AMINO ACID, and SOYBEAN MEAL.

AMP See also ADENOSINE MONOPHOSPHATE (AMP).

Amphibolic Pathway A metabolic pathway used in both catabolism and anabolism. See also ANABOLISM and CATABOLISM.

Amphipathic Molecules Molecules bearing both polar and nonpolar domains (within the same molecule). Some examples of amphipathic molecules are wetting agents (SDS), and membrane lipids such as lecithin. See also MICELLE, REVERSE MICELLE (RM), and POLARITY (CHEMICAL).

Amphiphilic Molecules Also known collectively as amphiphiles. Molecules possessing distinct regions of hydrophobic ("water hating") and hydrophilic ("water loving") character within the same molecule. When dissolved in water above a certain concentration (known as the CMC), these molecules are capable of forming high molecular weight aggregates, or micelles. See also CRITICAL MICELLE CONCENTRATION, HYDROPHOBIC, HYDROPHILIC, MICELLE, and REVERSE MICELLE (RM).

Amphoteric Compound A compound capable of both donating and accepting protons, and thus able to act chemically as either an acid or a base.

Amplification The production of additional copies of a chromosomal sequence, found as either intrachromosomal or extrachromosomal DNA. See also IN VITRO SELECTION.

Amyloid β Protein (AβP) A small protein that forms plaques in the brains and in the brain blood vessels of victims of Alzheimer's disease. AβP forms cation-selective ion channels in lipid bilayers (e.g., membranes surrounding cells). This ion channel formation disrupts calcium homeostasis, allowing (destructive) high concentrations of calcium ions in brain cells. See also PROTEIN, AMYLOID β PROTEIN PRECURSOR (AβPP), and ALZHEIMER'S DISEASE.

Amyloid β Protein Precursor (AβPP) A (collective) set of protein molecules, from which are derived Amyloid β Protein (AβP). See also PROTEIN and AMYLOID β PROTEIN (AβP).

Amyloid Placques See also AMYLOID β PROTEIN (AβP).

Amylopectin The form of starch that consists of multi-branched polymers, containing approximately 100,000 glucose units per molecule (polysaccharide). See also POLYMER, GLUCOSE, and POLYSACCHARIDES.

Amylose The form of starch that consists of unbranched polymers, containing approximately 4,000 glucose units per molecule (polysaccharide). It is present in potatoes at 23–29% content (variation is thought to be caused by different growing conditions). See also POLYMER, GLUCOSE, and POLYSACCHARIDES.

Anabolism The phase of intermediary metabolism concerned with the energy-requiring biosynthesis of cell components from smaller precursor molecules. See also CATABOLISM, ASSIMILATION, and METABOLISM.

Anaerobe An organism that lives in the absence of oxygen and generally cannot grow in the presence of oxygen. The catabolic metabolism of anaerobic microorganisms reduces a variety of organic and inorganic compounds in order to survive (e.g., carbon dioxide, sulfate, nitrate, fumarate, iron, manganese); and anaerobes produce a large number of end products of metabolism (e.g., acetic acid, propionic acid, lactic acid, ethanol, methane, etc.). See also CATABOLISM, METABOLISM, METABOLITE, REDUCTION (IN A CHEMICAL REACTION), and ANAEROBIC.

Anaerobic An environment without air or oxygen. See also ANAEROBE.

Analogue (also Analog) A compound (or molecule) that is a (chemical) structural derivative of a "parent" compound. The word is also used to describe a molecule which may be structurally similar (but not identical) to another, and which exhibits many or some of the same biological functions of the other.

For example, the large class of antibiotics known as the sulfa drugs are all analogues of the original synthetic chemical drug (known as Prontosil, cures streptococcal infections) discovered by the German biologist Gerhart Domagk. Herr Domagk's and other discoveries made possible a program of further chemical syntheses based upon the original (sulfanilamide) molecular structure, which resulted in the large number of sulfonamide (also called "sulfa") drugs that are available today. All of the sulfa drugs which were patterned after the original sulfanilamide molecular structure may be called sulfanilamide analogues.

Today, analogues are known by man for various vitamins, amino acids, purines, sugars, growth factors, and many other chemical compounds. Research chemists produce analogues of various molecules in order to ascertain the biological role of, or importance of, certain structures (within the molecule) to the molecule's function within a living organism. See also BIOMIMETIC MATERIALS, RATIONAL DRUG DESIGN, HETEROLOGY, GIBBERELLINS, and QUANTITATIVE STRUCTURE-ACTIVITY RELATIONSHIP (QSAR).

ANDA (to FDA) Abbreviated New Drug Application (to the U.S. Food and Drug Administration). See also NDA (TO FDA), "TREATMENT" IND REGULATIONS, and FOOD AND DRUG ADMINISTRATION (FDA).

Angiogenesis Formation/development of new blood vessels in the body. Discovered to be triggered and stimulated by Angiogenic Growth Factors, in the early 1980's. Angiogenesis is required for malignant tumors to metastasize (spread throughout the body), because it provides the (newly created) blood supply that tumors require. Angiogenesis is also crucial to the development of glaucoma and macular degeneration (major cause of blindness). The drug Thalidomide is a potent inhibitor of angiogenesis. See also ANGIOGENIC GROWTH FACTORS, TUMOR, CANCER, ANTIANGIOGENESIS, and CHIRAL COMPOUND.

Angiogenesis Factors See ANGIOGENIC GROWTH FACTORS.

Angiogenic Growth Factors Proteins that stimulate formation of blood vessels (e.g., in tissue being formed by the body to repair wounds). See also FILLER, EPITHELIAL CELLS, FIBROBLAST GROWTH FACTOR (FGF), MITOGEN, ANGIOGENIN, ENDOTHELIAL CELLS, TRANSFORMING GROWTH FACTOR-ALPHA (TGF-ALPHA), TRANSFORMING GROWTH FACTOR-BETA (TGF-BETA), PLATELET-DERIVED GROWTH FACTOR (PDGF), and ANGIOGENESIS.

Angiogenin One of the human angiogenic growth factors, it possesses potent angiogenic (formation of blood vessels) activity. In addition to stimulating (normal) blood vessel formation, angiogenin levels are correlated with placenta formation and tumor growth (tumors require new blood vessels). See also ANGIOGENIC GROWTH FACTORS, ANGIOGENESIS, TUMOR, and GROWTH FACTOR.

Angstrom (Å) 10^{-8} cm (3.937×10^{-9} inch).

Anion See ION.

Anneal The process by which the complementary base pairs in the strands of DNA combine. See also BASE PAIR (bp) and DEOXYRIBONUCLEIC ACID (DNA).

Antagonists Molecules that bind to certain proteins (e.g., receptors, enzymes) at a specific (active) site on that protein. The binding suppresses or inhibits the activity (function) of that protein. See also

RECEPTORS, ACTIVE SITE, CONFORMATION, AGONISTS, ENZYME, and ALLOSTERIC ENZYMES.

Anterior Pituitary Gland See PITUITARY GLAND.

Anti-Idiotypes Antibodies to antibodies. In other words, if a human antibody is injected into rabbits, the rabbit immune systems will recognize the human antibodies as foreign (regardless of the fact that they are antibodies) and produce antibodies against them. To the rabbit the foreign antibodies represent just another invader or nonself to be targeted and destroyed. Anti-idiotypes mimic antigens in that they are shaped to fit into the antibody's binding site (in lock-and-key fashion). As such, anti-idiotypes can be used to create vaccines that stimulate production of antibodies to the antigen (that the anti-idiotype mimics). This confers disease resistance (to the pathogen associated with that antigen) without the risk that a vaccine using attenuated pathogens entails (i.e., that the pathogen "revives" to cause the disease). See also ANTIBODY, MONOCLONAL ANTIBODY (MAb), ANTIGEN, IDIOTYPE, PATHOGEN, and ATTENUATED (PATHOGENS).

Anti-Interferon An antibody to interferon. Used for the purification of interferons. See also ANTIBODY, INTERFERONS, and AFFINITY CHROMATOGRAPHY.

Anti-Oncogenes See ONCOGENES and ANTISENSE (DNA SEQUENCE).

Antiangiogenesis Refers to any compound that works to prevent angiogenesis (formation/development of new blood vessels). Because angiogenesis is required for malignant tumors to grow and/or metastasize (spread), antiangiogenesis was proposed as a means to combat cancer, by Judah Folkman in 1970. Because angiogenesis is required for embryonic development, anti-angiogenic drugs inhibit proper development/growth of infants in the womb.

Drugs that have been found to possess antiangiogenic properties include fumagillin, ovalicin, and Thalidomide. See ANGIOGENESIS, ANGIOGENIC GROWTH FACTORS, TUMOR, and CANCER.

Antibiosis Refers to the processes via which one organism produces a substance that is toxic or repellent to another organism (e.g., a parasite) that is attacking the first organism. For example, certain varieties of corn/maize (*Zea mays L.*) produce chemical substances in their roots that are toxic to the corn rootworm. See ANTIBIOTIC, *BACILLUS THURINGIENSIS (B.t.)*, CORN and CORN ROOTWORM.

Antibiotic Organic compound that is naturally formed and secreted by various species of microorganisms and plants. It has a defensive function and is often toxic to other species (e.g., penicillin, originally produced by bread mold, is toxic to numerous human pathogens). Antibiotics act by inhibiting protein and nucleic acid (DNA and RNA) biosynthesis, hence killing the cells involved. Inorganic molecules may also have antibiotic properties. See also PATHOGEN, MICROORGANISM, PROTEIN, NUCLEIC ACIDS, PENICILLIN G, SYMBIOTIC, GRAM STAIN, GRAM-NEGATIVE, GRAM-POSITIVE, CELL, ANTIBIOSIS, and AUREOFACIN.

Antibody Also called immunoglobulin, Ig. A large defense protein that consists of two classes of polypeptide chains, light (L) chains and heavy (H) chains. A single antibody molecule consists of two identical copies of the L chain and two of the H chain. They are synthesized (i.e., made) by the immune system (B lymphocytes) of the organism. The antibody is composed of four proteins linked together to form a Y-shaped bundle of proteins (looks somewhat like a slingshot or two hockey sticks taped together at the handles). The amino acid sequence that makes up the stem (heavy chains) of the Y (i.e., the handles of the taped together hockey sticks) is similar for all antibodies. The stem is known as the Fc region of the antibody and it does not bind to antigen, but does have other regulatory functions.

The two arms of the Y are each made up of two side-by-side proteins called light chains and heavy chains (i.e., proteins are chains of amino acids), with identical antigen-binding (ab) sites on the tips of each "arm." The antibody is thus bivalent in that it has two binding sites for antigen. Taken together, the two arms of the Y are known as the Fab portions of the antibody molecule. The Fab portions can be cleaved from the antibody molecule with papain (an enzyme that is also used as a meat tenderizer) or the Fab portions can be produced via genetically engineered *Escherichia coli (E. coli)* bacteria.

When a foreign molecule (e.g., a bacterium, virus, etc.) enters the body, B lymphocytes are stimulated into becoming rapidly dividing blast cells, which mature into antibody-producing plasma cells. The plasma cells are triggered by the foreign molecule's epitope(s) [i.e., group or groups of specific atoms (also known as a hapten), that are recognized to be foreign by the body's immune system] into producing antibody molecules possessing antigen-binding (ab) sites (also called combining sites or determinants).

These fit into the foreign molecule's epitope. Thus, via the tips of its arms, the antibody molecule binds specifically to the foreign entity (antigen) that has entered the body. By this process it inactivates that foreign molecule or marks it for eventual destruction by other immune system cells.

System marking of the foreign molecule (e.g., pathogen or toxin) for destruction is accomplished by the fact that the stem of the Y (i.e., the Fc) fragment hangs free from the combined antibody-antigen clump, thereby providing a receptor for phagocytes, which roam throughout the body ingesting and subsequently destroying such "marked" foreign molecules. There are five classes of immunoglobulin: IgG, IgM, IgD, IgA, and IgE.

See also HUMORAL IMMUNITY, IMMUNOGLOBULIN (IgA, IgE, IgG, and IgM), PROTEIN, POLYPEPTIDE (PROTEIN), AMINO ACID, B LYMPHOCYTES, BLAST CELL, ANTIGEN, HAPTEN, EPITOPE, COMBINING SITE, DOMAIN (OF A PROTEIN), SEQUENCE (OF A PROTEIN MOLECULE), *ESCHERICHIA COLI (E. COLI)*, PATHOGEN, TOXIN, PHAGOCYTE, MARCOPHATE, MICROPHAGE, MONOCYTES, T CELLS, POLYMORPHONUCLEAR LEUKOCYTES, CELLULAR IMMUNE RESPONSE, POLYMORPHONUCLEAR GRANULOCYTES, GENETIC ENGINEERING, ENGINEERED ANTIBODIES, and RECEPTORS.

Antibody Affinity Chromatography A type of chromatography in which antibodies are immobilized onto the column material. The antibodies bind to their target molecules while the other components in the solution are not retained. In this way a separation (purification) is achieved. See also ANTIBODY, CHROMATOGRAPHY, and AFFINITY CHROMATOGRAPHY.

Anticodon A specific sequence of three nucleotides in a transfer RNA (tRNA), complementary to a codon (also three nucleotides) for an amino acid in a messenger RNA.

Antigen Also called an immunogen. Any large molecule or small organism whose entry into the body provokes synthesis of an antibody or immunoglobin (i.e., an immune system response). See also HAPTEN, ANTIBODY, EPITOPE, CELLULAR IMMUNE RESPONSE, and HUMORAL IMMUNITY.

Antigenic Determinant See HAPTEN, EPITOPE, and SUPER ANTIGENS.

Antihemophilic Factor VIII Also known as Factor VIII or Antihemophilic Globulin (AHG). See FACTOR VIII.

Antihemophilic Globulin Also known as Factor VIII or Antihemophilic Factor VIII. See FACTOR VIII.

Antioxidants Phytochemicals that act to prevent lipids from breaking down (e.g., to carcinogenic compounds). Synthetic analogues have also been manufactured (e.g., synthetic vitamins, etc.) which perform similar antioxidant function. See PHYTOCHEMICALS, LIPIDS, CARCINOGEN, and ANALOGUES.

Antiparallel Describes molecules that are parallel but point in opposite directions. The strands of the DNA double helix are antiparallel. See also DOUBLE HELIX.

Antisense (DNA sequence) A strand of DNA that has the same sequence as (i.e., is complementary to) messenger RNA. In genetic targeting using antisense molecules to block "bad" genes), antisense molecules are used to bind to a "bad" gene (e.g., an oncogene) messenger RNA (mRNA), thus cancelling the (cancer-causing) message of the gene and preventing cells from following its (tumor growth) instructions. Another example would be the use of antisense DNA to block the gene that codes for production of polygalacturonase (an enzyme that causes ripe fruit to soften). See also DEOXYRIBONUCLEIC ACID (DNA), CODING SEQUENCE, COMPLEMENTARY DNA (c-DNA), MESSENGER RNA (mRNA), GENETIC TARGETING, CANCER, POLYGALACTURONASE, ONCOGENES, SENSE, and GENE SILENCING.

Antisense RNA See ANTISENSE (DNA SEQUENCE).

Antithrombogenous Polymers Synthetic polymers (i.e., plastics) used to make medical devices that will be in contact with a patient's blood (e.g., catheters) and thus must not initiate the coagulation process as synthetic polymers usually do. The natural anticoagulant heparin is incorporated into the polymer, and is gradually released into the bloodstream by the polymer, thus preventing blood coagulation on the surface of the polymer.

Antitoxin See POLYCLONAL ANTIBODIES (USED IN HUMANS).

AP Atrial peptide. See also ATRIAL PEPTIDES.

Aplastic Anemia An autoimmune disease of the bone marrow. See also AUTOIMMUNE DISEASE.

APO-1/Fas See CD 95 PROTEIN.

Apoenzyme The protein part of a holoenzyme. Many (but not all) enzymes are composed of functional "pieces." For example, a protein piece (chain) and another piece that is an organic and/or inorganic molecule. This other piece is known as a cofactor and it may be removed from the enzyme under certain conditions. When this is done, the resulting inactive enzyme is known as an apoenzyme. The inactive apoenzyme becomes functionally active again if it is allowed to reombine with its cofactor. See also COFACTOR, ENZYME, and HOLOENZYME.

Apomixis A method of reproduction used by scientists to propagate (hybrid) plants without having to utilize sexual fertilization. By combining apomixis with tissue culture technology, cai Detian, Ma Piugfu, and Yao Jialin were able to thus propagate rice varieties in 1994. By "fixing" hybrid dominance, the need for (sexual) breeding is eliminated and the hybrid vigor is passed down via the seed from generation to generation. See also ASEXUAL, GERM CELL, HYBRID VIGOR, TISSUE CULTURE, HYBRIDIZATION (PLANT GENETICS), and F1 HYBRIDS.

Apoptosis Also called "Programmed cell death," it is a series of programmed steps that cause a cell to die via "self digestion" without rupturing and releasing intracellular contents (e.g., nucleus, chromosomes, refractile bodies, etc.) into the local (i.e., surrounding tissue) environment. Manifestations of cell apoptosis include shrinking of the cell's cytoplasm and chromatin condensation. See CD 95 PROTEIN, SIGNAL TRANSDUCTION, SIGNALLING, REFRACTILE BIODIES (RB), NUCLEUS CHROMOSOMES, CHROMATIN, CYTOPLASM, *FUSARIUM*, and p53 GENE.

Approvable Letter (from the FDA) One of the final steps in the Food and Drug Administration's (FDA) review process for new pharmaceuticals.

The letter precedes final FDA clearance for marketing of the new compound. See also FOOD AND DRUG ADMINISTRATION (FDA), IND, and IND EXEMPTION.

Aptamers Oligonucleotide molecules that bind (i.e., "stick to") other, specific molecules (e.g., proteins). Aptamer is from the Latin aptus ("to fit"). For example, in 1992, Louis Bock and John Toole isolated aptamers that bind and inhibit the blood-coagulation enzyme, thrombin. Since thrombin is crucial to the formation of blood clots (coagulation), such aptamers may be useful for anticoagulant therapy (e.g., to prevent blood clots following surgery or heart attacks). See also ENZYME, OLIGONUCLEOTIDE, INHIBITION, THROMBIN, THROMBUS, and THROMBOSIS.

Arabidopsis thaliana A small weed plant possessing 70,000 kilobase pairs in its genome, with very little repetitive DNA. This makes it an ideal model for studying plant genetics. At least two genetic maps have been created for *Arabidopsis thaliana* (one using yeast artificial chromosomes). Because of this a large base of knowledge about it has been accumulated by the scientific community.

Arabidopsis thaliana was first genetically engineered in 1986. In 1994, researchers succeeded in transferring genes for polyhydroxylbutylate ("biodegradable plastic") production into *Arabidopsis thaliana*. Because production of polyhydroxylbutylate (PHB) requires simultaneous expression of three genes (i.e., the PHB production process is "polygenic")—yet researchers have only been able to insert a maximum of two genes—they have to insert two genes into one plant and one gene into a second plant, then finally get the (total) three genes into (offspring) plants via traditional breeding. See also BRASSICA, GENE, EXPRESS, BASE PAIR (bp), KILOBASE PAIRS (Kbp), GENOME, GENETIC CODE, GENETIC MAP, GENETICS, TRAIT, POLYGENIC (TRAIT, PRODUCT, ETC.), DEOXYRIBONUCLEIC ACID (DNA), POLYHYDROXYLBUTYLATE (PHB), and YEAST ARTIFICIAL CHROMOSOMES (YAC).

Archaea Single-celled life forms that can live at extreme ocean depths (i.e., high pressure) and in the absence of oxygen. Enzymes robust (i.e., sturdy) enough for industrial process utilization have been isolated by scientists from *Archaea*. See also ENZYME, EXTREMOZYMES, CELL, ANAEROBE, and ANAEROBIC.

Arginine (arg) An amino acid, commonly abbreviated arg. In dry, bulk form arginine is colorless, crystalline, and water soluble. It is an essential amino acid of the α-ketoglutaric acid family. See also AMINO ACID, ESSENTIAL AMINO ACIDS, and NITRIC OXIDE SYNTHASE.

ARS Element A sequence of DNA that will support autonomous replication (sequence, ARS).

Ascites Liquid accumulations in the peritoneal cavity. Used as an input in one of the methods for producing monoclonal antibodies. See also

MONOCLONAL ANTIBODIES (MAb), PERITONEAL CAVITY/MEMBRANE, and ANTIBODY.

-ase The three-letter suffix that is added to a (root) word to denote an enzyme. For example, the stomachs of reindeer contain *lichenase*, an enzyme that enables reindeer to digest lichen that the reindeer consume as a source of winter food. See also ENZYME, PROTEASE, OXYGENASE, HUMAN PROTEIN KINASE C, HUMAN SUPEROXIDE DISMUTASE (hSOD), POLYMERASE, ATPase, ATP SYNTHASE, and REGULATORY ENZYME.

Asexual Denotes fertilization and/or reproduction by *in vitro* means. Without sex. See also *IN VITRO*, APOMIXIS, and GERM CELL.

Asparagine (asp) An amino acid, commonly abbreviated asp. In dry, bulk form asparagine appears as a white, crystalline solid. It is found in high amounts in many plants. See also AMINO ACID.

Aspartic Acid a dicarboxylic amino acid found in plants and animals, especially in molasses from young sugarcane and sugar beets. See also AMINO ACID.

Aspergillus flavus See AFLATOXIN, PEROXIDASE, and BETA CAROTENE.

Assay A test (specific technique) that measures a response to a test substance or the efficacy (effectiveness) of the test substance. See also IMMUNOASSAY, BIOASSAY, and HYBRIDIZATION SURFACES.

Assimilation The formation of "self" cellular material from small molecules derived from food. See also INSULIN-LIKE GROWTH FACTOR-1 (IGF-1), RIBOSOMES, and MESSENGER RNA (mRNA).

Association of Biotechnology Companies (ABC) An American trade association of companies involved in biotechnology and services to biotechnology companies (e.g., accounting, law, etc.). Formed in 1984, the ABC tended to consist of the smaller firms involved in biotechnology (and service firms that worked for *all* biotechnology companies). In 1993, the Association of Biotechnology Companies (ABC) was merged with the Industrial Biotechnology Association (IBA) to form the Biotechnology Industry Organization (BIO). See also INDUSTRIAL BIOTECHNOLOGY ASSOCIATION (IBA), BIOTECHNOLOGY INDUSTRY ORGANIZATION (BIO), and BIOTECHNOLOGY.

AT-III A human blood factor that promotes clotting. A deficiency of AT-III can be inherited, resulting from certain surgical procedures, certain illnesses, and sometimes use of certain oral contraceptives. See also FACTOR VIII.

ATCC See AMERICAN TYPE CULTURE COLLECTION (ATCC), TYPE SPECIMEN, and ACCESSION.

Atomic Weight The total mass of an atom, it is equal to the sum of the isotope's number of protons and neutrons (in the atom's nucleus). The atomic weights of the earth's elements are based on the assignment of exactly 12.000 as the atomic weight of the carbon-12 isotope (variation of atom). The atomic (weight) theory was established as a framework in

1869 by Meyer and Mendeléev, but standard precise values were not adopted internationally until an "international commission on atomic weights" was formed in 1899 in response to an initiative by the German Chemical Society. An element's atomic weight does not come out to a whole number (with the exception of carbon) because of the existence of isotopes which differ slightly with respect to the number of neutrons each contains. See also MOLECULAR WEIGHT and ISOTOPE.

ATP See ADENOSINE TRIPHOSPHATE.

ATP Synthase An enzyme complex that forms ATP from ADP and phosphate during oxidative phosphorylation in the inner mitochondrial membrane (in animals), in chloroplasts (in plants), and in cell membranes (in bacteria). This is an energy-producing reaction in that ATP is a high-energy compound used by cells to maintain their living condition. See ENZYME, CHLOROPLASTS, ADENOSINE TRIPHOSPHATE (ATP), ADENOSINE DIPHOSPHATE (ADP), and MITOCHONDRIA.

ATP Synthetase See ATP SYNTHASE.

ATPase Adenosine triphosphatase, an enzyme that hydrolyzes (clips the bond between two phosphates in) ATP to yield ADP, phosphate, and energy. The reaction is usually coupled to an energy-requiring process. ATP is hydrolyzed in the act of shivering and the energy produced is converted into heat to increase body temperature. This type of heat production involves what is known as a futile cycle because the energy is converted to (and wasted as) heat rather than used in motion, etc. See also ATP SYNTHASE, ENZYME, ADENOSINE TRIPHOSPHATE (ATP), ADENOSINE DIPHOSPHATE (ADP), FUTILE CYCLE, HYDROLYSIS, and HYDROLYZE.

Atrial Natriuretic Factor An atrial peptide hormone that may regulate blood pressure and electrolyte balance within the body. An example is a peptide hormone. See also HORMONE, ATRIAL PEPTIDES, and PEPTIDE.

Atrial Peptides (e.g., AP III) Endocrine components (proteins) that act to regulate blood pressure, as well as water and electrolyte homeostasis within the body. Atrial peptides are made by the heart in response to elevated blood pressure levels; and they stimulate the kidneys to excrete water and sodium into the urine, thus lowering blood pressure. They also slow the heartbeat. An example is a peptide hormone. See also ENDOCRINE HORMONES, HOMEOSTASIS, and ELECTROLYTE.

Attenuated (pathogens) Inactivated, rendered harmless (e.g., killed viruses used to make a vaccine). Some of the ways in which viruses and other pathogens may be attenuated are by heat, chemical, or radiation treatment. See also PATHOGEN.

Attenuation (of RNA) Premature termination of an elongating RNA chain. See also RIBONUCLEIC ACID (RNA).

Aureofacin An antifungal antibiotic produced by a strain of *Streptomyces aureofaciens.* At least one company has incorporated the gene for this antibiotic (which acts against wheat take-all disease) into a *Pseudomonas fluorescens,* to be used to confer resistance to wheat take-all disease. This is done by allowing the bacteria to colonize the wheat's roots. In this way the plant obtains the benefits of the antibiotic because the bacteria become a part of the plant. See also *PSEUDOMONAS FLUORESCENS, ENDOPHYTE, ANTIBIOTIC,* and *BACILLUS THURINGIENSIS (B.t.).*

Autogenous Control The action of a gene product (a molecule) that either inhibits (negative autogenous control) or activates (positive autogenous control) expression of the gene that codes for it. The presence of the product either causes or stops its own production. See also GENE.

Autoimmune Disease A disease in which the body produces an immunogenic (i.e., immune system) response to some constituent of its own tissue. In other words the immune system loses its ability to recognize some tissue or system within the body as "self" and targets and attacks it as if it were foreign. Autoimmune diseases can be classified into those in which predominantly one organ is affected (e.g., hemolytic anemia and chronic thyroiditis), and those in which the autoimmune disease process is diffused through many tissues (e.g., multiple sclerosis, systemic lupus erythematosus, and rheumatoid arthritis).

For example, multiple sclerosis is thought to be caused by T cells attacking the sheaths that surround the nerve fibers of the brain and spinal cord. This results in loss of coordination, weakness, and blurred vision. See also THYMUS, SUPERANTIGENS, T CELLS, and TUMOR NECROSIS FACTOR (TNF).

Autoradiography A technique to detect radioactively labeled molecules by creating an image on photographic film. The slab of gel or other material in which the molecules are held (suspended) is placed on top of a piece of photographic film. The two are then securely fastened together such that movement is eliminated and the film is exposed for a period of time. The exposed (to the radiation) film is subsequently developed and the radioactive area is seen as a dark (black) area. Among other uses, autoradiography has been used to track the spread of (radioactively labeled) viruses in a living plant. After treatment (i.e., the radioactive labeling process), the whole plant (in a slab) is placed on top of a piece of photographic film. When the film is subsequently developed, the "picture" seen is of a plant, with darker areas indicating regions of greater virus concentration. See also LABEL (RADIOACTIVE) and VIRUS.

Autosomes All chromosomes except the sex chromosomes. A diploid cell has two copies of each autosome.

Autotroph An organism that can live on very simple carbon and nitrogen sources, such as carbon dioxide and ammonia. See also HETEROTROPH.

Auxotroph Auxotrophic mutant. A mutant defective in the synthesis of a given biomolecule. The biomolecule must be supplied to the organism if normal growth is to be achieved.

Avidity (of an antibody) The "tightness of fit" between a given antibody's combining site and the antigenic determinant that it combines with. The firmness of the combination of antigen with antibody. See also ANTIGENIC DETERMINANT, ANTIBODY, ANTIGEN, COMBINING SITE, POLYCLONAL RESPONSE (OF IMMUNE SYSTEM TO A GIVEN PATHOGEN), and CATALYTIC ANTIBODY.

A_w See WATER ACTIVITY.

Azadirachtin The pharmacophore (i.e., active ingredient) in secretions of the tropical neem tree, which resists insect depradations. See also PHARMACOPHORE and NEEM TREE.

Azurophil-Derived Bactericidal Factor (ADBF) Potent antimicrobial protein produced by neutrophils (a type of white blood cell). See also LEUKOCYTES (WHITE BLOOD CELLS).

B

***B* Sitostanol** See BETA SITOSTANOL.

B Cells B lymphocytes. See also LYMPHOCYTE and B LYMPHOCYTES.

B Lymphocytes A class of white blood cells originating in the bone marrow and found in blood, spleen, and lymph nodes. They are the precursors of (blood) plasma cells that secrete antibodies directed against invading antigens. See also ANTIGEN, ANTIBODY, BLAST CELL, and LYMPHOCYTE.

B-DNA A helical form of DNA. B-DNA can be formed by adding back water to (dehydrated) A-DNA. B-DNA is the form of DNA of which James Watson and Francis Crick first constructed their model in 1953. It is found in fibers of very high (92 percent) relative humidity and in solutions of low ionic strength. This corresponds to the form of DNA that is thought to prevail in the living cell. See also DEOXYRIBONUCLEIC ACID (DNA), A-DNA, and ION.

Bacillus Rod-shaped bacteria.

Bacillus subtilis *(B. subtilis)* A (rod-shaped) aerobic bacterium commonly used as a host in recombinant DNA experiments. See also HOST VECTOR (HV) SYSTEM, DEOXYRIBONUCLEIC ACID (DNA).

Bacillus thuringiensis *(B.t.)* Discovered by bacteriologist Ishiwata Shigetane on a diseased silkworm in 1901. Later discovered on a dead Mediterranean flour moth, and first named *Bacillus thuringiensis*, by Ernst Berliner in 1915.

Today, *Bacillus thuringiensis* refers to a group of rod-shaped soil bacteria found all over the earth, that produce "cry" proteins which are indigestible by—yet still "bind" to—specific insects' gut (i.e., stomach) lining receptors, so those "cry" proteins are toxic to certain classes of insects (corn borers, corn rootworms, mosquitoes, black flies, some types of beetles, etc.), but which are harmless to all mammals. At least 20,000 strains of *Bacillus thuringiensis* are known.

Genes that code for the production of these "cry" proteins that are toxic to insects have been inserted by scientists since 1989 into vectors (i.e., viruses, other bacteria, and other microorganisms) in order to confer insect resistance to certain agricultural plants (e.g., via expression of those *B.t.* proteins by one or more tissues of the transgenic plant). For example, the *B.t.* strain known as *B.t. kurstaki,* which is fatal when ingested by the European corn borer was first (genetically) inserted into a corn plant (via vector) in 1991. *B.t. kurstaki* kills borers via perforation of that insect's gut by proteins that are coded-for by the *B.t. kurstaki* gene. The vectors as listed above are entities that can take up and carry the DNA into plant or other cells. Vectors are DNA-carrying vehicles. See also ENDOPHYTE, CORN, GENE, *PSEUDOMONAS FLUORESCENS, AGROBACTERIUM TUMEFACIENS,* AUREOFACIN, EUROPEAN CORN BORER (ECB), COWPEA TRYPSIN INHIBITOR (CpTI), PROTEIN, "SHOTGUN" METHOD [TO INTRODUCE FOREIGN (NEW) GENES INTO PLANT CELLS], CODING SEQUENCE, *FUSARIUM,* VECTOR, EXPRESS, GENETIC ENGINEERING, "EXPLOSION" METHOD [TO INTRODUCE FOREIGN (NEW) GENES INTO PLANT CELLS], BIOLISTIC® GENE GUN, CRY PROTEINS, CRY1A (b) PROTEIN, CRY1A (c) PROTEIN, CRY9C PROTEIN, *B.t. KURSTAKI*, *B.t. TENEBRIONIS*, and *B.t. ISRAELENSIS*.

Back Mutation Reverse the effect of a mutation that had inactivated a gene; thus it restores wild phenotype. See also PHENOTYPE and MUTATION.

Bacteria Any of a large group of microscopic organisms having round, rod-like, spiral, or filamentous unicellular or noncellular bodies that are often aggregated into colonies, are enclosed by a cell wall or membrane (procaryotes), and lack fully differentiated nuclei. Bacteria may exist as free-living organisms in soil, water, and organic matter, or as parasites in the live bodies of plants and animals. See also BACTERIOLOGY.

Bactericide See MICROBICIDE.

Bacteriocins Proteins produced by many types of bacteria, that are toxic (primarily) to other closely-related strains of the particular bacteria that produce those proteins. Bacteriocins hold promise (after genetic engineering of the DNA responsible for their production) for future possible use as natural food preservatives (i.e., acting against bacteria that cause food spoilage). See also PROTEIN, BACTERIA, BACTERIOLOGY, STRAIN, TOXIN, GENETIC ENGINEERING, DEOXYRIBONUCLEIC ACID (DNA), CODING SEQUENCE, and COLICINS.

Bacteriology The science and study of bacteria, a specialized branch of microbiology. The bacteria constitute a useful and essential group in the biological community. Although some bacteria prey on higher forms of life, relatively few are pathogens (disease-causing organisms). Life on earth depends on the activity of bacteria to mineralize organic compounds and to capture the free nitrogen molecules in the air for use by plants. Also, bacteria are important industrially for the conversion of raw materials into products such as organic chemicals, antibiotics, cheeses, etc. Genetically engineered bacteria are starting to be used to produce high value added pharmaceuticals and specialty chemicals. See also *ESCHERICHIA COLI (E. COLI)*.

Bacteriophage Discovered in 1917 by Felix d'Herelle (fr. "bacteria eaters"), a bacteriophage is a virus that attaches to, injects its DNA into, and multiplies inside bacteria; which causes bacteria to die. Often abbreviated as simply phage. Another name for virus. As an example, bacteriophage lambda is commonly used as a vector in rDNA experiments in *Escherichia coli* and attaches to a specific receptor, which in the bacteria also normally functions in sugar transport across the cell wall. Viruses come in many shapes and sizes. See also *ESCHERICHIA COLI (E. COLI)*, RECEPTORS, VIRUS, TRANSDUCTION, TRANSFECTION, and LAMBDA PHAGE.

Bacterium See BACTERIA.

Baculovirus A virus that infects lepidopteran insects (e.g., cotton bollworm or gypsy moth larva). Baculoviruses can be modified via genetic engineering to insert new genes into the larva, causing those larva to then produce proteins desired by man (e.g., pharmaceuticals). See also VIRUS, GENETIC ENGINEERING, GENE, PROTEIN, BACULOVIRUS EXPRESSION VECTORS (BEVs).

Baculovirus Expression Vectors (BEVs) Vectors (used by researchers to carry new genes into cells) in which the agent is a baculovirus (i.e., a virus that infects certain types of insects only). These could conceivably be use to make a genetically engineered "insecticide" that is specific to a targeted insect (i.e., wouldn't harm anything but that insect). For example, a BEV might be used to cause a cotton bollworm *adult* protein to be expressed when the bollworm is a *juvenile*, thus killing the bollworm before it has a chance to damage a cotton crop. See also BACULOVIRUS, VIRUS, VECTOR, GENE, PROTEIN, CELL, and GENETIC ENGINEERING.

BAR Gene See PAT GENE.

Base A substance with a pH in the range 7–14 which will react with an acid to form a salt. Bases normally taste bitter and feel slippery to the touch. See also ACID.

Base Excision Sequence Scanning (BESS) A method that can be utilized to detect a "point mutation" in DNA (via rapid DNA sequence scanning). See BASE PAIR (bp), NUCLEOTIDE, DEOXYRIBONUCLEIC ACID (DNA),

MUTATION, POINT MUTATION, EXCISION, SEQUENCING (OF DNA MOLECULES), and SEQUENCE (OF A DNA MOLECULE).

Base Pair (bp) Two nucleotides that are in different nucleic acid chains and whose bases pair (interact) by hydrogen bonding. In DNA, the nucleotide bases are adenine (which pairs with thymine) and guanine (which pairs with cytosine). See also DEOXYRIBONUCLEIC ACID (DNA), GENETIC CODE, and INFORMATIONAL MOLECULES.

Basic Fibroblast Growth Factor (BFGF) See FIBROBLAST GROWTH FACTOR (FGF).

Basophilic Staining strongly with basic dye. For example, basophil leukocytes are polymorphonuclear leukocytes which stain strongly with (take up a lot of) basic dyes. See also POLYMORPHONUCLEAR LEUKOCYTES (PMN).

Basophilis Also called basophilic leukocytes. A type of white blood cell that synthesizes and stores histamine and also contains heparin. When two IgE molecules of the same antibody "dock" at adjacent receptor sites on a basophil cell, the two IgE molecules capture an allergen between them. A chemical signal is sent to the basophil causing the basophil cell to release histamine, serotonin, bradykinin, and "slow-reacting-substance." Release of these chemicals into the body causes the blood vessels to become more permeable which consequently causes the nose to run. These chemicals also cause smooth muscle contraction, resulting in sneezing, coughing, wheezing, etc. See also MAST CELLS, ANTIGEN, ANTIBODY, HISTAMINE, WHITE BLOOD CELLS, BASOPHILIC, and POLYMORPHONUCLEAR LEUKOCYTES (PMN).

BBB See BLOOD-BRAIN BARRIER.

Bce4 The name of a promoter (region of DNA) that controls/enhances an oilseed plant's gene(s) that code for components (e.g., fatty acids, amino acids, etc.) of that plant's seeds. The Bce4 promoter causes such genes to be expressed during one of the earliest stages of canola plant's seed production, for instance. See also PROMOTER, DEOXYRIBONUCLEIC ACID (DNA), GENE, POLYGENIC (TRAIT, PRODUCT, ETC.), PLASTID, EXPRESS, CANOLA, SOYBEAN PLANT, and TRANSCRIPTION.

BESS Method See BASE EXCISION SEQUENCE SCANNING (BESS).

BESS T-Scan Method See BASE EXCISION SEQUENCE SCANNING (BESS).

Best Linear Unbiased Prediction (BLUP) A statistical (data) technique that is utilized by livestock breeders to determine the breeding (genetic trait) value of animals in a breeding program. See also GENETICS, TRAIT, PHENOTYPE, GENOTYPE, and EXPECTED PROGENY DIFFERENCES (EPD).

Beta Carotene A phytochemical (vitamin precursor) that is naturally produced in the endosperm portion of the corn (maize) kernel. If the kernel seed coat is torn (e.g., via insect chewing), the beta carotene inhibits growth of *Aspergillus flavus* fungi in the endosperm region of the kernel. See AFLATOXIN, FUNGUS, OH43, PHYTOCHEMICAL, and CAROTENOIDS.

Beta Cells Insulin-producing cells in the pancreas. If these cells are destroyed, childhood (also known as early-onset or Type I) diabetes results. See also ISLETS OF LANGERHANS.

Beta Conformation An extended, zig-zag arrangement of a polypeptide (molecule) chain. See also POLYPEPTIDE (PROTEIN).

Beta Interferon One of the interferons, it is a protein that was approved by America's Food & Drug Administration (FDA) in 1993 to be used to treat multiple sclerosis (MS) disease. See INTERFERONS, FOOD AND DRUG ADMINISTRATION (FDA), and PROTEIN.

Beta Sitostanol (β sitostanol) See SITOSTANOL.

BEVs See BACULOVIRUS and BACULOVIRUS EXPRESSION VECTORS.

BFGF Basic Fibroblast Growth Factor. See FIBROBLAST GROWTH FACTOR (FGF).

BGYF See BRIGHT GREENISH-YELLOW FLUORESCENCE.

BIO See BIOTECHNOLOGY INDUSTRY ORGANIZATION (BIO).

Bioassay Determination of the relative strength or bioactivity of a substance (e.g., a drug). A biological system (such as living cells, organs, tissues, or whole animals) is exposed to the substance in question and the effect on the living test system is measured. See also BIOLOGICAL ACTIVITY and ASSAY.

Biochemistry The study of chemical processes that comprise living things (systems). The chemistry of life and living matter. Despite the dramatic differences in the appearances of living things, the basic chemistry of all organisms is strikingly similar. Even tiny one-celled creatures carry out essentially the same chemical reactions that each cell of a complex organism (such as man) carries out. See also MOLECULAR BIOLOGY and MOLECULAR DIVERSITY.

Biochip An electronic device that uses biological molecules as the framework for other molecules that act as semiconductors, and functions as an integrated circuit. The future working parts of the science of bioelectronics, biochips may consist of two- or three-dimensional arrays of organic molecules used as switching or memory elements. If biochip technology proves to be feasible, one application will be to shrink currently existing biosensors in size. This would enable the biosensors to be implanted in the body or in organs and tissues for the sake of monitoring and controlling certain bodily functions. A future possibility is to try to provide sight for the blind using protein-covered electrodes implanted in the eyes. See also BIOELECTRONICS, BIONICS, and BIOSENSORS (ELECTRONIC).

Biocide Any chemical or chemical compound that is toxic to living things (systems). Literally "biokiller" or killer of biological systems. Includes insecticides, bactericides, fungicides, etc.

Biodegradable Describes any material that can be broken down by biological action (e.g., dissimilation, digestion, denitrification, etc.). The breakdown of material (chemicals) by microorganisms (bacteria, fungus, etc.). See also GLYCOLYSIS, METABOLISM, and NITRIFICATION.

Biodesulfurization The removal of organic and inorganic sulfur (a pollution source) from coal by bacterial and soil microorganisms. See also BIOLEACHING, BIORECOVERY, and BIOSORBENT.

Bioelectronics Also called biomolecular electronics. It is the field where biotechnology is crossed with electronics. The branch of biotechnology that deals with the electroactive properties of biological materials, systems, and processes together with their exploitation in electronic devices. Bioelectronics will attempt to replace traditional semiconductor materials (e.g., silicon or gallium arsenide) with organic materials such as proteins (biochips). See also BIOCHIP, BIOSENSORS (ELECTRONIC), and BIONICS.

Biogenesis The theory that living organisms are produced only by other living organisms. That is, the theory of generation from preexisting life. It is the opposite of abiogenesis, or spontaneous generation.

Biogeochemistry A branch of geochemistry that is concerned with biological materials and their relation to earth's chemicals in an area.

Bioinformatics See GENOMICS, FUNCTIONAL GENOMICS, and STRUCTURAL GENOMICS.

Bioleaching The biomediated recovery of precious metals from their ores. In the recovery of gold, for example, the microorganism *T. ferroxidans* may be used to cause the gold to leach out of the ore so it may then be concentrated and smelted. Aluminum may be similarly bioleached from clay ores, using heterotropic bacteria and fungi. See also BIORECOVERY, BIOGEOCHEMISTRY, BACTERIA, and BIOSORBENTS.

Biolistic® Gene Gun The word "biolistic" was coined from the words "biological" and "ballistic" (pertaining to a projectile fired from a gun). Used to shoot pellets that are coated with genes (e.g., for desired traits) into plant seeds or plant tissues, in order to get those plants to then express the new genes. The gun uses an actual explosive (.22 caliber blank) to propel the material. Steam may also be used as the propellant. The Biolistic Gene Gun was invented in 1983–1984 at Cornell University by John Sanford, Edward Wolf, and Nelson Allen. It and its registered trademark are now owned by E. I. du Pont de Nemours and Company. See "SHOTGUN" METHOD [TO INTRODUCE FOREIGN (NEW) GENES INTO PLANT CELLS], GENETIC ENGINEERING, GENE, and BIOSEEDS.

Biological Activity The effect (change in metabolic activity upon living cells) caused by specific compounds or agents. For example, the drug aspirin causes the blood to thin, that is, to clot less easily. See also BIOASSAY, PHARMACOPHORE, and RETINOIDS.

Biological Oxygen Demand (BOD) The oxygen used in meeting the metabolic needs of aerobic organisms in water containing organic compounds. Numerically, it is expressed in terms of the oxygen consumed in water at a temperature of 68°F (20°C) during a five-day period. The BOD is used as an indication of the degree of water pollution. See also METABOLISM.

Bioluminescence The enzyme-catalyzed production of light by living organisms, typically during mating or hunting. This word literally means "living light." Bioluminescence was first identified/analyzed in 1947, by William McElroy. Bioluminescence results when the enzyme luciferase comes into contact with adenosine triphosphate (ATP), inside the photophores (organs which emit the light) of the organism. Such production of light by living organisms is exemplified by fireflies, South America's railroad worm, and by many deep ocean marine organisms.

Bioluminescence has been utilized by man as a genetic marker (e.g., to cause a genetically engineered plant to glow as evidence that a gene was successfully transferred into that plant). Another use of bioluminescence by man is for the rapid detection of foodborne pathogenic bacteria (e.g., in a food processing factory). One rapid-test for bacteria uses two chemical reagents that first break down bacteria cell membranes, then cause the ATP from those broken cells to luminesce. Another rapid-test uses a electrophoresis to first separate the sequences of bacteria's DNA (following its extraction from cell and enzymatic fragmentation), cause those separated sequences to luminesce, then a camera is used to record the sequence-pattern light emission and compare that pattern to patterns-of-pathogenic-bacteria previously stored in a database. See also ENZYME, MARKER (GENETIC MARKER), BACTERIA, TOXIN, PATHOGENIC, *ESCHERICHIA COLIFORM* 0157:H7 *(E. COLI* 0157:H7*)*, CELL, ADENOSINE TRIPHOSPHATE, GENETIC ENGINEERING, ELECTROPHORESIS, POLYACRYLAMIDE GEL ELECTROPHORESIS (PAGE), SEQUENCE (OF A DNA MOLECULE), and RESTRICTION ENDONUCLEASES.

Biomass All organic matter grown by the photosynthetic conversion of solar energy (e.g., plants), and organic matter from animals.

Biomimetic Materials Synthetic (i.e., man-made) molecules or systems that are analogues of natural (i.e., made by living organisms) materials. For instance, molecules have been synthesized by man that act chemically like natural proteins, but are not as easily degraded by the digestive system (as are those natural protein molecules). Other systems such as reverse micelles and/or liposomes exhibit certain properties that mimic certain aspects of living systems. See also PROTEIN, DIGESTION

(WITHIN ORGANISMS), REVERSE MICELLE (RM), LIPOSOMES, ANALOGUE, and BIOPOLYMER.

Biomolecular Electronics See BIOELECTRONICS.

Bionics An interscience discipline for constructing artificial systems that resemble or have the characteristics of living systems.

Biophysics An area of scientific study in which physical principles, physical methods, and physical instrumentation are used to study living systems or systems related to life. It overlaps with biophysical chemistry, which is more specialized in scope since it is concerned with the physical study of chemically isolated substances found in living organisms.

Biopolymer A high molecular weight organic compound found in nature, whose structure can be represented by a repeated small unit [i.e., monomer (links)]. Common biopolymers include cellulose (long-chain sugars found in most plants and the main constituent of dried woods, jute, flax, hemp, cotton, etc.) and proteins in general and specifically collagen and gelatin. See also MOLECULAR WEIGHT, PROTEIN, and POLYMER.

Biorecovery The use of microorganisms (including bacteria, fungi, and algae) in the recovery of (collecting of) various metals and/or organic compounds from ores or garbage (other matrices). See also BIOLEACHING, CONSORTIA, and BIOSORBENTS.

Biosafety See CONVENTION ON BIOLOGICAL DIVERSITY (CBD).

Biosafety Protocol See CONVENTION ON BIOLOGICAL DIVERSITY (CBD), INTERNATIONAL PLANT PROTECTION CONVENTION (IPPC).

Bioseeds Plant seeds produced via genetic engineering of existing plants. See also GENETIC ENGINEERING, BIOLISTIC® GENE GUN, HERBICIDE-TOLERANT CROP, PAT GENE, EPSP SYNTHASE, ALS GENE, CP4 EPSPS, GLYPHOSATE OXIDASE, CHOLESTEROL OXIDASE, HIGH-LYSINE CORN, HIGH-METHIONINE CORN, HIGH-PHYTASE CORN (OR HIGH-PHYTASE SOYBEANS), HIGH-STEARATE SOYBEANS, LOW-STACHYOSE SOYBEANS, LOX NULL, PLANT'S NOVEL TRAIT (PNT), "SHOTGUN" METHOD (TO INTRODUCE FOREIGN NEW GENES INTO PLANT CELLS), *BACILLUS THURINGIENSIS (B.t.)*, *B.t. KURSTAKI*, *B.t. TENEBRIONIS*, *B.t. ISRAELENSIS*, CRY PROTEINS, CRY1A (b) PROTEIN, CRY1A (c) PROTEIN, and CRY9C PROTEIN.

Biosensors (chemical) Chemically based devices that are able to detect and/or measure the presence of certain molecules (e.g., DNA, antigens, pesticides, etc.). These devices are currently created in the following forms:

(1) A two-part diagnostic test that can detect the presence of trace amounts of specific chemicals (e.g., pesticides). The (chemical) biosensor consists of an immobilized enzyme (to bind the trace chemical) combined with a color reagent (to indicate visually the presence of the trace chemical).

(2) A one-part test that can detect specific DNA segments in complex ("dirty," multiple component) samples. The biosensor consists of 13-nm gold particles onto which are attached numerous nucleotide "molecular chains." Each "nucleotide chain" contains twenty-eight nucleotides. The thirteen nucleotides that are closest to each gold particle serve as a "spacer," and solutions containing such (spaced) randomly distributed gold particles appear red in color when illuminated by light.

The fifteen nucleotides that are farthest from each gold particle are chosen to be complementary to, and thus bind to (complementary), nucleotide sequences in the target (e.g., DNA) molecule. In the presence of the specific target molecule, a closely linked network of gold particles and double-stranded nucleotide molecular chains forms (overcoming the 13-nucleotide "spacer" which previously held apart the gold particles). When double-stranded chains form (i.e., target molecule is present), the distance between gold particles becomes less than the size of those particles, which makes the solution containing (bound) particles appear blue in color when illuminated by light.

See also ENZYME, IMMUNOASSAY, NANOCRYSTAL MOLECULES, NANOTECHNOLOGY, DEOXYRIBONUCLEIC ACID (DNA), NANOMETERS (nm), ANTIGEN, SEQUENCE (OF A DNA MOLECULE), NUCLEOTIDE, POLYMER, COMPLEMENTARY DNA, DOUBLE HELIX, DUPLEX, and SELF-ASSEMBLY (OF A LARGE MOLECULAR STRUCTURE).

Biosensors (electronic) Electronic sensors that are able to detect and measure the presence of biomolecules such as sugars or DNA segments. Currently created by:

(a) Fusing organic matter (e.g., enzymes, antibodies, receptors, or nucleic acids) to tiny electrodes; yielding devices that convert natural chemical reactions into electric current to measure blood levels of certain chemicals (e.g., glucose or insulin), control functions in an artificial organ, monitor some industrial processes, act as a robot's "nose," etc.

(b) Fusing organic matter (e.g., segment of DNA, antibody, enzyme, etc.) onto the surfaces of etched silicon wafers; yielding devices that convert supramolecular interactions [e.g., nucleotide hybridization, enzyme-substrate binding, lectin-carbohydrate (sugar) interactions, antibody-antigen binding, host-guest complexation, etc.] into electric current via a charge-coupled device (CCD) detector that measures the shift in interference pattern caused by change in refractive index that results when (sensed) molecule tightly binds to the fused organic matter. For such an etched-silicon-wafer biosensor, the nucleotide hybridization (binding) enables the detection of femtomolar (10^{-15} mole or 0.000000000000001) concentrations of DNA. If the (sensed) DNA segment is not complementary to the fused DNA segment, there is no significant change in the interference pattern.

A major future goal is to build future generations of biosensors directly into computer chips. (Researchers have discovered that proteins can replace certain metals in semiconductors.) This would enable low-cost mass production via processes similar to those now used for existing semiconductor chips, with circuits built right into the sensor to process data picked up by the biological matter on the chip. See also BIOELECTRONICS, ENZYME, GENOSENSORS, RECEPTORS, ANTIBODY, INSULIN, COMBINATORIAL CHEMISTRY, SUBSTRATE (CHEMICAL), LECTINS, SUGAR MOELCULES, CARBOHYDRATES (SACCHARIDES), GLUCOSE (GLc), DEOXYRIBONUCLEIC ACID (DNA), NUCLEOTIDE, HYBRIDIZATION (MOLECULAR GENETICS), HYBRIDIZATION SURFACES, ANTIGEN, COMPLEMENTARY DNA (c-DNA), GENE, NANOTECHNOLOGY, and TEMPLATE.

Biosilk A biomimetic, man-made fiber produced by: A. sequencing the "dragline silk" protein that is produced by the orb-weaving spider; B. synthesizing genes to code for the "dragline silk" protein (components); C. expressing those genes in a suitable host (i.e., yeast, bacteria) to cause production of the protein(s); D. dissolving the protein in a solvent, and then "spinning" the protein into fiber form by passing the liquid (dissolved protein) through a small orifice, followed by drying to remove the solvent. See also BIOMIMETIC MATERIALS, BIOPOLYMER, PROTEIN, SEQUENCING (OF PROTEIN MOLECULES), GENE, GENE MACHINE, SYNTHESIZING (OF DNA MOLECULES), DEOXYRIBONUCLEIC ACID (DNA), EXPRESS, and SUPERCRITICAL CARBON DIOXIDE.

Biosorbents Microorganisms which, either by themselves or in conjunction with a support/substrate system (e.g., inert granules) effect the extraction (e.g., from ore) and/or concentration of desired (precious) metals or organic compounds by means of selective retention of those entities. Retention of organic compounds (e.g., gasoline) may be for the purpose of cleaning polluted soil. See also BIORECOVERY, BIOLEACHING, and CONSORTIA.

Biosphere All the living matter on or in the earth, the oceans and seas, and the atmosphere. The area of the planet in which life is found to occur.

Biosynthesis Production of a chemical compound or entity by a living organism.

Biotechnology The means or way of manipulating life forms (organisms) to provide desirable products for man's use. For example, beekeeping and cattle breeding could be considered to be biotechnology-related endeavors. The word *biotechnology* was coined in 1919 by Karl Ereky, to apply to the interaction of biology with human technology.

However, usage of the word biotechnology in the United States has come to mean all parts of an industry that knowingly create, develop, and market a variety of products through the willful manipulation, on a molecular level, of life forms or utilization of knowledge pertaining to living systems.

A common misconception is that biotechnology refers only to recombinant DNA (rDNA) work. However, recombinant DNA is only one of the many techniques used to derive products from organisms, plants, and parts of both for the biotechnology industry. A list of areas covered by the term biotechnology would more properly include: recombinant DNA, plant tissue culture, rDNA or gene splicing, enzyme systems, plant breeding, meristem culture, mammalian cell culture, immunology, molecular biology, fermentation, and others. See also GENETIC ENGINEERING, BIORECOVERY, RECOMBINANT DNA, RECOMBINATION, DEOXYRIBONUCLEIC ACID (DNA), BIOLEACHING, GENE SPLICING, MAMMALIAN CELL CULTURE, and FERMENTATION.

Biotechnology Industry Organization (BIO) An American trade association composed of companies and individuals involved in biotechnology and in services to biotechnology companies (e.g., accounting, law, etc.). Formed in 1993, the BIO was created by the merger of its two predecessor trade associations; the Association of Biotechnology Companies (ABC) and the Industrial Biotechnology Association (IBA). The BIO works with the government and the public to promote safe and rational advancement of genetic engineering and biotechnology. See also BIOTECHNOLOGY, ASSOCIATION OF BIOTECHNOLOGY COMPANIES (ABC), INDUSTRIAL BIOTECHNOLOGY ASSOCIATION (IBA), JAPAN BIO-INDUSTRY ASSOCIATION, and SENIOR ADVISORY GROUP ON BIOTECHNOLOGY (SAGB).

Biotransformation (of a biosynthesized product) See POST-TRANSLATIONAL MODIFICATION OF PROTEIN.

Biotransformation (of an introduced compound) See the *biological* portion of definition of PERSISTENCE.

Black-layered (corn) An indicator of a corn plant's maturity. It refers to a distinctive dark line that forms in each corn kernel at maturity. See CORN.

Black-lined (corn) See BLACK-LAYERED (CORN).

Blast Cell A large, rapidly dividing cell that develops from a B cell (B lymphocyte) in response to an antigenic stimulus. The blast cell then becomes an antibody-producing plasma cell. See also ANTIGEN, ANTIBODY, B LYMPHOCYTES, and LYMPHOCYTE.

Blast Transformation The process via which a B cell (B lymphocyte) becomes a blast cell. See also ANTIBODY, B LYMPHOCYTES, and BLAST CELL.

Blood-Brain Barrier (BBB) The specialized layer of endothelial cells that line all blood vessels in the brain. The BBB prevents most organisms (e.g., bacteria) and toxins from entering the brain via the bloodstream. However, the BBB does allow oxygen and needed nutrients (e.g., iron tryptophan, etc.) to enter the brain from the bloodstream. For example, receptors that line BBB cell surfaces (on the bloodstream side of the BBB) "latch onto" transferrin molecules (which contain iron molecules) as those transferrin molecules pass by in the bloodstream. These TRANSFERRIN RECEPTORS first bind to the (passing) transferrin molecules, transport those transferrin molecules through the BBB via a process called vaginosis, then release those transferrin molecules (in order to supply needed iron to the brain cells). Factors such as aging, trauma, stroke, multiple sclerosis, and some infections will cause an increase in the permeability of the BBB. See also ENDOTHELIAL CELLS, TOXIN, TRANSFERRIN, TRANSFERRIN RECEPTOR, CHELATING AGENT, RECEPTORS, VAGINOSIS, HEME, BACTERIA, TRYPTOPHAN (trp), and SEROTONIN.

Blood Clotting See FIBRIN.

Blood Derivatives Manufacturing Association A trade organization of firms involved in producing pharmaceuticals from collected blood. See also SERUM, BUFFY COAT (CELLS), and SEROLOGY.

Blood Plasma See PLASMA.

Blood Platelets See PLATELETS.

Blood Serum See SERUM.

Blunt-End DNA A segment of DNA that has both strands terminating at the same base pair location, that is, fully base-paired DNA. No sticky ends. See also STICKY ENDS.

Blunt-End Ligation A method of joining blunt-ended DNA fragments using the enzyme T4 ligase which can join fully base-paired, double-stranded DNA. See also LIGASE, DEOXYRIBONUCLEIC ACID (DNA), BASE PAIR (bp), and BLUNT-END DNA.

BLUP See BEST LINEAR UNBIASED PREDICTION.

BOD See BIOLOGICAL OXYGEN DEMAND.

Boletic Acid See FUMARIC ACID.

Bollworms See *HELIOTHIS VIRESCENS, HELICOVERPA ZEA, PECTINOPHORA GOSSYPIELLA,* and *B.t. KURSTAKI.*

Bone Morphogenetic Proteins (BMP) A family of proteinaceous growth factors (nine identified as of 1994) for bone tissue formation (e.g., at the site where a bone has been broken). BMPs stimulate a "recruitment" of bone-forming cells (e.g., to the site of bone injury) which first form cartilage, then that cartilage is mineralized to form bone. See also GROWTH FACTOR, PERIODONTIUM, and PROTEIN.

Bovine Somatotropin (BST) Also called bovine growth hormone. A protein hormone, produced in a cow's pituitary gland, that increases the efficiency of the cow in converting its feed into milk. Increases milk production in cows, and promotes cell growth in healing tissues of all ages of cattle. Promotes body growth of young cattle. See also PROTEIN, GROWTH HORMONE (GH), HORMONE, SOMATOMEDINS, and SPECIES SPECIFIC.

bp Common abbreviation for base pair. See also BASE PAIR (bp).

Brassica A fast-growing category of the mustard plant family, which also produces sulfur-based gases (a natural defense against certain fungi and insect pests). For example, Australian CSIRO scientists discovered in 1994 that sulfur-based isothiocyanates emitted by *Brassica* actively combat Wheat Take-All Disease (a fungal disease that attacks the roots of the wheat plant). See also *ARABIDOPSIS THALIANA,* WHEAT TAKE-ALL DISEASE, and FUNGUS.

Brassica campestre See *BRASSICA*.

Brassica campestris See CANOLA, *BRASSICA*.

Brassica napus See CANOLA, *BRASSICA*.

Bright Greenish-Yellow Fluorescence (BGYF) An indication of the presence of fungus (e.g., in a sample of grain), when light of an appropriate wavelength is shone on sample. For example, when the fungus *Aspergillus flavus* infects cottonseed during boll development on the cotton plant, the resultant seed (when harvested) shows BGYF on its lint and linters. That fungus gains entry into the bolls typically via holes made by the pink bollworm (*Pectinophora gossypiella*). See MYCOTOXINS, AFLATOXIN, FUNGUS, PECTINOPHORA GOSSYPIELLA, and FLUORESCENCE.

Broad Spectrum See GRAM STAIN.

Broth A fluid culture medium (for growing microorganisms). See also MEDIUM and CULTURE MEDIUM.

BSE Bovine spongiform encephalopathy. A neurodegenerative disease of cattle. See PRION.

BSP Biosafety protocol. See CONVENTION ON BIOLOGICAL DIVERSITY (CBD).

BST See BOVINE SOMATOTROPIN (BST).

B.t See *BACILLUS THURINGIENSIS (B.t.)*.

B.t.k. See *B.t. KURSTAKI*.

B.t. israelensis One of the approximately 30 subspecies groupings within the approximately 20,000 different strains of the soil bacteria known (collectively) as *Bacillus thuringiensis (B.t.)*. When eaten (e.g., due to presence on food), the protoxin proteins produced by *B.t. israelensis* are toxic to mosquitoes and black fly (Diptera) larvae. See *BACILLUS THURINGIENSIS (B.t.)*, and PROTOXIN.

B.t. kurstaki One of the approximately 30 subspecies groupings within the approximately 20,000 different strains of the soil bacteria known (collectively) as *Bacillus thuringiensis (B.t.)*. When eaten (e.g., as part of a genetically engineered plant), the protoxin proteins produced by *B.t.*

kurstaki are toxic to certain caterpillars (Lepidoptera larvae), such as the European corn borer (pyralis). See also *BACILLUS THURINGIENSIS (B.t.)*, PROTOXIN, CRY1A (b) PROTEIN, and EUROPEAN CORN BORER.

B.t. tenebrionis One of the approximately 30 subspecies groupings within the approximately 20,000 different strains of the soil bacteria known (collectively) as *Bacillus thuringiensis (B.t.)*. When eaten (e.g., as part of a genetically engineered plant), the protoxin proteins produced by *B.t. tenebrionis* are toxic to certain insects. See *BACILLUS THURINGIENSIS (B.t.)*, and PROTOXIN.

Buffy Coat (cells) The layer of white blood cells (leukocytes) that separates out when blood is subjected to centrifugation. See also ULTRACENTRIFUGE, LEUKOCYTES (WHITE BLOOD CELLS), PLASMA, and BLOOD DERIVATIVES MANUFACTURING ASSOCIATION.

Bundesgesundheitsamt (BGA) German Federal Health Organization. The German Government agency that must approve new pharmaceutical products for sale within Germany, it is the equivalent of the U.S. Food and Drug Administration (FDA). See also FOOD AND DRUG ADMINISTRATION (FDA), KOSEISHO, COMMITTEE FOR PROPRIETARY MEDICINAL PRODUCTS (CPMP), COMMITTEE ON SAFETY IN MEDICINES, MEDICINES CONTROL AGENCY (MCA), and EUROPEAN MEDICINES EVALUATION AGENCY (EMEA).

C

C Value The total amount of DNA in a haploid genome. See also DEOXYRIBONUCLEIC ACID (DNA), HAPLOID, and GENOME.

c-DNA See COMPLEMENTARY DNA.

C-DNA Also known as copy DNA. A helical form of DNA. It occurs when DNA fibers are maintained in 66 percent relative humidity in the presence of lithium ions. It has fewer base pairs per turn than B-DNA. See also B-DNA, DEOXYRIBONUCLEIC ACID (DNA), and COMPLEMENTARY DNA (c-DNA).

Cadherins A class of adhesion molecules that function as cellular adhesion receptors. See ADHESION MOLECULE and RECEPTORS.

Calcium Channel-Blockers Drugs (e.g., verapamil or nifedipine) that are used to slow down calcium movement through cell membranes. This leads to dilation of the blood vessels and reduces the heart's workload. Blood vessels need calcium to contract (causing flow constriction and hence an increase in blood pressure), so the drug-induced shortage of available calcium causes the body's blood vessels to remain dilated (which results in lower blood pressure).

Research in 1996 indicated the possibility that certain types of calcium channel-blockers might lead to increased rates of some cancers. If so, this is likely due to the drug preventing enough calcium availability for normal apoptosis in body cells. See also CELL, CANCER, MEMBRANE TRANSPORT, and APOPTOSIS.

Calcium Oxalate A crystalline salt that is normally deposited in the cells of many plants. In many animals, calcium oxalate is excreted in the urine, or retained by the animal's body in the form of urinary calculi. See OXALATE and CELL.

Callipyge (means "beautiful buttocks" in the Greek language) An inherited trait in livestock (e.g., sheep) that results in thicker, meatier hindquarters. First identified as a genetic trait in 1983, this desirable trait results in a higher meat yield per animal. See also TRAIT, GENOTYPE, PHENOTYPE, and WILD TYPE.

Callus An undifferentiated cluster of plant cells that is a first step in regeneration of plants from tissue culture. See also SOMOCLONAL VARIATION.

Calorie The amount of heat (energy) required to raise the temperature of one gram of water from 14.5°C (58°F) to 15.5°C (60°F) at a constant pressure of one standard atmosphere.

CaMV See CAULIFLOWER MOSAIC VIRUS 35S PROMOTER.

CaMV 35S See CAULIFLOWER MOSAIC VIRUS 35S PROMOTER.

Canavanine An uncommon amino acid. It is used in biology as an arginine (another amino acid) analogue. It is a potent growth inhibitor of many organisms. See also AMINO ACID and BIOMIMETIC MATERIALS

Cancer The name given to a group of diseases that are characterized by uncontrolled cellular growth (e.g., formation of tumor). See also ONCOGENES, TUMOR-SUPPRESSOR GENES, TUMOR, TELOMERES, RETINOIDS, TELOMERASE, and NEOPLASTIC GROWTH.

CANDA Computer Assisted New Drug Application. An application to the U.S. Food and Drug Administration (FDA) seeking approval of a drug that has undergone Phase 2 and Phase 3 clinical trials. A CANDA is submitted in the form of computer-readable (e.g., clinical) data that provides the FDA with a sophisticated database that allows the FDA reviewers to evaluate (e.g., statistically) the data themselves, directly. See also NDA (TO FDA), NDA (TO KOSEISHO), FOOD AND DRUG ADMINISTRATION (FDA), MAA, and PHASE I CLINICAL TESTING.

Canola *Brassica napus* or *Brassica campestris* strains of the rapeseed plant, which were developed by plant breeders after the 1960's. This was because oil produced from rapeseed grown prior to 1971 contained 30%–60% erucic acid (high dietary levels of which were associated with cardiac lesions in experimental animals via toxicology testing).

By 1974, canola varieties producing oil containing less than 5% erucic acid constituted virtually all of that year's Canadian rapeseed crop, and Canadian breeders continued to develop new canola varieties with ever-lower erucic acid content.

In 1982, Canada filed with the U.S. Food and Drug Administration (FDA) to have low-erucic-acid rapeseed (LEAR) oil affirmed to be GRAS (Generally Recognized As Safe) which the FDA did. LEAR was one of the first foodstuffs to be determined to be "substantially equivalent" under the OECD-defined criteria for "substantial equivalence" because LEAR was shown (in OECD petition) to be very similar to, and composed of the same basic components as, traditional rapeseed oil (and other commonly consumed vegetable oils) except for a lower level of erucic acid (the component of concern, per above). See also STRAIN, FATTY ACID, OLEIC ACID, GRAS LIST, and ORGANIZATION FOR ECONOMIC COOPERATION AND DEVELOPMENT (OECD).

CAP Catabolite gene-activator protein, also known as CRP, catabolite regulator protein (or cyclic AMP receptor protein). The protein mediates the action of cyclic AMP (cAMP) on transcription in that cAMP and CAP must first combine. The cAMP-CAP complex then binds to the promoter regions of *Escherichia coli* and stimulates transcription of its operon. Since a cell component increases rather than inhibits transcription, this type of regulation of gene expression is called positive transcriptional control. In contrast see CATABOLITE REPRESSION. See also *ESCHERICHIA COLI (E. COLI)*, TRANSCRIPTION, and OPERON.

Capsid The external protein coat of a virus particle that surrounds the nucleic acid. The individual proteins that make up the capsid are called capsomers or protein subunits. It has been discovered that resistance to certain viral diseases may be imparted to some plants by inserting the gene for production of the protein coat into the plants. See also TOBACCO MOSAIC VIRUS (TMV), VIRUS, and PROTEIN.

Capsule An envelope surrounding many types of microorganisms. The capsule is usually composed of polysaccharides, polypeptides, or polysaccharide-protein complexes. These materials are arranged in a compact manner around the cell surface. Capsules are not absolutely essential cellular components. See MICROORGANISM, POLYSACCHARIDES, POLYPEPTIDE (PROTEIN), PROTEIN, CELL, GRAM NEGATIVE (G–), MANNANOLIGOSACCHARIDES (MOS), and GRAM POSITIVE (G+).

CARB See CENTER FOR ADVANCED RESEARCH IN BIOTECHNOLOGY (CARB).

Carbetimer An antineoplastic (i.e., anticancer) low molecular weight polymer that acts against several types of cancer tumors, perhaps via stimulation of the patient's immune system. It has minimal toxicity.

Carbohydrate Engineering The selective, deliberate alteration/creation of carbohydrates (and the oligosaccharide side chains of glycoprotein molecules) by man. See also GLUCONEOGENESIS, GLYCOBIOLOGY, GLYCOFORM, GLYCOLIPID, GLYCOLYSIS, GLYCOPROTEIN, GLYCOSIDASES, RESTRICTION ENDOGLYCOSIDASES, GLYCOSIDE, and GLYCOSYLATION (TO GLYCOSYLATE).

Carbohydrates (saccharides) A large class of carbon-hydrogen-oxygen compounds. Monosaccharides are called simple sugars, of which the most abundant is D-glucose. It is both the major fuel for most organisms and constitutes the basic building block of the most abundant polysaccharides, such as starch and cellulose. While starch is a fuel source, cellulose is the primary structural material of plants. Carbohydrates are produced by photosynthesis in plants. Most, but not all, carbohydrates are represented chemically by the formula $Cx(H_2O)n$, where *n* is three or higher. On the basis of their chemical structures, carbohydrates are classified as polyhydroxy aldehydes, polyhydroxy ketones, and their derivatives. See also GLUCOSE (GLc), GLYCOGEN, MONOSACCHARIDES, OLIGOSACCHARIDES, and POLYSACCHARIDES.

Carcinogen A cancer-causing agent. See also PROTO-ONCOGENES.

Carotenoids A general term for a group of plant-produced pigments ranging in color from yellow to red and brown. The carotenes and the xanthophylls, orange to yellow in color, are the most common. Carotenoids are responsible for the coloration of certain plants (e.g., the carrot) and of some animals (e.g., the lobster). The carotenoid pigments are transferred to animals as an element in their foods. Carotenoids are composed of isoprene units (usually eight) which may be modified by the addition of other chemical groups on the molecule. The carotenes are of importance to higher animals because they are utilized in the formation of vitamin A. See also VITAMIN and BETA CAROTENE.

Cartilage-Inducing Factors A and B Compounds produced by the body which also have immunosuppressive activity. See also IMMUNOSUPPRESSIVE.

Cassette A "package" of genetic material (containing more than one gene) that is inserted into the genome of a cell via gene splicing techniques. May include promoter(s), leader sequence, termination codon, etc. See also GENE SPLICING, LEADER SEQUENCE, PROMOTER, GENETIC CODE, TERMINATION CODON (SEQUENCE), GENETIC ENGINEERING, TRANSGENE, and GENOME.

Catabolism Energy-yielding pathway. The phase of metabolism involved in the energy-yielding degradation of nutrient (food) molecules. See also DISSIMILATION and METABOLISM.

Catabolite Activator Protein See CAP.

Catabolite Repression Common in bacteria. The decreased expression of catabolic enzymes as brought about by a catabolite such as glucose. For example, glucose is the preferred fuel source for certain bacteria and when it is present in the culture medium it represses the formation of enzymes that are required for the utilization of other fuel sugars, such as for example β-galactosidase. Since glucose or other catabolites (other molecules derived from glucose) cause the repression, it is known as catabolite repression. See also CAP, OPERON, GLUCOSE (GLc), and ADENOSINE MONOPHOSPHATE (AMP).

Catalase An enzyme that catalyzes the very rapid decomposition of hydrogen peroxide to water and oxygen. Catalase is in the group of enzymes known as metalloenzymes because it requires the presence of a metal in order to be catalytically active. The metal (known as a cofactor) is, in the case of catalase, iron. Found in both plants and animals. See also HYDROLYSIS, HUMAN SUPEROXIDE DISMUTASE (hSOD), PEG-SOD (POLYETHYLENE GLYCOL SUPEROXIDE DISMUTASE).

Catalyst Any substance (entity), either of protein or of nonproteinaceous nature, that increases the rate of a chemical reaction, without being consumed itself in the reaction. In the biosciences, the term "enzyme" is used for a proteinaceous catalyst. Enzymes catalyze biological reactions. See also ENZYME, CATALYTIC SITE, ACTIVE SITE, CATALYTIC ANTIBODY, and SEMISYNTHETIC CATALYTIC ANTIBODY.

Catalytic Antibody An antibody that is produced (e.g., via monoclonal antibody techniques) in response to a carefully selected antigen (e.g., target molecule in bloodstream, or molecule involved in chemical reaction of interest) which itself catalyzes the "splitting" of a molecule in the bloodstream (e.g., heroin into two harmless small molecules) or mimics:

(1) Restriction endonucleases that cleave (cut) proteins or DNA molecules precisely at specific locations on those molecules
(2) Restriction endoglycosidases that are capable of cleaving oligosaccharides or polysaccharide molecules precisely at specific locations on those molecules
(3) Transition state chemical complex in the chemical reaction that is to be catalyzed—resultant antibody acts both as an antibody (to the selected transition-state-complex antigen) and as a catalyst (for the chemical reaction possessing that selected transition state chemical complex)

This catalyst (enzyme) thus possesses the remarkable specificity of an antibody (i.e., specific only to the desired transition-state reactant) which holds the potential to yield chemical reaction products of greater purity than those achieved via current (less specific) catalysts. Because the immune system will (in theory) produce an antibody to virtually every molecule of sufficient size to be detected by the immune system (i.e., 6 to 34 Angstroms), it should be possible to raise catalytic antibodies for a large number of industrial chemical reactions that are currently catalyzed via conventional (less specific) catalysts. See also OLIGOSACCHARIDES, CATALYST, ANTIBODY, RESTRICTION ENDONUCLEASES, RESTRICTION ENDOGLYCOSIDASES, MONOCLONAL ANTIBODIES (MAb), ANTIGEN, TRANSITION STATE (IN A CHEMICAL REACTION), PROTEIN, ACTIVATION ENERGY, SEMISYNTHETIC CATALYTIC ANTIBODY, ANGSTROM (Å), and ABZYMES.

Catalytic Site The site (geometric area) on an enzyme molecule (or other catalyst) that is actually involved in the catalytic process. The catalytic site usually consists of a small portion of the total area of the enzyme. See also CATALYST, ENZYME, ACTIVE SITE, and CATALYTIC ANTIBODY.

Catecholamines Hormones (such as adrenalin) that are amino derivatives of a base structure known as catechol. Catecholamines are released into the bloodstream by exercise, and act as natural tranquilizers. See also ENDORPHINS and HORMONE.

Cation See ION, CHELATION, and CHELATING AGENT.

Cauliflower Mosaic Virus 35S Promoter (CaMV 35S) A promoter (sequence of DNA) that is often utilized in genetic engineering to control expression of (inserted) gene; i.e., synthesis of desired protein in a plant. See also VIRUS, PROMOTER, DEOXYRIBONUCLEIC ACID (DNA), GENE, GENETIC ENGINEERING, and PROTEIN.

CBD See CONVENTION ON BIOLOGICAL DIVERSITY (CBD).

CCC DNA A covalently linked circular DNA molecule, such as a plasmid. See also DEOXYRIBONUCLEIC ACID (DNA) and PLASMID.

CD4 EPSPS See EPSP SYNTHASE and CP4 EPSPS.

CD4 EPSP Synthase See EPSP SYNTHASE and CP4 EPSPS.

CD4-PE40 An experimental drug discovered in 1988 by Ira Pastan and Bernard Moss that has indicated potential to combat acquired immune deficiency syndrome (AIDS). CD4-PE40 is a conjugated protein consisting of a CD4 protein (molecule) attached to *Pseudomonas* exotoxin (a substance produced by *Pseudomonas* bacteria that is toxic to certain living cells). The gp 120 glycoprotein on the surface of the HIV (i.e., AIDS) virus attaches preferentially to the CD4 portion of this immunoconjugate, and the virus is inactivated by the *Pseudomonas* exotoxin portion of this immunoconjugate. See also PROTEIN, CD4 PROTEIN, GP120 PROTEIN, SOLUBLE CD4, IMMUNOTOXIN, CONJUGATED PROTEIN, ACQUIRED IMMUNE DEFICIENCY SYNDROME (AIDS), HUMAN IMMUNODEFICIENCY VIRUS (TYPE 1 and TYPE 2), RICIN, and ABRIN.

CD4 Protein An adhesion molecule (protein) imbedded in the outer wall (envelope) of human immune system and brain cells that functions as the receptor (door to entry into the cell) for the HIV (AIDS) virus. The gp120 envelope glycoprotein of the HIV (i.e., AIDS virus) directly interacts with the CD4 protein on the surface of helper T cells to enable the virus to invade the helper T cells. See also T CELL RECEPTORS, ADHESION MOLECULES, GP120 PROTEIN, and SOLUBLE CD4.

CD44 Protein One of the adhesion molecules (embedded in the surface of the linings of blood vessels) that assists the neutrophils on their journey from the bloodstream through the walls of blood vessels (e.g., to combat pathogens into adjacent tissues). Tumor cells also exploit CD44 molecules in order to metastasize (spread throughout the body's tissue from a single beginning tumor) via a similar (tumor cell)-through-blood vessel-wall adhesion molecule mechanism. See also ADHESION MOLECULES, CD4 PROTEIN, PROTEIN, NEUTROPHILS, PATHOGEN, TUMOR, CANCER, and SOLUBLE CD4.

CD95 Protein Also called APO-1/Fas, it is a transmembrane protein (embedded within the surface membrane of the cell) that transmits apoptosis ("programmed" cell death) "signal" into cells. Transduction of that apoptosis signal occurs when certain ligands or antigens (i.e., the APO-1/Fas antigen) bind to the extracellular (i.e., portion outside of cell membrane) part (i.e., receptor) of the CD95 protein. See APOPTOSIS, PROTEIN, CELL, SIGNAL TRANSDUCTION, SIGNALLING, NUCLEAR RECEPTORS, ANTIGEN, RECEPTOR, and FUSARIUM.

Cecrophins (lytic proteins) Proteins produced by certain white blood cells [called cytotoxic T lymphocytes (CTL) or killer T cells]. The proteins allow lysis (i.e., bursting) of infected cells. Cecrophins are amphopathic (i.e., contain both a hydrophobic region and a hydrophilic region); and work by "worming" the hydrophobic portion into the cell membrane (so the hydrophobic portion of the cecrophin molecule is out of the water). This creates a transmembrane pore (i.e., a hole in the membrane) which is lined with the cecrophin's hydrophilic portion. Membranes function simply to separate various components. This separation is required for life to exist. When holes are introduced into cell membranes, water rushes into the targeted cell due to differences in osmotic pressure and the cell ruptures (explodes). T cecrophins are only able to lyse (i.e., burst) infected cells because only "sick" cells have a weakened cytoskeleton (located just inside the cell membrane), which cannot prevent the contents of the cell from spilling out through the pores (created by cecrophins). See also HELPER T CELLS (T4 CELLS), PATHOGEN, COMPLEMENT (COMPONENT OF IMMUNE SYSTEM), HYDROPHOBIC, HYDROPHILIC, COMPLEMENT CASCADE, LYSE, and LYSIS.

Cell The fundamental self-containing unit of life. The living tissue of every multicelled organism is composed of these fundamental living units. Certain organisms may consist of only one cell, such as yeast or bacteria, protozoa, some algae, and gametes (the reproductive stages) of higher organisms (see MICROBIOLOGY). Larger organisms are subdivided into organs that are relatively autonomous but cooperate in the functioning of that plant or animal. Unicellular (i.e., single-cell) organisms perform all life functions within the one cell. In a higher organism (i.e., a multicellular organism), entire populations of cells (i.e., an organ) may be designated a particular specialized task (e.g., the heart to facilitate circulation). The cells of muscle tissue are specialized for movement and those of bone and connective tissue, for structural support. See also GAMETE, GERM CELL, and OOCYTES.

Cell Culture The *in vitro* (i.e., outside of body, in a test tube) propagation of cells isolated from living organisms. See also MAMMALIAN CELL CULTURE, DISSOCIATING ENZYMES, and HARVESTING ENZYMES.

Cell Differentiation The process whereby descendants of a common parental cell achieve and maintain specialization of structure and function. In humans, for instance, all the different types of cells (e.g., muscle cells, bone cells, etc.) differentiate from the simple sperm and egg. In humans, the various blood cell types (e.g., red blood cells, white blood cells, etc.) differentiate from stem cells in the bone marrow. Cell differentiation is caused/triggered/assisted by colony stimulating factors (CSFs), growth factors (GFs), and certain other proteins. See also STEM CELLS, STEM CELL ONE, "HEDGEHOG" CELL-DIFFERENTIATION PROTEINS, ERYTHROCYTES, and LEUKOCYTES.

Cell-Differentiation Proteins The various growth factors and other proteins which cause/assist in cell differentiation. See also CELL DIFFERENTIATION.

Cell Fusion The combining of cell contents of two or more cells to become a single cell. Fertilization is such a process (fusing of gametes' cells). See also GAMETE.

Cell Recognition See ADHESION MOLECULE, SIGNAL TRANSDUCTION, and RECEPTORS.

Cell-Mediated Immunity See CELLULAR IMMUNE RESPONSE.

Cellular Adhesion Molecule See ADHESION MOLECULE.

Cellular Adhesion Receptors See ADHESION MOLECULE, RECEPTORS, INTEGRINS, SELECTINS, and CADHERINS.

Cellular Affinity Tendency of cells to adhere specifically to cells of the same type. This property is lost in some cancer cells. See also CELL, ADHESION MOLECULE, and CELL DIFFERENTIATION.

Cellular Immune Response Also called cell-mediated immunity. The immune response that is carried out by specialized cells, in contrast to the response carried out by soluble antibodies. The specialized cells that make up this group include cytotoxic T lymphocytes (CTL), helper T lymphocytes, macrophages, and monocytes. This system works in concert with the humoral immune response. See also HUMORAL IMMUNITY, T CELLS, T CELL RECEPTORS, PHAGOCYTE, HELPER T CELLS (T4 CELLS), CYTOKINES, and MACROPHAGE.

Cellulase The enzyme that digests cellulose to simple sugars such as glucose. See also ENZYME and DIGESTION (WITHIN CHEMICAL PRODUCTION PLANTS).

Cellulose A polymer of glucose units found in all plant matter; the most abundant biological compound on earth. See also CARBOHYDRATES (SACCHARIDES), GLUCOSE (GLC), and VAN DER WAALS FORCES.

Center for Advanced Research in Biotechnology (CARB) A protein engineering research consortium that was established in Rockville, Maryland during 1989 by the U.S. Government, the University of Maryland, and local government. See also PROTEIN ENGINEERING.

Central Dogma The historical organizing principle of molecular genetics; it states that genetic information flows from DNA to RNA to protein—or, stated in another way: DNA makes RNA which makes protein. This principle was first stated by Watson and Crick. It is, however, not rigorously accurate as illustrated by the facts that:

(a) The enzyme reverse transcriptase produces ("makes") DNA using an RNA template.
(b) Prions do not contain any DNA.

See also MOLECULAR GENETICS, COMPLEMENTARY DNA (c-DNA), ENZYME, REPLICATION (OF VIRUS), TRANSCRIPTION, TRANSLATION, DEOXYRIBONUCLEIC ACID (DNA), RIBONUCLEIC ACID (RNA), MESSENGER RNA (mRNA), PRION, and TEMPLATE.

Centrifuge A machine that is used to separate heavier from lighter molecules and cellular components and structures. See also ULTRACENTRIFUGE.

Centromere A constricted region of a chromosome that includes the site of attachment to the mitotic or meiotic spindle. See also CHROMOSOMES, MEIOSIS, CHROMATIN, MITOSIS, KARYOTYPE, and KARYOTYPER.

Cerebrose See GALACTOSE (gal).

CFTR See CYSTIC FIBROSIS TRANSMEMBRANE REGULATOR PROTEIN.

C GIAR See CONSULTATIVE GROUP ON INTERNATIONAL AGRICULTURAL RESEARCH.

cGMP Current Good Manufacturing Practices. The set of current, up-to-date methodologies, practices, and procedures mandated by the Food and Drug Administration (FDA) which are to be followed in the testing and manufacture of pharmaceuticals. The set of rules and regulations promulgated and enforced by the FDA to ensure the manufacture of safe clinical supplies. The cGMP guidelines are more fine-tuned and up to date (technologically speaking) than the more general GMP. See also PHASE I CLINICAL TESTING, IND, and GOOD MANUFACTURING PRACTICES (GMP).

Chakrabarty Decision *Diamond* vs. *Chakrabarty*, U.S. Department of Commerce, 1980; a landmark case in which the U.S. Supreme Court held that the inventor of a new microorganism whose invention otherwise met the legal requirements for obtaining a patent, could not be denied a patent solely because the invention was alive. It essentially allowed the patenting of life forms. See also U.S. PATENT AND TRADEMARK OFFICE (USPTO), and MICROORGANISM.

Channel-Blockers See CALCIUM CHANNEL-BLOCKERS.

Chaperones Protein molecules inside living cells that assist with correct protein folding as the protein molecule emerges from the cell's ribosome. Examples of such chaperone molecules include heat-shock protein 70 and heat-shock protein 40. See also PROTEIN FOLDING, HEAT-SHOCK PROTEINS, PROTEIN, RIBOSOMES, and CELL.

Chaperonins See CHAPERONES, MOLECULAR CHAPERONES, and PROTEIN FOLDING.

CHD Coronary heart disease. See LOW-DENSITY LIPOPROTEINS (LDLP).

Chelating Agent A molecule capable of binding metal atoms. The chelating agent/metal complex is held together by coordination bonds which have a strong polar character. One example of a common chelating agent is ethylenediamine tetraacetate (EDTA) which tightly and reversibly binds Mg^{2+} and other divalent cations (positively charged ions). If a chelate is allowed to bind to metal ions required for enzyme activity, the enzyme will be inactivated (inhibited). Cobalamin (vitamin B_{12}), EDTA and the iron-porphyrin complex of heme (which provides the red color of blood) are other examples of chelates. See also EDTA, CHELATION, HEME, and TRANSFERRIN.

Chelation The binding of metal cations (metal atoms or molecules possessing a positive electrical charge) by atoms possessing unshared electrons (thus the electrons can be "donated" to a bond with a cation). The binding of the metal (cation) to the (electron-excess) chelator atom (ligand) results in formation of a chelator/metal cation complex. The intra-atom bonds thus formed are given the name of coordination bonds.

The properties of the chelator/metal cation complex frequently differ markedly from the "parent" cation. Both carboxylate and amino (molecular) groups readily bind metal cations. One of the most widely used chelators is EDTA (ethylenediamine tetraacetate). It has a strong affinity for metal cations possessing two (bi) or more positive (electrical) charges. Each EDTA molecule binds one metal cation. The EDTA molecule can be visualized as a "hand" (having only four fingers) which grasps the metal cation. Some enzymes (which require metal cations for their activity) are inactivated by EDTA (and other chelators) in that the chelators preferentially remove the metal from the enzyme. See also ION, EDTA, LIGAND (IN BIOCHEMISTRY), CARBOHYDRATES (SACCHARIDES), ENZYME, HEME, CHELATING AGENT, and TRANSFERRIN.

Chemometrics An empirical methodology utilized to (inexpensively) infer a chemical quantity/value from (indirect) measurement(s) of other physical/chemical values (which can be obtained inexpensively).

The term chemometrics was coined in 1975 by Bruce Kowalski. One example of the use of chemometrics is to infer the TME(N) or "true metabolizable energy" of high-oil corn from that corn's protein and oil (fat) content. See HIGH-OIL CORN, TME (N), PROTEIN, and FATS.

Chemopharmacology Therapy (to cure disease) by chemically synthesized drugs. See also PHARMACOLOGY and CISPLATIN.

Chemotaxis Sensing of, and movement toward or away from a specific chemical agent by living, freely moving cells.

Chimera An organism consisting of tissues or parts of diverse genetic constitution. An example of a chimera would be a centaur; the half-man, half-goat figure of Greek mythology. The word "chimera" is from the mythological creature by that name which possessed the head of a lion, the body of a goat, and the tail of a serpent. The word chimera is very general and may be applied to any number of entities. For example, chimeric antibodies may be produced by cell cultures in which the variable, antigen-binding regions are of murine (mouse) origin while the rest of the molecule is of human origin. It is hoped that this combination will lead to an antibody which, when injected, would not give rise to a lesser immune response by the host. See also CHIMERIC DNA and CHIMERIC PROTEINS.

Chimeric DNA Recombinant DNA containing spliced genes from two different species. See also RECOMBINANT DNA (rDNA), and GENETIC ENGINEERING.

Chimeric Proteins Fused proteins from different species, produced from the chimeric DNA template. See also CHIMERA, CHIMERIC DNA, DEOXYRIBONUCLEIC ACID (DNA), and ENGINEERED ANTIBODIES.

Chiral Compound A chemical compound that contains an asymmetrical center and is capable of occurring in two nonsuperimposable mirror images. This phenomenon was first described by Louis Pasteur. "Chiral" is a word that is derived from the Greek *cheir* (meaning "hand").

For example, human hands may be used to illustrate chirality in that when the left and right hands are held one on top of the other, one thumb sticks out on one side while the other thumb sticks out on the other side. The point is that the same number and type of fingers and thumbs exist in both hands, but their arrangement in space may be different. So it is with the arrangement of a given molecule's (e.g., a drug's) atoms in three-dimensional space.

Approximately 40 percent of drugs on the market today consist of chiral compounds. In many chiral drugs, only one type of the molecule is beneficially biologically active (i.e., acts beneficially to control disease, reduce pain, etc.), while the other type of the drug molecule is either inactive or else causes undesired impacts (called "side effects" of the drug mixture). For example, one enantioner of the drug thalidomide is a potent angiogenesis inhibitor, but the other enantiomer causes birth defects in babies of pregnant women taking it. See also STEREOISOMERS, ANGIOGENESIS, OPTICAL ACTIVITY, ENANTIOMERS, and *cis/trans* ISOMERISM.

Chitin A water-insoluble polysaccharide polymer composed of *N*-acetyl-D-glucosamine units which forms the exoskeletons of arthropods (insects) and crustacea. Shellac is produced from chitin. See also POLYSACCHARIDES and POLYMER.

Chitinase An enzyme that degrades chitin. It is produced by certain fungi and actinomycetes that destroy the eggs (shells) of harmful roundworms. See also CHITIN, ENZYME, STRESS PROTEINS, FUNGUS, and AFLATOXIN.

Chloroplast Transit Peptide (CTP) A transit peptide that, when fused to a protein, acts to transport that protein into chloroplast(s) in a plant. Once (both are) inside the chloroplast, the transit peptide is cleaved off the protein and that protein is then free (to do the task it was designed for).

For example, the CP4 EPSPS enzyme in genetically engineered glyphosate-resistant soybean [*Glycine max (L) Merrill*] is transported into the soybean plant's chloroplasts by the CTP known as "N-terminal petunia chloroplast transit peptide." After (both) reach the chloroplast, the CTP is cleaved and degraded, so the CP4 EPSPS is then free to do its task (i.e., confer resistance to glyphosate). See PEPTIDE, CHLOROPLASTS, GATED TRANSPORT (OF A PROTEIN), VESICULAR TRANSPORT (OF A PROTEIN), TRANSIT PEPTIDE, FUSION PROTEIN, PROTEIN, SOYBEAN PLANT, CP4 EPSPS, EPSP SYNTHASE, and HERBICIDE-TOLERANT CROP.

Chloroplasts Specialized chlorophyll-containing photosynthetic organelles in eucaryotic cells. See also EUCARYOTE, ORGANELLES, CELL, CHLOROPLAST, and TRANSIT PEPTIDE.

Cholesterol An essential material for creation of cell membranes, and a "building block" for certain hormones and acids used by the body. However, deposition of cholesterol on the interior walls of blood vessels results in atherosclerosis, an often fatal disease. See also HIGH-DENSITY LIPOPROTEINS (HDLPS), LOW-DENSITY LIPOPROTEINS (LDLPS), CELL, HORMONE, SITOSTANOL, FRUCTOSE OLIGOSACCHARIDES, and CHOLESTEROL OXIDASE.

Cholesterol Oxidase An enzyme that catalyzes the breakdown of cholesterol molecules (causing oxygen consumption in the breakdown process). Because cholesterol molecules are essential for creation and maintenance of cell membranes and some hormones, an excess of cholesterol oxidase can be harmful.

When the gene (that codes) for cholesterol oxidase is inserted into the genome of the corn (maize) plant, it may enable that plant to resist many of the worm pests (e.g., corn earworm, European corn borer, corn rootworm, black cutworm, armyworm, etc.) that attack corn (maize) in the field.

When the gene (that codes) for cholesterol oxidase is inserted into the cotton plant, it may enable that plant to resist weevils that attack cotton in the field. See ENZYME, GENE, GENETIC ENGINEERING, GENOME, CORN, *HELICOVERPA ZEA (H. ZEA)*.

Choline A nutrient that takes part in many of the metabolism processes in the human body. Naturally present in egg yolks, organ meats, dairy products, soybean lecithin, spinach, and nuts. Choline promotes the synthesis of high-density lipoproteins (i.e., HDLP, also known as "good" cholesterol) by the liver.

One active metabolite of choline is Platelet Activating Factor (PAF), which is involved in the body's hormonal and reproductive functions. Choline is so important in proper infant development/growth that it is included in manufactured infant formula at the rate of at least 7 mg per 100 kcal. See LECITHIN, METABOLISM, METABOLITE, HIGH-DENSITY LIPOPROTEINS (HDLP), HORMONE, and SOYBEAN OIL.

Chromatids Copies of a chromosome produced by replication within a living eucaryotic cell during the prophase (i.e., the first stage of mitosis). They are compact cylinders consisting of DNA coiled around flexible rods of histone protein. See also CHROMATIN, EUCARYOTE, MITOSIS, CHROMOSOMES, REPLICATION (OF VIRUS), HISTONES, and PROTEIN.

Chromatin From the Greek word for color. Named by Walter Flemming in 1879, due to the fact that chromatin's band-like structures stained darkly, chromatin is the complex of DNA and (histone) protein of which the chromosomes are composed. Consisting of fibrous swirls of unraveled DNA molecules in the nucleus of the interphase (i.e., the prolonged period of cell growth between cell division phases) eucaryote cell. Chromatin DNA gradually coils itself around flexible rods of histone protein during the prophase (i.e., the first stage of mitosis), forming two parallel compact cylinders (called chromatids) connected by a knot-like structure (called a centromere) at their middles. In appearance they are sort of like two rolls of carpeting standing side-by-side that are tied together with rope at their middles.

These (recently replicated) cylinders (that are joined at their middles) are homologous chromosomes (i.e., the genes of the two chromosomes are linked in the same linear order within the DNA strands of both chromosomes). While they are still joined at their middles, these paired chromosomes appear X-shaped when photographed by a karyotyper to produce a karyotype.

Chromatin is usually not visible during the interphase of a cell, but can be made more visible during all phases by reaction with basic stains (dyes) specific for DNA. See also BASOPHILIC, DEOXYRIBONUCLEIC ACID (DNA), PROTEIN, HISTONES, CHROMATIDS, CHROMOSOMES, MITOSIS, REPLICATION (OF VIRUS), CENTROMERE, KARYOTYPE, EUCARYOTE, and KARYOTYPER.

Chromatography A process by which complex mixtures of different molecules may be separated from each other. This is accomplished by subjecting the mixture to many repeated partitionings between a flowing phase and a stationary phase. Chromatography constitutes one of, if not *the* most fundamental separation techniques used in the biochemistry/biotechnology arena to date. See also POLYACRYLAMIDE GEL ELECTROPHORESIS (PAGE), SUBSTRATE (IN CHROMATOGRAPHY), AFFINITY CHROMATOGRAPHY, BIOTECHNOLOGY, AGAROSE, and GEL FILTRATION.

Chromosome Map See LINKAGE MAP.

Chromosomes Discrete units of the genome carrying many genes, consisting of (histone) proteins and a very long molecule of DNA. Found in the nucleus of every plant and animal cell. See also GENOME, GENE, GENETIC CODE, CHROMATIN, CHROMATIDS, KARYOTYPE, and KARYOTYPER.

Chymosin Also known as rennin. It is an enzyme used to make cheeses (from milk). Chymosin occurs naturally in the stomachs of calves, and is one of the oldest commercially used enzymes. Chymosin (rennin) is chemically similar to renin, an enzyme that plays an important role in regulating blood pressure in humans. See also RENIN.

Cilia Protein-based structures that occur in certain cells of both the plant and animal world. Cilia are very tiny hair-like structures and occur in large numbers on the outside of certain cells. In higher organisms such as man, they usually function to move extracellular material along the cell surface. An example is the "sweeping-out-of-foreign matter" action of cilia in the bronchial tubes in which very small particles are moved into the throat to be expelled or swallowed. Lower organisms may use cilia for locomotion (swimming). Cilia are used in the swimming motion of bacteria towards sources of nutrients in a process called chemotaxis. Cilia are shorter and occur in larger numbers per cell than flagella. Singular: cilium. See also CHEMOTAXIS and FLAGELLA.

Ciliary Neurotrophic Factor (CNTF) A human protein that has been shown to help the survival of those cells in the nervous system that act to convey sensation and control the function of muscles and organs. CNTF was approved by the U.S. FDA to treat amyotrophic lateral sclerosis (also known as Lou Gehrig's disease) in 1992. Amyotrophic lateral sclerosis causes a victim's muscles to degenerate severely, and it affects approximately 30,000 people per year in the United States. CNTF *might* prove useful for treating Alzheimer's Disease and/or other human neurological diseases. See also PROTEIN, CELL, NERVE GROWTH FACTOR (NGF), and FOOD AND DRUG ADMINISTRATION (FDA).

***cis*-Acting Protein** A *cis*-acting protein has the exceptional property of acting only on the molecule of DNA from which it was expressed. See also *trans*-ACTING PROTEIN and DEOXYRIBONUCLEIC ACID (DNA).

Cisplatin A drug that is used in chemotherapy regimens against certain types of cancer tumors. Cisplatin works against (tumor) cells by binding to the cell's DNA and generating intrastrand cross-links (between the two strands of the DNA molecule). These intrastrand cross-links prevent replication, and cause cell death. See also CHEMOPHARMACOLOGY, CANCER, DEOXYRIBONUCLEIC ACID (DNA), REPLICATION FORK, and REPLICATION (OF DNA).

***cis/trans* Isomerism** A type of geometrical isomerism found in alkenic systems in which it is possible for each of the doubly bonded carbons to carry two different atoms or groups. Two similar atoms or groups may be on the same side (i.e., *cis*) or on opposite sides (i.e., *trans*) of a plane bisecting the alkenic carbons and perpendicular to the plane of the alkenic systems. See also ISOMER, CHIRAL COMPOUND, and *TRANS* FATTY ACIDS.

***cis/trans* Test** Assays (determines) the effect of relative configuration on expression of two (gene) mutations. In a double heterozygote, two mutations in the same gene show mutant phenotype in *trans* configuration, wild (phenotype) in *cis* configuration. The phenotypic distinction is referred to as the position effect. See also GENE, PHENOTYPE, *cis*-ACTING PROTEIN, POSITION EFFECT, HETEROZYGOTE, and MUTATION.

Cistron Synonymous with gene. See also GENE.

Citrate Synthase The enzyme that is utilized (e.g., by plants) to synthesize (i.e., create) citric acid. See ENZYME and CITRIC ACID.

Citrate Synthase Gene A gene that codes for (i.e., causes to be produced by an organism possessing that gene) the enzyme known as citrate synthase. See GENE, ENZYME, EXPRESS, CITRATE SYNTHASE, and CITRIC ACID.

Citrate Synthase (CSb) Gene A bacterial gene that is utilized by certain bacteria (e.g., *Pseudomonas*) to code for (i.e., cause to be produced by bacterium possessing that gene) the enzyme known as citrate synthase. That enzyme is utilized to synthesize (i.e., create) citric acid.

In 1996, Luis Herrera-Estrella discovered that inserting the CSb gene from *Pseudomonas aeruginosa* into certain plants caused those plants to produce up to ten times more citrate in their roots, and to release up to four times more citric acid from those roots into the surrounding soil (thus decreasing aluminum toxicity, which slows plant growth and decreases crop yields). See GENE, ENZYME, EXPRESS, CITRATE SYNTHASE, and CITRIC ACID.

Citric Acid a tricarboxylic acid occurring in plants, especially citrus fruits. It is used as a flavoring agent, as an antioxidant in foods, and as a sequestering agent. The commercially produced form of citric acid melts at 153°C (307°F). Citric acid is found in all cells, its central role is in the metabolic process.

Some plants naturally release citric acid from their roots into the surrounding soil, in order for that citric acid to chemically "bind" aluminum ions that are present in some soils. Such aluminum, which slows plant growth and decreases crop yields, is present to a certain degree (which causes at least some crop yield reduction) in approximately one-third of the world's arable land. For example, 70% of the agricultural land in the country of Colombia possesses harmful amounts/conditions of aluminum to damage crops.

Corn (maize) yields are reduced up to 80% by such aluminum in soils. Soybeans, cotton, and field bean yields are also reduced. See also METABOLISM, ACID, CELL, CITRATE SYNTHASE, CITRATE SYNTHASE GENE, CITRATE SYNTHASE (Csb) GENE, CITRIC ACID CYCLE, METABOLITE, CELL, ION, SOYBEAN PLANT, and CORN.

Citric Acid Cycle Also known as the tricarboxylic acid cycle [TCA cycle because the citric acid molecule contains three (tri) carboxyl (acid) groups]. Also known as the Krebs cycle after H. A. Krebs, who first postulated the existence of the cycle in 1937 under its original name of "citric acid cycle." A cyclic sequence of chemical reactions that occurs in almost all aerobic (air-requiring) organisms. A system of enzymatic reactions in which acetyl residues are oxidized to carbon dioxide and hydrogen atoms, and in which formation of citrate is the first step. See also CITRIC ACID, CITRATE SYNTHASE, CITRATE SYNTHASE GENE, CITRATE SYNTHASE (Csb) GENE, ACID, AEROBIC, METABOLISM, ENZYME, and OXIDATION.

CKR-5 Proteins See HUMAN IMMUNODEFICIENCY VIRUS TYPES 1 & 2, RECEPTORS, and PROTEIN.

Clades The taxonomic sub-groups within cladistics. See also CLADISTICS.

Cladistics Initially popularized by Willi Hennig's 1950 book entitled *Phylogenetic Systematics,* cladistics is a system of taxonomic classification of organisms (and/or their specimens) that is based upon (determined) similar lines of selected shared traits. See CLADES, TYPE SPECIMEN, GENETICS, AMERICAN TYPE CULTURE COLLECTION (ATCC), and TRAIT.

Clinical Trial One of the final stages in the collection of data (for drug approval prior to commercialization) in which the new drug is tested in human subjects. Used to collect data on effectiveness, safety, and required dosage. See also PHASE I CLINICAL TESTING, FOOD AND DRUG ADMINISTRATION (FDA), KOSEISHO, BUNDESGESUNDHEITSAMT (BGA), COMMITTEE ON SAFETY IN MEDICINES, AND COMMITTEE FOR PROPRIETARY MEDICINAL PRODUCTS (CPMP).

Clone (a molecule) To create copies of a given molecule via various methods. See also POLYMERASE CHAIN REACTION (PCR), MONOCLONAL ANTIBODY (MAb), COCLONING (OF MOLECULES), and ANTIBODY.

Clone (an organism) A group of individual organisms (or cells) produced from one individual cell through asexual processes that do not involve the interchange or combination of genetic material. As a result, members of a clone have identical genetic compositions. For example, protozoa and bacteria frequently reproduce asexually (i.e., without sex) by a process called binary fission. In binary fission a single-celled organism undergoes cell division. The result is two cells with identical genetic composition. When these two identical cells undergo division, the result is four cells with identical genetic composition. These identical offspring are all members of a clone. The word "clone" may be used either as a noun or a verb.

Clostridium A genus of bacteria. Most are obligate anaerobes, and form endospores. See also ANAEROBE and ENDOSPORE.

CMC See CRITICAL MICELLE CONCENTRATION.

CMV See CUCUMBER MOSAIC VIRUS.

CNTF See CILIARY NEUROTROPHIC FACTOR.

CoA See COENZYME A.

Coccus A spherical-shaped bacterium. See also *BACILLUS*.

Cocloning (of molecules) The additional (accidental) cloning (i.e., copying) of extra molecular fragments, other than the desired one, that sometimes occurs when a scientist is attempting to clone a molecule. See also CLONE (A MOLECULE), POLYMERASE CHAIN REACTION (PCR), and Q-BETA REPLICASE TECHNIQUE.

Codex Alimentarius See CODEX ALIMENTARIUS COMMISSION.

Codex Alimentarius Commission An international regulatory body that is part of the United Nations' Food and Agriculture Organization (FAO), it is one of the three international SPS (sanitary and phytosanitary) standard-setting organizations that is recognized by the World Trade Organization (WTO). It was created in 1962 by the UN's FAO and the World Health Organization (WHO). It has 147 member nations.

In the Latin language, *Codex Alimentarius* means "food law" or "food code." The Codex Alimentarius Commission is responsible for execution of the Joint FAO/WHO Food Standards Program. The Codex Alimentarius standards are a set of international food mandates that have been adopted by the Commission. The Commission is composed of delegates from member country governmental agencies. The Codex Secretariat is headquartered in Rome, Italy.

The Commission periodically determines, then publishes a list of food ingredients and maximum allowable levels that it deems safe for human consumption (known as the *Codex Alimentarius)*. See also MAXIMUM

RESIDUE LEVEL (MRL), SPS, INTERNATIONAL PLANT PROTECTION CONVENTION (IPPC), INTERNATIONAL OFFICE OF EPIZOOTICS (OIE), and WORLD TRADE ORGANIZATION (WTO).

Coding Sequence The region of a gene (DNA) that encodes the amino acid sequence of a protein. See also GENETIC CODE, INFORMATIONAL MOLECULES, GENE, and MESSENGER RNA (mRNA).

Codon A triplet of nucleotides [three nucleic acid units (residues) in a row] that code for an amino acid (triplet code) or a termination signal. See also GENETIC CODE, TERMINATION CODON (SEQUENCE), AMINO ACID, NUCLEOTIDE, INFORMATIONAL MOLECULES, and MESSENGER RNA (mRNA).

Coenzyme A nonproteinaceous organic molecule required for the action of certain enzymes. The coenzyme contains as part of its structure one of the vitamins. This is why vitamins are so critically important to living organisms. Sometimes the same coenzyme is required by different enzymes that are involved in the catalysis of different reactions. By analogy, a coenzyme is like a part of a car such as a tire which can be identified in and of itself and which can, furthermore, be removed from the car. The car (enzyme), however, must of necessity have the tire in order to carry out its prescribed function. Coenzymes have been classified into two large groups: fat soluble and water soluble. Examples of a few water-soluble vitamins are: thiamin, biotin, folic acid, vitamin C, and vitamin B_{12}. Examples of fat-soluble vitamins are: vitamins A, D, E, and K. See also ENZYME, CATALYST, HOLOENZYME, VITAMIN, and POLYPEPTIDE (PROTEIN).

Coenzyme A A water-soluble vitamin known as pantothenic acid. A coenzyme in all living cells. It is required by certain condensing enzymes and functions in acyl-group transfer and in fatty-acid metabolism. Abbreviated CoA. See also ENZYME.

Cofactor A nonprotein component required by some enzymes for activity. The cofactor may be a metal ion or an organic molecule called a coenzyme. The term "cofactor" is a general term. Cofactors are generally heat stable. See also COENZYME, HOLOENZYME, and MOLECULAR WEIGHT.

Cofactor Recycle The regeneration of spent cofactor by an auxiliary reaction such that it may be reused many times over by a cofactor-requiring enzyme during a reaction. See also COFACTOR, HOLOENZYME, and ENZYME.

Cohesive Termini See STICKY ENDS.

Cold Hardening A process of acclimatization in which certain organisms produce specific proteins that protect them from freezing to death during the winter. Among other organisms, the common housefly, the fruit fly, and "no-see-em's" (i.e., *Culicoides variipennis)* can produce these proteins (e.g., during the gradually decreasing temperatures of a typical fall season in North America). The amount of such proteins produced within their bodies is proportional to the severity and duration of the cold experienced.

For example, prior to cold hardening, *Culicoides variipennis* insects usually die after exposure for two hours to a temperature of 14°F (–10°C). If those insects are first exposed for one hour to a temperature of 41°F (5°C), approximately 98 percent of these insects can then survive exposure for three days to a temperature of 14°F (–10°C). See also ACCLIMATIZATION, PROTEIN, LOW-TILLAGE CROP PRODUCTION, NO-TILLAGE CROP PRODUCTION, and *DROSOPHILA*.

Colicins Proteins produced by *Escherichia coli (E. coli)*, that are toxic (primarily) to other closely-related strains of bacteria. The particular *E. coli* that produce a given colicin are generally unaffected by the colicin that they produce. See also BACTERIOCIN, BACTERIOLOGY, STRAIN, BACTERIA, PROTEIN, TOXIN, and *ESCHERICHIA COLI (E. COLI)*.

Collagen The major structural protein in connective tissue. It is instrumental in wound healing [stimulated by fibroblast growth factor (FGF), platelet-derived growth factor, and insulin-like growth factor-1]. See also PROTEIN, FIBROBLAST GROWTH FACTOR (FGF), PLATELET-DERIVED GROWTH FACTOR (PDGF), and INSULIN-LIKE GROWTH FACTOR-1 (IGF-1).

Collagenase An enzyme that catalyzes the cleavage of collagen. One example of this is when bacteria in the mouth cause production of collagenase that then cleaves (i.e., breaks down) the collagen that holds teeth in place. Some cancers use collagenase to break down connective tissues in the body they inhabit, to enable the cancers to form the (new) blood vessels that nourish those cancers and help those cancers to spread through the body. Collagenase may also be responsible indirectly for certain autoimmune diseases such as arthritis, via breaking down the protective proteoglycan coat that covers cartilage in the body. See also STROMELYSIN (MMP-3), PROTEOLYTIC ENZYMES, ENZYME, COLLAGEN, CANCER, and AUTOIMMUNE DISEASE.

Colony A growth of a group of microorganisms derived from one original organism. After a sufficient growth period, the growth is visible to the eye without magnification.

Colony Hybridization A technique using *in situ* hybridization to identify bacterial colonies carrying inserted DNA that is homologous with some particular sequence (probe). See also DNA PROBE, HOMOLOGY, *IN SITU*, and REGULATORY SEQUENCE.

Colony Stimulating Factors (CSFs) Specific glycoprotein growth factors required for the proliferation and differentiation of hematopoietic progenitor cells. Different CSFs stimulate the growth of different cells. See also MACROPHAGE COLONY STIMULATING FACTOR (M-CSF), GRANULOCYTE COLONY STIMULATING FACTOR (G-CSF), GRANULOCYTE-MACROPHAGE COLONY STIMULATING FACTOR (GM-CSF), EPIDERMAL

GROWTH FACTOR (EGF), FIBROBLAST GROWTH FACTOR (FGF), HEMATOLOGIC GROWTH FACTOR (HGF), INSULIN-LIKE GROWTH FACTOR-1 (IGF-1), MEGAKARYOCYTE STIMULATING FACTOR (MSF), NERVE GROWTH FACTOR (NGF), PLATELET-DERIVED GROWTH FACTOR (PDGF), TRANSFORMING GROWTH FACTOR-ALPHA (TGF-ALPHA), and TRANSFORMING GROWTH FACTOR-BETA (TGF-BETA).

Combinatorial Biology A term used to describe the set of DNA technologies that are utilized to generate a large number of samples of new chemicals (metabolites) via creation of non-natural metabolic pathways. This collection of samples thus generated, is called a "library," and the samples are then tested for potential use (e.g., for therapeutic effect, in the case of pharmaceutical). These technologies enable greater efficiency in a pharmaceutical researcher's screening process for drug discovery. See also COMBINATORIAL CHEMISTRY, TARGET, MOLECULAR DIVERSITY, METABOLISM, INTERMEDIARY METABOLISM, METABOLITE, and RECEPTORS.

Combinatorial Chemistry A term used to describe the set of technologies that are utilized to generate a large number of samples of (new) chemicals, which are then tested for potential use (e.g., for therapeutic effect, in the case of pharmaceutical). These large numbers of chemical samples, thus generated, are called a "library" and are screened (e.g., for therapeutic effect) via a variety of laboratory, biosensor, computational, or animal tests. For a library that is used for new drug screening, high diversity in molecular structure among the chemicals in the library is desired, to increase the efficiency of the screening process. One method used to measure diversity of the molecular structure among samples in a library is called "molecular fingerprinting." If two samples are identical in molecular structure, the "fingerprint" coefficient is 1.0. If two samples are totally dissimilar in molecular structure, the coefficient is 0. The diversity of a library is measured by comparing each sample's molecular structure to that of all the others in the library. See also COMBINATORIAL BIOLOGY, TARGET, MOLECULAR DIVERSITY, RECEPTORS, and BIOSENSORS (ELECTRONIC).

Combining Site The site on an antibody molecule that locks (binds) onto an epitope (hapten). See also ANTIBODY, EPITOPE, ENGINEERED ANTIBODIES, HAPTEN, and CATALYTIC ANTIBODY.

Commission of Biomolecular Engineering An agency of the French government, established to oversee and regulate all genetic engineering activities in the country of France. See GENETIC ENGINEERING, RECOMBINANT DNA ADVISORY COMMITTEE (RAC), ZKBS (CENTRAL COMMISSION ON BIOLOGICAL ACTIVITY), INDIAN DEPARTMENT OF BIOTECHNOLOGY, and GENE TECHNOLOGY OFFICE.

Committee for Proprietary Medicinal Products (CPMP) The European Union's (EU's) scientific advisory organization dealing with new human pharmaceuticals approval. Its recommendations (e.g., to either approve or not approve a new product) are usually adopted by the European Medicines Evaluation Agency (EMEA), to which the CPMP reports.

Within 60 days of a CPMP "approval for recommendation" being adopted by the EMEA, each of the EU's member countries must advise the EMEA of its progress toward a regulatory decision on that pharmaceutical's submission for approvals. See also FOOD AND DRUG ADMINISTRATION (FDA), KOSEISHO, EUROPEAN MEDICINES EVALUATION AGENCY (EMEA), COMMITTEE ON SAFETY IN MEDICINES, and BUNDESGESUNDHEITSAMT (BGA).

Committee on Safety in Medicines The British Government agency that must approve new pharmaceutical products for sale within the United Kingdom. In concert with the Medicines Control Agency (MCA), it regulates all pharmaceutical products in the United Kingdom. It is the equivalent of the U.S. Food and Drug Administration. See also FOOD AND DRUG ADMINISTRATION (FDA), MEDICINES CONTROL AGENCY (MCA), COMMITTEE FOR PROPRIETARY MEDICINAL PRODUCTS (CPMP), KOSEISHO, NDA (TO KOSEISHO), IND, BUNDESGESUNDHEITSAMT (BGA), and EUROPEAN MEDICINES EVALUATION AGENCY.

Committee for Veterinary Medicinal Products (CVMP) The European Union's (EU's) scientific advisory organization dealing with approvals of new medicinal products intended use in animals. Its recommendations (e.g., to either approve, or not approve a new product) are usually adopted by the European Medicines Evaluation Agency (EMEA). See also COMMITTEE FOR PROPRIETARY MEDICINAL PRODUCTS (CPMP), FOOD AND DRUG ADMINISTRATION (FDA), KOSEISHO, COMMITTEE ON SAFETY IN MEDICINES, MEDICINES CONTROL AGENCY (MCA), EUROPEAN MEDICINES EVALUATION AGENCY (EMEA), and BUNDESGESUNDHEITSAMT (BGA).

Competence Factor See PLATELET-DERIVED GROWTH FACTOR (PDGF).

Complement (component of immune system) A group of more than 15 soluble proteins found in blood serum that interacts in a sequential fashion, in which a precurser molecule is converted into an active enzyme. Each enzyme uses the next molecule in the system as a substrate and converts it into its active (enzyme) form. This cascade of events and reactions leads ultimately to the formation of an attack complex that forms a transmembrane channel in the cell membrane. It is the presence of the channel that leads to lysis (rupturing) of the cell. See also COMPLEMENT CASCADE, CECROPHINS (LYTIC PROTEINS), HUMORAL IMMUNITY, LYSE, and LYSIS.

Complement Cascade The precisely regulated, sequential interaction of proteins (in the blood) that is triggered by a complex of antibody and antigen to cause lysis of infected cells. The triggering of lysis by multivalent antibody-antigen complexes is mediated by the classical pathway, beginning with the activation of C1, the first component (protein) of the pathway. This activation step, in which C1 undergoes conversion from a zymogen to an active protease, results in sequential cleavage of the C4, C2, C3, and C5 components (proteins). C5b, a fragment of C5, then joins C6, C7, and C8 to penetrate the (cell) membrane bearing the antigen. Finally, the binding of some 16 molecules of C9 to this "bridgehead" produces large pores in the (cell) membrane, which cause the lysis and destruction of the target cell. See also ANTIBODY, ANTIGEN, LYSIS, COMPLEMENT (COMPONENT OF IMMUNE SYSTEM), ZYMOGENS, HUMORAL SYSTEM, and CECROPHINS (LYTIC PROTEINS).

Complementary DNA (c-DNA) A single-stranded DNA that is complementary to a strand of RNA. The DNA is synthesized *in vitro* by an enzyme known as reverse transcriptase. It is a DNA copy of mRNA (messenger RNA) and this "rebukes" the Central Dogma. See also c-DNA, DEOXYRIBONUCLEIC ACID (DNA), MESSENGER RNA (mRNA), and CENTRAL DOGMA.

Compound Q See TRICHOSANTHIN.

Computer Assisted New Drug Application (also called Computer Assisted NDA) See CANDA.

Configuration The three-dimensional arrangement in space of substituent groups in stereoisomers.

Conformation The three-dimensional arrangement of substituent groups in a protein or other molecular structure that is free to assume different positions. The geometric form or shape of a protein in three-dimensional space. See also NATIVE CONFORMATION, TERTIARY STRUCTURE, EFFECTOR, and PROTEIN FOLDING.

Conjugate A molecule created by fusing together (e.g., via recombination or chemically) two unlike (different) molecules. The purpose of this is to create a molecule in which one of the original molecules has one function, for example, a toxic, cell-killing function, while the other original molecule has another function, such as targeting the toxin to a specific site which might include cancerous cells. For example, molecules of interleukin-2 (IL-2) have been fused with molecules of diphtheria toxin to create a conjugate that does the following:

(1) It enters leukemia and lymphoma cells. Because these two types of cancer cells possess IL-2 receptors on their surfaces, the IL-2 (targeting function) binds to that receptor and is internalized.

(2) The diphtheria toxin (killing function) then shuts down protein synthesis within the cancer cells.

(3) It then kills the cancerous cells.

This type of approach is widespread and there are many different types of conjugates. One type of conjugate consists of enzymes used in the treatment of certain molecular diseases attached covalently to polyethylene glycol (PEG). In this case the PEG greatly diminishes both the immunogenicity (the ability to induce an immune reaction) and the antigenicity (the ability to react with preformed antibodies). Antibodies may be used as vectors to carry both relatively small molecules of destructive chemicals or proteins to specific sites (cells) within the body. Antibodies may be coupled to enzymes, toxins, and/or ribosome-inhibiting proteins, as well as to radioisotopes. These conjugates are known collectively as immunoconjugates. See also IMMUNOCONJUGATE, CONJUGATED PROTEIN, FUSION PROTEIN, RECOMBINATION, TOXIN, INTERLEUKIN-2 (IL-2), RICIN, ABRIN, RECEPTORS, RIBOSOMES, and MESSENGER RNA (mRNA).

Conjugated Protein A protein containing a metal or an organic prosthetic group, or both. For example, a glycoprotein is a conjugated protein bearing at least one oligosaccharide group. See also PROSTHETIC GROUP, GLYCOPROTEIN, PROTEIN, OLIGOSACCHARIDES, CONJUGATE, and CD4-PE40.

Conjugation A process akin to sexual reproduction occurring in bacteria; mating in bacteria. A process that involves cell-to-cell contact and the one-way transfer of DNA from the donor to the recipient. In contrast to some other DNA-transfer processes of bacteria, conjugation may involve the transfer of large portions of the genome. The discovery caused considerable controversy at the time. See also TRANSFORMATION, TRANSDUCTION, DEOXYRIBONUCLEIC ACID (DNA), GENOME, and SEXUAL CONJUGATION.

Consensus Sequence The nucleotide sequence (within a DNA molecule) which gives the *most common* nucleotide at each position (along that sequence of that DNA molecule), for those instances (in certain organisms) where a (usually small) number of variations in nucleotide sequences can occur (e.g., for a given nucleotide sequence such as a promoter sequence). See also NUCLEOTIDE, DEOXYRIBONUCLEIC ACID (DNA), SEQUENCE (OF A DNA MOLECULE), GENETIC CODE, GENE, and PROMOTER.

Conserved A term used to describe a domain (region) of a molecule on the surface of a rapidly mutating microorganism (e.g., the AIDS virus) that remains the same in all, or most, variations of that microorganism. If that *conserved* region is suitable to act as an antigen (hapten, epitope), it may be possible to create a successful vaccine against that microorganism that would otherwise be unsuccessful due to the fact that the rapid mutation would cause it (the AIDS virus) to appear to be "different" than the one (antigen) the vaccine was designed against. See also DOMAIN (OF A PROTEIN), GP 120 PROTEIN, SUPERANTIGENS, MUTATION, ACQUIRED IMMUNE DEFICIENCY SYNDROME (AIDS), ANTIGEN, HAPTEN, EPITOPE, VIRUS, and HIV-1 and HIV-2.

Consortia Microorganisms that interact with each other (or at least "coexist peacefully") when growing together. An example of such interaction/coexistence would be bioleaching. See also BIOLEACHING, BIORECOVERY, BIODESULFURIZATION, and BIOSORBENTS.

Constitutive Enzymes Enzymes that are part of the basic, permanent enzymatic machinery of the cell. They are formed at a constant rate and in constant amounts regardless of the metabolic state of the organism. For example, enzymes that function in the production of cell-usable energy (such as ATP) might be good candidates. And this, in fact, is the case with the enzymes of the glycolytic sequence, which is the most ancient energy-yielding catabolic pathway. See also ENZYME and METABOLISM.

Constitutive Genes Expressed as a function of the interaction of RNA polymerase with the promoter, without additional regulation. They are sometimes also called "household genes" in the context of describing functions expressed in all cells at a low level. See also GENE, RNA POLYMERASE, and PROMOTER.

Constitutive Heterochromatin The inert state of permanently nonexpressed sequences, usually satellite DNA. See also EXPRESS, CODING SEQUENCE, DEOXYRIBONUCLEIC ACID (DNA), and CHROMATIN.

Constitutive Mutations Mutations (unplanned changes) that cause genes that are nonconstitutive (have controlled protein expression) to become constitutive (in which state the protein is expressed all of the time). See also CONSTITUTIVE GENES, MUTATION, and REGULATORY SEQUENCE.

Construct See CASSETTE and TRANSGENE.

Consultative Group on International Agricultural Research (CGIAR) An organization that is cosponsored by the Rome-based United Nations Food and Agriculture Organization (FAO), the United Nations Development Programme (UNDP), and the World Bank. The CGIAR is an association of 43 public and private donors that jointly support seventeen international agricultural research centers that are located primarily in developing countries. Twelve of the research centers have collectively assembled 500,000 different preserved samples of major food, forage and forest plant species into a gene bank. This, the world's largest internationally held collection of genetic resources, was legally placed under the auspices of the FAO in 1994 in order "to hold the collection in trust for the international community." Since 1970, CGIAR's collection has supported research efforts to develop better varieties of staple foods consumed primarily in developing countries of the world. See also AMERICAN TYPE CULTURE COLLECTION (ATCC), TYPE SPECIMEN.

Contaminant By definition, any unwanted or undesired organism, compound, or molecule present in a controlled environment. Unwanted presence of an entity in an otherwise clean or pure environment.

Continuous Perfusion A type of cell culture in which the cells (either mammalian or otherwise) are immobilized in a part of the system, and nutrients/oxygen are allowed to flow through the stationary cells, thus effecting nutrient/waste exchange. Ideally the system incorporates features that retard the activity of proteolytic enzymes, and reduce the need for anti-infective agents (e.g., antibiotics) and fetal bovine serum, which are required by most other cell culture systems. Continuous perfusion is used because, among other things, it eliminates the need to separate the cells from the culture medium when fresh medium is exchanged for old. See also MAMMALIAN CELL CULTURE, ENZYME, and PROTEOLYTIC ENZYMES.

Convention on Biological Diversity (CBD) The international treaty governing the conservation and use of biological resources around the world, that was signed by more than 150 countries at the 1992 United Nations Conference on Environment and Development.

Article 19.4 of the CBD called for the establishment of a "protocol on biosafety" to govern the trans-national-boundary movement of non-indigenous living organisms. See also CONSULTATIVE GROUP ON INTERNATIONAL AGRICULTURAL RESEARCH (CGIAR), and INTERNATIONAL PLANT PROTECTION CONVENTION (IPPC).

Convergent Improvement See TRANSGRESSIVE SEGREGATION.

Coordination Chemistry See CHELATION.

Copy DNA (C-DNA) See C-DNA.

Copy Number The number of molecules (copies) of an individual plasmid or plastid that is typically present in a single (e.g., bacterial for *plasmid*, plant for *plastid*) cell. Each plasmid has a characteristic copy number value ranging from 1 to 50 or more. Higher copy numbers result in a higher yield of the protein encoded for by the plasmid gene in each cell. See also PLASMID, PLASTID, PROTEIN, GENE, EXTRANUCLEAR GENES, GENETIC CODE, and MULTI-COPY PLASMIDS.

Corepressor A small molecule that combines with the repressor to trigger repression (the shutting down) of transcription. See also TRANSCRIPTION.

Corn The domesticated plant *Zea mays L.* also known as maize.

A green, leafy (grain) plant that is one of the world's largest providers of edible starch and fructose (sugar) for mankind's use. This summer annual plant varies in height from two feet (0.5 meter) to more than twenty feet (6 meters) tall. The seeds (kernels) are borne in cobs, ranging in size from two feet long to smaller than a man's thumb.

Grown widely in the world's temperate zones, corn is grown as far north as latitude 58° in Canada and Russia; and as far south as latitude 40° in the Southern Hemisphere.

During the 1980's, scientists were able to insert genes from *Bacillus thuringiensis (B.t.)* bacteria into the corn plant, to make that plant resistant to certain insects. During the 1990's, scientists were able to insert genes into the corn plant, to make it tolerant to certain herbicides; and to cause the corn plant to produce monoclonal antibodies (MAb).

Some of the major economic pests of corn include the European corn borer (*Ostrinia nubialis*), corn earworm (*Helicoverpa zea*), corn rootworm (*Diabrotica virgifera virgifera*). See HYBRIDIZATION (PLANT GENETICS), *BACILLUS THURINGIENSIS (B.t.)*, PROTEIN, STRESS PROTEINS, CRY PROTEINS, CRY1A (b) PROTEIN, CRY1A (c) PROTEIN, CRY9C PROTEIN, "STACKED" GENES, OPAGUE-2, HIGH-METHIONINE CORN, HIGH-LYSINE CORN, *B.t. KURSTAKI*, VALUE-ENHANCED GRAINS, *HELICOVERPA ZEA (H. ZEA)*, CHLOROPLAST TRANSIT PEPTIDE (CTP), HERBICIDE-TOLERANT CROP, EUROPEAN CORN BORER (ECB), AFLATOXIN, *FUSARIUM*, CORN ROOTWORM, TRANSPOSON, GLUTAMATE DEHYDROGENASE, BLACK-LAYERED (CORN), and MONOCLONAL ANTIBODIES (MAb).

Corn Earworm See *HELICOVERPA ZEA (H. ZEA)* and CORN.

Corn Rootworm Also known as the western corn rootworm, it is the larva stage of the corn rootworm beetle (*Diabrotica virgifera virgifera*), which historically has laid its eggs on corn/maize (*Zea mays L.*) plants. When they hatch, the larva must feed on the roots of the corn/maize plant in order to live. Some strains of *Bacillus thuringiensis (B.t.)* have proven to be effective against the corn rootworm, when sprayed onto them or genetically engineered into the corn/maize plant.

In the mid-1990's, a new genetic variant (known as the "eastern phenotype") was discovered in America. It prefers to lay its eggs on soybean plants instead of corn plants. See CORN, PHENOTYPE, SOYBEAN PLANT, STRAIN, *BACILLUS THURINGIENSIS (B.t.)*, and GENETIC ENGINEERING.

Corticotropin See ACTH.

Cowpea Mosaic Virus (CpMV) A virus that infects cowpea plants (which are known as black-eyed peas in the United States), but does not infect animals. Researchers have discovered how to cause CpMV to express certain animal virus proteins (i.e., antigens) on its surface, via genetic engineering. These virus antigens hold potential to replace the antigens currently used in vaccines, which are fraught with problems due to their production in animal cells, bacterial cells, or yeast cells. In addition, CpMV acts as an intrinsic natural adjuvant to the (animal virus) antigens, since it provokes an immune response itself. See also VIRUS, COWPEA TRYPSIN INHIBITOR (CpTI), EXPRESS, PROTEIN, ADJUVANT (TO A PHARMACEUTICAL), IMMUNE RESPONSE, and ANTIGEN.

Cowpea Trypsin Inhibitor (CpTI) A chemical that is naturally coded for by a certain cowpea plant gene. It kills certain insect larvae by inhibiting digestion of ingested trypsin by the larvae, thereby starving the larvae to death. See also TRYPSIN, GENE, and CODING SEQUENCE.

COX Cyclooxygenase. A human protein (enzyme) that helps to protect the inner lining of the stomach. Aspirin and some other pain-relieving drugs can block the protective action of cyclooxygenase. See ENZYME and PROTEIN.

CP4 EPSP Synthase See CP4 EPSPS.

CP4 EPSPS The enzyme 5-enolpyruvyl-shikimate-3-phosphate synthase, which is naturally produced by an *Agrobacterium* species (strain CP4) of soil bacteria. CP4 EPSPS is essential for the functioning of that bacterium's metabolism biochemical pathway. CP4 EPSPS happens to be unaffected by glyphosate-containing herbicides, so introduction of the CP4 EPSPS gene into crop plants (e.g., soybeans) makes those plants essentially impervious to glyphosate-containing herbicides. See also ENZYME, METABOLISM, GENE, GENETIC ENGINEERING, EPSP SYNTHASE, SOYBEAN PLANT, GLYPHOSATE OXIDASE, BACTERIA, CHLOROPLAST TRANSIT PEPTIDE (CTP), and HERBICIDE-TOLERANT CROP.

CPMP See COMMITTEE FOR PROPRIETARY MEDICINAL PRODUCTS (CPMP).

CpMV See COWPEA MOSAIC VIRUS.

CpTI See COWPEA TRYPSIN INHIBITOR

Critical Micelle Concentration Also known as the CMC of a surfactant. It is the lowest surfactant concentration at which micelles are formed. That is, the CMC represents that concentration of surfactant at which the individual surfactant molecules aggregate into distinct, high molecular weight spherical entities called micelles. Or from another viewpoint, it represents the concentration of a surfactant, above which micelles or reverse micelles will spontaneously form through the process of self-aggregation (self-assembly). See also MICELLE and REVERSE MICELLE (RM).

Cross Reaction When an antibody molecule (against one antigen) can combine with (bind to) a different (second) antigen. This sometimes occurs because the second antigen's molecular structure (shape) is very similar to that of the first antigen. See also ANTIBODY and ANTIGEN.

Crossing Over The reciprocal exchange of material between chromosomes that occurs during meiosis. The event is responsible for genetic recombination. The process involves the natural breaking of chromosomes, the exchange of chromosome pieces, and the reuniting of DNA molecules. See also LINKAGE, DEOXYRIBONUCLEIC ACID (DNA), CHROMOSOMES, and RECOMBINATION.

Crown Gall See *AGROBACTERIUM TUMEFACIENS.*

CRP See CAP.

Cry Proteins A class of proteins produced by *Bacillus thuringiensis (B.t.)* bacteria (or plants into which a *B.t.* gene has been inserted). Cry proteins are toxic to certain categories of insects (corn borers, mosquitoes, black flies, some types of beetles, etc.), but harmless to mammals and beneficial insects. See also *BACILLUS THURINGIENSIS (B.t.)*, PROTEIN, BACTERIA, GENE, PROTOXIN, and CORN.

Cry1A (b) Protein One of the "cry" proteins, it is a protoxin that—when eaten by certain insects (e.g., Lepidoptera larvae such as the European corn borer)—is toxic to those insects. However if eaten by a mammal, the Cry1A (b) is digested within one minute, harmlessly. See CRY PROTEINS, PROTEIN, *B.T. KURSTAKI*, PROTOXIN, and EUROPEAN CORN BORER (ECB).

Cry1A (c) Protein One of the "cry" proteins. See CRY PROTEINS.

Cry9C Protein One of the "cry" proteins, it is a protoxin that—when eaten by European corn borer, southwestern cornborer, black cutworm, and some species of armyworm—is toxic to those insects. However, if eaten by a mammal, the cry9C is nontoxic. See CRY PROTEINS, *BACILLUS THURINGIENSIS (B.t.)*, PROTOXIN, PROTEIN, and EUROPEAN CORN BORER (ECB)

CSF See COLONY STIMULATING FACTORS.

CTAB See HEXADECYLTRIMETHYLAMMONIUM BROMIDE.

CTP See CHLOROPLAST TRANSIT PEPTIDE.

Culture Any population of cells (e.g., bacteria, algae, protozoa, virus, yeasts, plant cells, mammalian cells, etc.) growing on, or in a medium that supports their growth. Typically used to refer to a population of the cells of a single species or a single strain. A medium which contains only one specific organism (e.g., *E. coli* bacteria) is known as a pure culture. A culture may be preserved (i.e., stored alive) via freezing, drying (in which the cells go dormant), subculturing on an agar medium, or other preservation methods. See also CULTURE MEDIUM, TYPE SPECIMEN, LYOPHILIZATION, AMERICAN TYPE CULTURE COLLECTION (ATCC), SPECIES, STRAIN, CELL CULTURE and MAMMALIAN CELL CULTURE.

Culture Medium Any nutrient system for the artificial cultivation of bacteria or other cells. It usually consists of a complex mixture of organic and inorganic materials. For example, the classic culture (growth) medium used for bacteria consists of nutrients (required by that bacteria) plus agar to solidify or semi-solidify the nutrient containing mass. See also MEDIUM, AGAR, CELL CULTURE, and MAMMALIAN CELL CULTURE.

Curing Agent A substance that increases the rate of loss of plasmids during bacterial growth. See also GROWTH (MICROBIAL) and PLASMID.

Current Good Manufacturing Practices See cGMP.

Cut An enzyme-induced, highly specific break in both strands of a DNA molecule (opposite one another). The enzymes involved are called restriction enzymes. See also RESTRICTION ENDONUCLEASES, ENZYME, and DEOXYRIBONUCLEIC ACID (DNA).

Cyclic AMP A molecule of AMP (adenosine monophosphate) in which the phosphate group is joined to both the 3′ and the 5′ positions of the ribose, forming a cyclic (ring) structure. When cAMP binds to CAP, the complex is a positive regulator of procaryotic transcription. See also ADENOSINE MONOPHOSPHATE (AMP), CAP, PROCARYOTES, and TRANSCRIPTION.

Cyclodextrin A macrocyclic (doughnut-shaped) carbohydrate ring produced enzymatically from starch. The external surface is hydrophobic while the interior is hydrophilic in nature. The hole of the doughnut is large enough to accommodate guest molecules. Uses include solubilization, separation, and stabilization of molecules in the interior cavity of or in association with the cyclodextrin molecules.

Cycloheximide Also called actidione. A chemical that inhibits protein synthesis by the 80S eucaryotic ribosomes; it does not, however, inhibit the 70S ribosomes of procaryotes. The chemical blocks peptide bond formation by binding to the large ribosomal subunits. See also PROTEIN and RIBOSOMES.

Cyclooxygenase See COX.

Cyclosporin An immune-system-supressing drug that was isolated from a mold in the mid-1970s by the Swiss firm F. Hoffmann-LaRoche & Co. AG. The drug is used to prevent (organ recipient's) immune system from "rejecting" a transplanted organ and typically must be taken by the organ recipient for the duration of his/her lifetime.

Cyclosporin's mechanism of action is to prevent the divalent calcium cation (Ca^{2+}) from entering T lymphocytes to activate certain genes within those T lymphocytes (that trigger the "rejection" process).

In 1996, Thomas Eisner reported that the mold *Tolypocladium inflatum*, from which cyclosporin is harvested, prefers a natural (wild) substrate of a deceased dung beetle. See also T-LYMPHOCYTES, FUNGUS, XENOGENEIC ORGANS, CATION, GENE, GRAFT-VERSUS-HOST DISEASE (GVHD), HUMAN LEUKOCYTE ANTIGENS (HLA), and MAJOR HISTOCOMPATIBILITY COMPLEX (MHC).

Cyclosporine See CYCLOSPORIN.

Cysteine (cys) An amino acid of molecular weight (mol wt) 121 Daltons. It is incorporated in many proteins. It possesses a sulfhydryl group (SH) that makes cysteine a mild reducing agent. Cysteine can cross-link with another cysteine located on the same or on a different polypeptide chain to form disulfide bridges. The "free" cysteine group is called a *thiol group*. See also AMINO ACID, CYSTINE, DISULFIDE BOND, POLYPEPTIDE (PROTEIN), and PROTEIN.

Cystic Fibrosis See CYSTIC FIBROSIS TRANSMEMBRANE REGULATOR PROTEIN (CFTR).

Cystic Fibrosis Transmembrane Regulator Protein (CFTR) A protein that regulates proper chloride ion transport across the cell membranes of human lung airway epithelial cells. When the gene that codes for CFTR protein is damaged or mutated, the (mutant) CFTR protein fails to function properly, which causes mucous (and bacteria) to accumulate in the lungs. This lung disease is known as Cystic Fibrosis. See also PROTEIN, GENE, ION, DEOXYRIBOCYCLEIC ACID (DNA), INFORMATIONAL MOLECULES, GENOME, GENETIC CODE, RIBOSOMES, and TRANSCRIPTION.

Cystine Two cysteine amino acids that are covalently linked via a disulfide bond. These units are important in biochemistry in that disulfide bridges represent one important way in which the conformation of a protein is maintained in the active form. Cystine bridges lock the structure of the proteins in which they occur in place by disallowing certain types of (molecule) chain movement. When the disulfide bond is with a "free" cysteine (i.e., one that is not a part of the same protein molecule's amino-acid backbone) the "free" cysteine is known as a *thiol group*. Cystine can be metabolized from methionine by certain animals (e.g., swine), but not vice versa. See also CYSTEINE (cys), AMINO ACID, CONFORMATION, PROTEIN, METHIONINE (met), METABOLISM, and DISULFIDE BOND.

Cytochrome Any of the complex protein respiratory pigments (enzymes) occurring within plant and animal cells. They usually occur in mitochondria, and function as electron carriers in biological oxidation. Cytochromes are involved in the "handing off" of electrons to each other in a stepwise fashion. In the process of "handing off" other events take place which result in the production of energy that the cell needs and is able to use.

Cytochrome P450 An enzyme that contains an iron-heme cofactor. It catalyzes many different biological hydroxylation reactions. Essentially, the enzyme renders fat-soluble (hydrophobic) molecules water soluble or more water soluble (by introduction of the hydrophilic hydroxyl group) so that the molecules may be removed (washed) from the body via the kidneys. This enzyme is being investigated for its potential as a catalyst in the hydroxylation of specific (valuable) industrial chemicals. See also CYTOCHROME, ENZYME, COFACTOR, HEME, HYDROXYLATION REACTION, and CYTOCHROME P4503A4.

Cytochrome P4503A4 An enzyme that, in humans, catalyzes reactions involved in the metabolism (breakdown) of certain pharmaceuticals. Those pharmaceuticals include some sedatives, antihypertensives, the antihistamine terfenadine, and the immunosuppressant cyclosporin. See ENZYME, CYTOCHROME P450, METABOLISM, HISTAMINE, CYCLOSPORIN, and CYTOCHROME.

Cytokines A large class of glycoproteins similar to lymphokines but produced by nonlymphocytic cells such as normal macrophages, fibroblasts, keratinocytes and a variety of transformed cell lines. They participate in regulating immunological and inflammatory processes, and can contribute to repair processes and to the regulation of normal cell growth and differentiation. Although cytokines are not produced by glands, they are hormone-like in their intercellular regulatory functions. They are active at very low concentrations and for the most part appear to function nonspecifically. For example, the cytokines stimulate the endothelial cells to express (synthesize and present) P-selectins and E-selectins on the internal surfaces (of blood vessels). These selectins protrude into the bloodstream, which causes passing white blood cells (leukocytes) to adhere to the selectins, then leave the bloodstream by "squeezing" between adjacent endothelial cells. Cytokines are exemplified by the interferons. See also INTERLEUKIN-1 (IL-1), LYMPHOKINES, INTERFERONS, GLYCOPROTEIN, PROTEIN, T CELLS, INTERLEUKIN-6 (IL-6), MACROPHAGE, LECTINS, FIBROBLASTS, HORMONE, ENDOTHELIAL CELLS, ENDOTHELIUM, SELECTINS, P-SELECTIN, ELAM-1, LEUKOCYTES, and ADHESION MOLECULES.

Cytolysis The dissolution of cells, particularly by destruction of their surface membranes. See also LYSIS, CECROPHINS (LYTIC PROTEINS), LYSOZYME, MAGAININS, COMPLEMENT (COMPONENT OF IMMUNE SYSTEM), and COMPLEMENT CASCADE.

Cytomegalovirus (CMV) A virus that infects 40–90 percent of American heterosexuals, and about 95 percent of homosexuals. CMV normally produces a latent (nonclinical, nonobvious) infection, but when AIDS or other events cause immume system suppression. CMV produces a febrile (fever-causing) illness that is usually mild in nature but can become retinitis (eye infection). CMV can be treated (to halt life- and sight-threatening infection) in immunocompromised patients (i.e., transplant patients and AIDS victims) with Ganciclovir, an antiviral compound developed by Syntex or Foscarnet, a compound developed by Astra Pharmaceuticals.

In 1996, Stephen E. Epstein found that latent CMV may cause changes in artery wall cells that aid clogging of arteries in adults (esp. following balloon angioplasty). See also VIRUS and ACQUIRED IMMUNE DEFICIENCY SYNDROME (AIDS).

Cytopathic Damaging to cells.

Cytoplasm The protoplasmic contents of the cell not including the nucleus. See also NUCLEUS, CELL, and PROTOPLASM.

Cytosine A pyrimidine occurring as a fundamental unit (one of the bases) of nucleic acids. See also NUCLEIC ACIDS.

Cytotoxic Poisonous to cells.

Cytotoxic Killer Lymphocyte See also CYTOTOXIC T CELLS.

Cytotoxic T Cells Also called killer T cells. T cells that have been created by stimulated helper T cells. The T refers to cells of the cellular system rather than to cells of the humoral system (B cells). Cytotoxic T cells detect and destroy infected body cells by use of a special type of protein. The protein attaches to the infected cell's membrane and forms holes in it. This allows the uncontrolled leakage of ions out of and water into the cell, causing cell death. In general, the loss of the integrity of the cell membrane leads to death. The cytotoxic T cells also transmit a signal to the (leaking) infected cells that causes the cell to "chew up" its DNA. This includes its own DNA as well as that of the virus. See also CECROPHINS (LYTIC PROTEINS), MAGAININS, INTERLEUKIN-4 (IL-4), HELPER T CELLS (T4 CELLS), VIRUS, T CELLS, SUPPRESSOR T CELLS, PROTEIN, INTERLEUKIN-2 (IL-2), and DEOXYRIBONUCLEIC ACID (DNA).

D

δ Endotoxins See DELTA ENDOTOXINS.

D Loop A region within mitochondrial DNA in which a short stretch of RNA is paired with one strand of DNA, displacing the original partner DNA strand in this region. The same term is used also describe the displacement of a region of one strand of duplex DNA by a single-stranded invader in the reaction catalyzed by RecA protein. See also DEOXYRIBONUCLEIC ACID (DNA).

Daidzen See ISOFLAVONES.

Daidzein See ISOFLAVONES.

Dalton A unit of mass very nearly equal to that of a hydrogen atom (precisely equal to 1.0000 on the atomic mass scale). Named after John Dalton (1766–1844) who developed the atomic theory of matter. It is 1.660×10^{-24} gram. See also KILODALTON (Kd).

Deamination The removal of amino groups from molecules.

Defective Virus A virus that, by itself, is unable to reproduce when infecting its host (cell), but that can grow in the presence of another virus. This other virus provides the necessary molecular machinery that the first virus lacks.

Degenerate Codons Two or more codons that code for the same amino acid. For example, isoleucine is specified by the AUU, AUC, and AUA triplets. Since in this case more than one triplet codes for isoleucine the codons are called degenerate. See also GENETIC CODE and CODON.

Dehydrogenases Enzymes that catalyze the removal of pairs of hydrogen atoms from their substrates. See also SUBSTRATE (CHEMICAL), GLUTAMATE DEHYDROGENASE, ENZYME, and DEHYDROGENATION.

Dehydrogenation The removal of hydrogen atoms from molecules. When those molecules are the components of vegetable oils/fats, this results in a lower content percentage of "saturated" fats. See also FATS, MONOUNSATURATED FATS, SATURATED FATTY ACIDS, and FATTY ACID.

Delaney Clause Formerly part of American federal law (1959 Delaney amendment to Food, Drug and Cosmetic Act), it was eliminated during 1996. The Delaney Clause had set a zero-risk tolerance level for carcinogenic pesticide residues in processed foods. See also CARCINOGEN.

Deletions Loss of a section of the genetic material from a chromosome. The size of a deleted material can vary from a single nucleotide to sections containing a number of genes. See also GENE and CHROMOSOMES.

Delta Endotoxins See CRY PROTEINS and PROTEIN.

Denaturation The loss of the native conformation of a macromolecule resulting, for instance, from heat, extreme pH (i.e., acidity or basicity) changes, chemical treatment, etc. It is accompanied by loss of biological activity. See also CONFORMATION, CONFIGURATION, and MACROMOLECULE.

Denatured DNA DNA that has been converted from double-stranded to single-stranded form by a denaturation process such as heating the DNA solution. In the case of heat denaturation, the solution becomes very gelatinous and viscous. See also DENATURATION, DEOXYRIBONUCLEIC ACID (DNA), and DUPLEX.

Denaturing Gradient Gel Electrophoresis See DENATURING POLYACRYLAMIDE GEL ELECTROPHORESIS.

Denaturing Polyacrylamide Gel Electrophoresis The use of PAGE (polyacrylamide gel electrophoresis) in order to separate and analyze DNA fragments (sequences) after that DNA is first denatured. This methodology can be utilized to scan DNA in order to detect point mutations. See POLYACRYLAMIDE GEL ELECTROPHORESIS (PAGE), POINT MUTATION, DENATURING GRADIENT GEL ELECTROPHORESIS, DEOXYRIBONUCLEIC ACID (DNA), DENATURED DNA, and BASE EXCISION SEQUENCE SCANNING.

Dendrimers Polymers (i.e., molecules composed of repeating atomic units within the molecule) that repeatedly branch (while "growing" due to addition of more atoms in a repeating pattern) until that branching is stopped by the physical constraint of contacting itself (i.e., having formed a complete, hollow sphere). Discovered during the 1970s by Donald Tomalia, dendrimers possess sites on their exterior surface to which genetic material (e.g., genes or other portions of DNA) can be "attached." Dendrimers bearing such genetic material have been shown to be able to successfully transfer that genetic material into more than thirty types of living animal cells. See also POLYMER, DENDRITIC POLYMERS, GENE, GENETIC ENGINEERING, GENE DELIVERY (GENE THERAPY), INFORMATIONAL MOLECULES, CODING SEQUENCE, TUMOR SUPPRESSOR GENES, DEOXYRIBONUCLEIC ACID (DNA), GENETIC TARGETING, and GENETICS.

Dendritic Langerhans Cells A type of cell, located in the mucous membranes of the mouth and genital areas, that permits the human immunodeficiency virus (i.e., the virus that causes AIDS) to enter and infect the body—even when there are no cuts or abrasions through those mucous membranes. See also HUMAN IMMUNODEFICIENCY VIRUS (TYPE 1 and TYPE 2), ACQUIRED IMMUNE DEFICIENCY SYNDROME (AIDS), ADHESION MOLECULES, and DENDRITIC POLYMERS.

Dendritic Polymers Polymers (i.e., molecules composed of repeating atomic units within the molecule) that repeatedly branch (while "growing" due to the addition of more atoms in a repeating pattern) until that branching is stopped (e.g., by physical constraints, for those polymers within living tissues). In the absence of physical constraints, dendritic polymers can continue branching (and growing) until they form a complete (hollow) sphere. See also POLYMER and DENDRIMERS.

Denitrification Reduction of nitrate to nitrites or into gaseous oxides of nitrogen, or even into free nitrogen by organisms. See also REDUCTION (IN A CHEMICAL REACTION)

Deoxynivalenol A mycotoxin, also known as DON. See DON and MYCOTOXINS.

Deoxyribonucleic Acid (DNA) Discovered by Frederick Miescher in 1869, it is the chemical basis for genes. The chemical building blocks (molecules) of which genes (i.e., paired nucleotide units that code for a protein to be produced by a cell's machinery, such as its ribosomes) are constructed. Every inherited characteristic has its origin somewhere in the code of the organism's complement of DNA. The code is made up of subunits, called nucleic acids. The sequence of the four nucleic acids is interpreted by certain molecular machines (systems) to produce the required proteins of which the organism is composed. The structure of the DNA molecule was elucidated in 1953 by James Watson, Francis Crick, and Maurice Wilkins. The DNA molecule is a linear polymer made up of deoxyribonucleotide repeating units (composed of the sugar 2-deoxyribose, phosphate, and a purine or pyrimidine base). The bases are linked by a phosphate group, joining the 3′ position of one sugar to the 5′ position of the next sugar. Most molecules are double-stranded and antiparallel, resulting in a right-handed helix structure that is held together by hydrogen bonds between a purine on one chain and pyrimidine on the other chain. DNA is the carrier of genetic information, which is encoded in the sequence of bases; it is present in chromosomes and chromosomal material of cell organelles such as mitochondria and chloroplasts, and also present in some viruses. See related terms under A-DNA, B-DNA, c-DNA, and Z-DNA. See also TRANSCRIPTION, ANTIPARALLEL, DOUBLE HELIX, MESSENGER RNA (mRNA), NUCLEOTIDE, PROTEIN, RIBOSOMES, GENETIC CODE, GENE, CHROMOSOMES, CHROMATIDS, and CHROMATIN.

Deprotection (of a peptide) See also HF CLEAVAGE.

Desferroxamine Manganese An iron chelating agent, (i.e., it chemically binds to iron atoms in the blood, thus trapping the iron atoms. The molecule also acts as an hSOD mimic by capturing harmful oxygen free radicals in the blood before they damage the walls of blood vessels. Recent research indicates that desferroxamine manganese may be useful in blocking the onset of cataracts. See also HUMAN SUPEROXIDE DISMUTASE (hSOD), XANTHINE OXIDASE, and LAZAROIDS.

Desulfovibrio A genus of bacteria that reduces sulfate to H_2S (hydrogen sulfide). Energy is obtained by oxidation of H_2 or organic molecules. Not a strict autotroph because CO_2 cannot be used as a sole carbon source. See also REDUCTION (IN A CHEMICAL REACTION) and AUTOTROPH.

Dextran A polysaccharide produced by yeasts and bacteria as an energy storage reservoir (analogous to fat in humans). Consists of glucose residues, joined almost exclusively by alpha-1,6 linkages. Occasional branches (in the molecule) are formed by alpha 1,2, alpha 1,3, or alpha 1,4 linkages. Which linkage is used depends on the species of yeast or bacteria producing the dextran. See also POLYSACCHARIDES.

Dextrorotary (D) Isomer A stereoisomer that rotates the plane of plane-polarized light to the right. Dextro = right. See also STEREOISOMERS, LEVOROTARY (L) ISOMER, and POLARIMETER.

DHA See DOCOSAHEXANOIC ACID.

Diadzein See DAIDZEIN and ISOFLAVONES.

Dialysis The separation of low molecular weight compounds from high molecular weight components in solution by diffusion through a semipermeable membrane. Frequently utilized to remove salts, and to remove biological effectors (such as nicotinamide adenine dinucleotides, nucleotide phosphates, etc.) from polymeric molecules such as protein, DNA, or RNA. Commonly used membranes have a molecular weight cutoff (threshold) of around 10,000 Daltons, but other membrane pore sizes are available. See also HOLLOW FIBER SEPARATION (OF PROTEINS) and ACTIVE TRANSPORT.

Diamond* vs. *Chakrabarty See also CHAKRABARTY DECISION.

Diastereoisomers Four variations of a given molecule, consisting of a pair of stereoisomers about a second asymmetric carbon atom for each of the two isomers of the first asymmetric carbon atom. See also STEREOISOMERS and CHIRAL COMPOUND.

Digestion (within chemical production plants) Breakdown of feedstocks by various processes (chemical, mechanical, and biological) to yield their desired building-block components for inclusion as raw materials in subsequent chemical or biological processes.

Digestion (within organisms) The enzyme-enhanced hydrolysis (breakdown) of major nutrients (food) in the gastrointestinal system to yield their building-block components (to the organism), such as amino acids, fatty acids, or other essential nutrients. See also HYDROLYSIS, FATS, AMINO ACIDS, ESSENTIAL AMINO ACIDS, ESSENTIAL NUTRIENTS, FATTY ACID, ESSENTIAL FATTY ACIDS, "IDEAL PROTEIN" CONCEPT, ENZYME, and ABSORPTION.

Diploid The state of a cell in which each of the chromosomes, except for the sex chromosomes, is always represented twice (46 chromosomes in humans). In contrast to the haploid state in which each chromosome is represented only once. See also DIPLOPHASE, CHROMOSOMES, HOMOZYGOUS, and TRIPLOID.

Diplophase A phase in the life cycle of an organism in which the cells of the organism have two copies of each gene. When this state exists the organism is said to be diploid. See also DIPLOID, GENE, HOMOZYGOUS, and CELL.

Disaccharides Carbohydrates consisting of two covalently linked monosaccharide units—hence "di" for "two." See also OLIGOSACCHARIDES, MONOSACCHARIDES, and POLYSACCHARIDES.

Dissimilation The breakdown of food material to yield energy and building blocks for cellular synthesis. See also DIGESTION (WITHIN ORGANISMS).

Dissociating Enzymes See also HARVESTING ENZYMES.

Distribution See "ADME" TESTS and PHARMACOKINETICS (PHARMACODYNAMICS).

Disulfide Bond An important type of covalent bond formed between two sulfur atoms of different cysteines in a protein. Disulfide bonds (linkages, bridges) contribute to holding proteins together and also help provide the internal structure (conformation) of the protein. See also PROTEIN, CYSTEINE (cys), and CYSTINE.

Diversity Biotechnology Consortium A nonprofit U.S. organization that was formed in August of 1994 by a group of research institutions and companies. The Consortium's first president is Stuart A. Kauffman of the Santa Fe Institute. The Consortium's purpose is to further the use of molecular diversity as a tool in drug design and in the study of mutating viruses. See also MOLECULAR DIVERSITY, RATIONAL DRUG DESIGN, MOLECULAR BIOLOGY, VIRUS, MUTATION, MUTANT, SITE-DIRECTED MUTAGENESIS, COMBINATORIAL CHEMISTRY, and COMBINATIONAL BIOLOGY.

DNA See also DEOXYRIBONUCLEIC ACID.

DNA Analysis See also DNA PROFILING.

DNA Chimera One DNA molecule composed of DNA from two different species. See also CHIMERA.

DNA Fingerprinting See DNA PROFILING.

DNA Ligase An enzyme that creates a phosphodiester bond between the 3′ end of one DNA segment and the 5′ end of another, while they are base-paired to a template strand. The enzyme seals (joins) the ends of single-stranded DNA in a duplex DNA chain. DNA ligase constitutes a part of the DNA repair mechanism available to the cell. See also NICK, LIGASE, DEOXYRIBONUCLEIC ACID (DNA), and DUPLEX.

DNA Marker See MARKER.

DNA Polymerase An enzyme that catalyzes the synthesis of DNA. It does this by catalyzing the addition of deoxyribonucleotide residues to the free 3′-hydroxyl end of a DNA chain, starting from a mixture of the appropriate triphosphorylated bases, which are dATP, dGTP, dCTP and dTTP. This chemical reaction is reversible and, hence, DNA polymerase also functions as an exonuclease. See also ENZYME, EXONUCLEASE, *TAQ* DNA POLYMERASE, DEOXYRIBONUCLEIC ACID (DNA), and SYNTHESIZING (OF DNA MOLECULES).

DNA Probe Also called gene probe or genetic probe. Short, specific (complementary to desired gene) artificially produced segments of DNA used to combine with and detect the presence of specific genes (or shorter DNA segments) within a chromosome. If a DNA probe of known composition and length is mingled with pieces of DNA (genes) from a chromosome, the probe will cling to its exact counterpart in the "chromosomal DNA pieces" (genes), forming a stable double-stranded hybrid. The presence of this (now) "labeled" probe is detected visually or with the aid of another detection instrument.

Because the composition of the DNA probes is known, scientists can riffle through a chromosome, spotting segments of DNA (i.e., genes) that seem to be linked to genetic diseases. See also MUSCULAR DYSTROPHY (MD), PROBE, POLYMERASE CHAIN REACTION (PCR), GENE, POLYMERASE CHAIN REACTION (PCR) TECHNIQUE, CHROMOSOMES, DOUBLE HELIX, DUPLEX, HYBRIDIZATION, HYBRIDIZATION SURFACES, DEOXYRIBONUCLEIC ACID (DNA), HOMEOBOX, and RAPID MICROBIAL DETECTION (RMD).

DNA Profiling Invented in 1985 by Alec Jeffreys, it is a technique used by forensic (i.e., crime-solving) chemists to match biological evidence (e.g., a blood stain) from a crime scene to the person (e.g., the assailant) involved in that particular crime. DNA profiling involves the use of RFLP (restriction fragment length polymorphism) analysis or ASO/PCR (allele-specific oligonucleotide/polymerase chain reaction) analysis to analyze the specific sequence of bases (i.e., nucleotides) in a piece of DNA taken from the biological evidence. Since the specific sequence of bases in DNA molecules is different for each individual (due to DNA polymorphism), a criminal's DNA can be matched to that of the evidence to prove guilt or innocence. Biological evidence may include among other things blood, hair, nail fragments, skin, and sperm. See also DEOXYRIBONUCLEIC ACID (DNA), RESTRICTION FRAGMENT LENGTH POLYMORPHISM (RFLP) TECHNIQUE, POLYMORPHISM (CHEMICAL), POLYMERASE CHAIN REACTION (PCR) TECHNIQUE,

ALLELE, NUCLEOTIDE, NUCLEIC ACIDS, OLIGOMER, GENETIC CODE, INFORMATIONAL MOLECULES, OLIGONUCLEOTIDE, and CODON.

DNA Synthesis See also SYNTHESIZING (OF DNA MOLECULES).

DNA Typing See also DNA PROFILING.

DNA Vaccines Products in which "naked" genes (i.e., pieces of bare DNA) are used to stimulate an immune response (e.g., either a cellular immune response, humoral immune response, or otherwise raising antibodies against the pathogen from which the naked genes have arisen/been derived). See also DEOXYRIBONUCLEIC ACID (DNA), IMMUNE RESPONSE, CELLULAR IMMUNE RESPONSE, HUMORAL IMMUNITY, ANTIBODY, "NAKED" GENE, PATHOGEN, and DNA VECTOR.

DNA Vector A vehicle (such as a virus) for transferring genetic information (DNA) from one cell to another. See also BACTERIOPHAGE, RETROVIRUSES, and VECTOR.

DNA-Dependent RNA Polymerase See RNA POLYMERASE.

DNA-RNA Hybrid A double helix that consists of one chain of DNA hydrogen bonded to a chain of RNA by means of complementary base pairs. See also HYBRIDIZATION and DOUBLE HELIX.

DNAse Deoxyribonuclease, an enzyme that degrades (cuts up) DNA. See ENZYME and DEOXYRIBONUCLEIC ACID (DNA).

Docosahexanoic Acid (DHA) One of the omega-3 (n-3) polyunsaturated fatty acids (PUFA), DHA is important in the development of the human infant's brain, spinal cord, and retina tissues. DHA aids optimal brain and nervous system development in human infants. Naturally present in human breast milk and fish oil. See POLYUNSATURATED FATTY ACIDS (PUFA).

Domain (of a chromosome) May refer either to a discrete structural entity defined as a region within which supercoiling is independent of other domains, or to an extensive region, including an expressed gene that has heightened sensitivity to degradation by the enzyme DNAse I. See also GENE, EXPRESS, and ENZYME.

Domain (of a protein) A discrete continuous part of the amino acid sequence that can be equated with a particular function. See also COMBINING SITE, EPITOPE, IDIOTYPE, PROTEIN, p53 PROTEIN, and MINIMIZED PROTEIN.

Dominant (gene) See DOMINANT ALLELE.

Dominant Allele Discovered by Gregor Mendel in the 1860s, it is a gene that produces the same phenotype when it is heterozygous as it does when it is homozygous (i.e., trait, or protein, is expressed even if only one copy of the gene is present in the genome). See also RECESSIVE ALLELE, HETEROZYGOTE, HOMOZYGOUS, PHENOTYPE, GENOTYPE, and GENOME.

DON Abbreviation for the mycotoxin deoxynivalenol, which is produced by *Fusarium* fungi DON. Also known as "vomitoxin," because it can cause some animals to vomit if they consume it. See MYCOTOXINS, DEOXYNIVALENOL, *FUSARIUM*, FUNGUS, and VOMITOXIN.

Donor Junction The junction between the left 5′ end of an exon and the right 3′ end of an intron. See also EXON and INTRON.

Double Helix The natural coiled conformation of two complementary, antiparallel DNA chains. This structure was first put forward by Watson and Crick in 1953. See also DEOXYRIBONUCLEIC ACID (DNA).

Down Promoter Mutations Those mutations that decrease the frequency of initiation of transcription. Down promoter mutations lead to the production of less mRNA than is the case in the nonmutated state. See also mRNA, MUTATION, and TRANSCRIPTION.

Drosophila The name of a type of fruit fly that is commonly used in genetics experiments; due to its short life cycle (14 days) and simple genome (four chromosome pairs). Because of this, a large base of knowledge about Drosophila genetics has been accumulated by the world's scientific community. See also GENETICS, GENOME, GENETIC CODE, GENETIC MAP, CHROMOSOMES, COLD HARDENING, and HOMEOBOX.

Duchenne Muscular Dystrophy (DMD) Gene See also MUSCULAR DYSTROPHY (MD).

Duplex The double-helical structure of DNA (deoxyribonucleic acid). See also DOUBLE HELIX and DEOXYRIBONUCLEIC ACID (DNA).

E

E-Selectin See ELAM-1.

Early Development This refers to the period of a phage infection before the start of DNA replication. See also PHAGE (BACTERIOPHAGE), BACTERIOPHAGE and DEOXYRIBONUCLEIC ACID (DNA).

Early vs. Late Genes Those genes transcribed early in a bacteriophage-mediated infection process as compared to those genes transcribed some time later. May require different "p factors" (sigma) for recognition of promotors. See also GENE and PROMOTER.

Early vs. Late Proteins During viral infection, viral-specific proteins are synthesized at characteristic times after infection. They are called "early" and "late." Often under positive control of bacterial and viral sigma factors. See also EARLY VS. LATE GENES and PROTEIN.

EAA See EXCITATORY AMINO ACIDS.

ECB See EUROPEAN CORN BORER (ECB).

E. coli See *Escherichia coliform (E. coli).*

***E. coli* 0157:H7** See *Escherichia coliform* 0157:H7 *(E. coli* 0157:H7*).*

Ecology The study of the interrelationships between organisms and their environment. See also HABITAT.

"Edible Vaccines" Edible substances, bearing antigens, that cause activation of an animal's immune system via that animal's GALT (gut-associated lymphoid tissues). These "edible vaccines" are derived from transgenic plants (e.g., grains, tubers, fruits, etc.) or eggs (i.e., via the activation of the hens' immune system to cause them to secrete desired molecules into the eggs they lay). See also GUT-ASSOCIATED LYMPHOID TISSUES (GALT), ANTIGEN, CELLULAR IMMUNE RESPONSE, MOLECULAR PHARMING™, AND HUMORAL IMMUNITY.

EDTA Ethylenediamine tetraacetate. An organic molecule which, due to the chemical groups it contains and their juxtaposition within that molecule, is able to chelate (bind) certain other molecules such as divalent metal cations. EDTA thus inhibits some enzymes requiring such ions for activity. See also CHELATION, COFACTOR, and CHELATING AGENT.

Effector A class of (usually small) molecules that regulates the activity of a specific protein (e.g., enzyme) molecule by binding to a specific site on the protein. Control of (existing) enzyme molecules may be achieved by combination of the effector with the enzyme. The effector molecule may either physically block the active site on the enzyme molecule, or alter the three-dimensional conformation of the enzyme molecule. That conformation change results in a change in the enzyme's catalytic activity.

Effector is a general term. Effector molecules may be activators (cause an *increase* in the enzyme's catalytic activity) or inhibitors (cause a *decrease* in the enzyme's catalytic activity).

A special class of effector, known as an allosteric effector, binds to enzyme molecule at a site other than the enzyme's active site (thereby activating or inhibiting). See also PROTEIN, ENZYME, CONFORMATION, ALLOSTERIC ENZYMES, ALLOSTERIC SITE, ACTIVE SITE, FEEDBACK INHIBITION, and CATALYTIC SITE.

EGF See EPIDERMAL GROWTH FACTOR (EGF).

EGF Receptor A protein embedded in the surface of the membranes of skin cells. The receptor consisting of (1) an outside (of the cell membrane) enzyme that recognizes epidermal growth factor (EGF) and binds to it, and (2) an enzyme on the inside of the cell membrane, which is of the tyrosine kinase class. When free EGF comes in contact with an EGF receptor, they bind (in a lock-and-key fashion), and then enter the cell (through the cell membrane) together (where EGF then stimulates growth). The EGF receptor (and receptors in general) is like a butler who allows the EGF (a guest) to enter the cell (home). See also ONCOGENES and PROTEIN.

EHEC See ENTEROHEMORRHAGIC *E. COLI.*

EIA See ENZYME IMMUNOASSAY.

ELAM-1 Also known as E-selectin, it is a selectin molecule that is synthesized by endothelial cells after (adjacent) tissue is infected. ELAM-1 molecules then help leukocytes to leave the bloodstream to fight the infection. See also SELECTINS, LECTINS, ADHESION MOLECULES, and LEUKOCYTES.

Elastase An enzyme secreted by neutrophils (white blood cells that engulf pathogens) which catalyzes the cleavage (breakdown) of specific proteins that function to provide elasticity to certain tissues. May be indirectly responsible for some autoimmune diseases, such as arthritis (which results from breakdown of cartilage tissue). Elastase may also be indirectly responsible for the emphysema (caused by loss of lung elasticity) that results from prolonged smoke inhalation. When α-1 antitrypsin (anti-elastase) efficacy is reduced (via smoke) the now-unrestrained excess elastase destroys alveolar walls in the lungs by digesting elastic fibers and other connective tissue proteins. See also LEUKOCYTES (WHITE BLOOD CELLS), NEUTROPHILS, and PROTEOLYTIC ENZYMES.

Electrolyte Any compound (e.g., salt, acid, base, etc.) which in aqueous solution dissociates into ions (charged atom-sized particles). Electrolytes may either be strong (completely or nearly completely dissociated) or weak (only partially dissociated). See also ION.

Electron Carrier A protein, such as a flavoprotein or a cytochrome, that can gain and lose electrons reversibly and functions in the transfer of electrons from one carrier to another until the electron is taken up by a final molecule or atom such as oxygen. See also PROTEIN and CYTOCHROME.

Electron Microscopy (EM) A technique for greatly magnifying and visualizing very small entities such as viruses and even large molecules. The technique uses beams of electrons instead of light rays. Because of the physics involved, beams of electrons permit much greater magnification than is possible with a light microscope. Electron microscopes have been used to examine the structures of viruses, bacteria, pollen grains, molecules etc.

Electropermeabilization See ELECTROPORATION.

Electrophoresis A technique for separating molecules based on the differential movement of charged particles through a matrix when subjected to an electric field. The term is usually applied to large ions of colloidal particles dispersed in water. The most important use of electrophoresis (currently) is in the analysis of proteins, and then a technique known as gel electrophoresis is used. Since the proportion of proteins varies widely in different diseases, electrophoresis can be used for diagnostic purposes.

Electrophoresis, through agarose or other gel matrices, is a common way to separate, identify, and purify plasmid DNA, DNA fragments resulting from digestion (of DNA) with restriction endonucleases, and RNA. Electrophoresis is also used to study bacteria and viruses, nucleic acids, and some types of molecules, including amino acids. See also PROTEIN, AMINO ACID, BIOLUMINESCENCE, POLYACRYLAMIDE GEL ELECTROPHORESIS (PAGE), CHROMATOGRAPHY, GEL, AGAROSE, PLASMID, DEOXYRIBONUCLEIC ACID (DNA), RESTRICTION ENDONUCLEASES, RIBONUCLEIC ACID (RNA), BACTERIA, and VIRUS.

Electroporation A process utilized to introduce a foreign gene into the genome of an organism. In 1995, the U.S. company Dekalb Genetics Corp. received a patent for producing genetically engineered corn via introduction of a foreign gene into corn cells via electroporation.

Electroporation, also called electroporesis, or electropermeabilization, uses a brief direct-current (dc) electrical pulse to cause formation of "micropores" (tiny holes) in the surface of cells or protoplasts suspended in a solution (water) containing DNA sequences (genes). After the gene(s) enter cell via the temporarily created micropores, the electrical pulse ceases, and the micropores close so that the gene(s) cannot depart the cell. The cell then incorporates (some) of the new genetic material (genes) into its genetic complement (genome), and produces whatever product (i.e., a protein) the newly introduced gene codes for. See also CODING SEQUENCE, GENETIC ENGINEERING, VECTOR, BIOLISTIC® GENE GUN, "EXPLOSION" METHOD [TO INTRODUCE FOREIGN (NEW) GENES INTO PLANT CELLS], *AGROBACTERIUM TUMEFACIENS*, GENE, GENOME, CELL, CORN, PROTOPLAST, DEOXYRIBONUCLEIC ACID (DNA), and PROTEIN.

Electroporesis See ELECTROPORATION.

ELISA (test for proteins) An enzyme-linked immunosorbent assay (hence the acronym) which can readily measure less than a nanogram (10^{-9}g) of a protein. This assay is more sensitive than simple immunoassay (tests) because one of the two antibodies used to bind and quantitate (measure) the protein's antigen, based on two concurrent epitopes within the protein, is attached to an enzyme. The enzyme can rapidly convert an added colorless substrate into a colored product, or a nonfluorescent substrate into an intensely fluorescent product (thus enabling finer quantitation). See also ABSORBANCE (A), IMMUNOASSAY, PROTEIN, ANTIGEN, ENZYME, NANOGRAM (ng), and FLUORESCENCE.

EMAS Eco-Management and Audit Scheme.

Embryology The study of the early stages in the development of an organism. In these stages a single highly specialized cell, the egg, is transformed into a complex many-celled organism resembling its parents. See CELL, ANTIANGIOGENESIS, and GAMETE.

EMEA See EUROPEAN MEDICINES EVALUATION AGENCY.

Emulsion A stable dispersion of one liquid in a second, immiscible (i.e., nonmixable) liquid. For example, milk is an emulsion of oil (fat) in water and latex paint is an emulsion of paint resin in water.

Enantiomers From the Greek word *enantios,* which means "opposite." Enantiomers are a pair of nonidentical, mirror image molecules. This means that both molecules are made up of the same atoms, that is, they have the same molecular formula, but the constituent groups that are attached to a carbon atom can be arranged in two different ways (forms) around the carbon atom. This gives rise to an asymmetric molecule that can exist in either of two mirror-image forms whose mirror images are not superimposable. A pair of these molecules is known as enantiomers. The four attached groups are all different from each other. See also RACEMATE, OPTICAL ACTIVITY, CHIRAL COMPOUND, and ENANTIOPURE.

Enantiopure Refers to a compound (e.g., a pharmaceutical) that consists of only *one* of that compound's two possible enantiomers. Sometimes expressed in relative terms. For example, 98% enantiopure would refer to a compound that consists of 98% (of) desired enantiomer. See also ENANTIOMERS, CHIRAL COMPOUND, RACEMATE, and OPTICAL ACTIVITY.

Endergonic Reaction A chemical reaction with a positive standard free energy change (i.e., an "uphill" reaction). An (heat) energy-requiring reaction. A nonspontaneous reaction at ambient temperature. See also EXERGONIC REACTION and FREE ENERGY.

Endocrine Glands Glands that secrete their products (hormones) into the blood, which then carries them to their specific target organs. For example, adrenalin, produced in the adrenal glands, is carried to the heart (and other muscles) when needed during periods of stress. The endocrine glands are: the pituitary, thyroids, adrenals, pancreas, ovaries (in females) and testes (in males). Endocrine glands are found in some invertebrates as well as in vertebrates. See also HORMONE and ENDOCRINE HORMONES.

Endocrine Hormones These are the products secreted by the endocrine glands. These help control long-term bodily processes, such as growth, lactation, sex cycles, and metabolic adjustment. The endocrine system and the nervous system are interdependent and often referred to collectively as the neuroendocrine system. For example, the juvenile hormone, found in insects and annelids, affects sexual maturation. There is currently great interest among scientists in the potential use of such hormones in the control of destructive insects. See also ENDOCRINE GLANDS, HORMONE, and PHEROMONES.

Endocrinology The branch of science that studies the endocrine glands, hormones, and hormone-like substances. See also ENDOCRINE GLANDS, HORMONE, and ENDOCRINE HORMONES.

Endocytosis Also called receptor-mediated endocytosis. The import of substances (e.g., hormones, viruses, and toxins) into a cell via specific receptor/ligand binding. The entity under consideration binds to a receptor(s) located in the plasma (cell) membrane, which then invaginates (infolds) hence taking up the entity via "endosomes" (formed by pinching-off of infold to form a "bag") into vesicles located within the cell. It is one route to deliver essential metabolites to cells (e.g., low-density lipoprotein), and it is a means to modulate the cell's responses to many protein hormones and growth factors (e.g., insulin, epidermal growth factor, and nerve growth factor). It is a route by which certain proteins targeted for destruction can be taken up and delivered to the cell's lysosomes. For example, phagocytic cells have receptors enabling them to take up antigen-antibody complexes for subsequent destruction by the phagocytic cell. This route is also a means exploited by certain viruses and toxins to gain entry into cells through the otherwise impervious cell membranes (e.g., the AIDS virus and the Semliki Forest Virus). Disorders of endocytosis can lead to disease states (e.g., high cholesterol levels in the blood of people whose low-density lipoprotein receptors are impaired). Drugs (e.g., certain painkillers) can be targeted to specific receptors via receptor mapping (RM) and receptor fitting (RF) for greater efficacy. See also INVASIN, ADHESION MOLECULE, CD4 PROTEIN, EXOCYTOSIS, T CELL RECEPTORS, SIGNAL TRANSDUCTION, VAGINOSIS, RECEPTORS, RECEPTOR FITTING (RF), HIGH-DENSITY LIPOPROTEINS (HDLP), LOW-DENSITY LIPOPROTEINS (LDLP), RECEPTOR MAPPING (RM), SIGNALLING, and NUCLEAR RECEPTORS.

Endoglycosidase An enzyme capable of hydrolyzing (i.e., breaking) interior bonds in the oligosaccharide molecular branches of a glycoprotein molecule. That is, the enzyme is capable of cutting a sugar-to-sugar bond anywhere within the sugar polymer molecule (depending, of course, on the specificity of the enzyme). This is in contrast to an exoglycosidase, which must cut away at the polymer from the outside, that is, from the free end, one unit (or section as the case may be) at a time. See also EXOGLYCOSIDASE, GLYCOPROTEIN, ENZYME, OLIGOSACCHARIDES, RESTRICTION ENDOGLYCOSIDASES, and HYDROXYLATION REACTION.

Endometrium The lining of the uterus.

Endonucleases A class of enzymes capable of hydrolyzing (breaking) the interior phosphodiester bonds of DNA or RNA chains. As opposed to cleavage (by exonucleases) at the terminal bonds (ends) of a chain. See also EXONUCLEASE (its opposite) and ENDOGLYCOSIDASE, above, because the idea is the same.

Endophyte A microrganism (i.e., fungus) that lives inside vascular tissues of plants. At least one company has incorporated the gene for a protein toxic to insects (taken from *Bacillus thuringiensis*) into an endophyte to confer insect resistance to the plant. See also *BACILLUS THURINGIENSIS (B.t.)* and FUNGUS.

Endoplasmic Reticulum (ER) A highly specialized, complex network of branching, intercommunicating tubules (surrounded by membranes) found in the cytoplasm of most animal and plant cells. The two types of ER recognized are: the rough ER and smooth ER. ER that is covered with many ribosomes is called rough and the ER without or with fewer ribosomes attached is called smooth. This nomenclature comes about because of the appearance of the ER under high magnification. The rough ER is very well developed to facilitate cells carrying on abundant protein synthesis.

Endorphins Hormones produced in the brain, which act as natural painkillers. For example, runners and long-distance walkers achieve something of a "high" due to endorphins released by the brain during long runs or walks. See also ENKEPHALINS, CATECHOLAMINES, and HORMONE.

Endosome See ENDOCYTOSIS.

Endospore A highly resistant, dormant inclusion body formed within certain bacteria. To kill spores, temperatures above boiling are usually needed. For this, pressure cookers and autoclaves are required. Endospores have survival value since the spore can remain for long periods of time in a nongrowing state and then, under appropriate conditions, can be induced to germinate and regenerate the original cell. Endospore formation may be viewed as being akin to hibernation, that is, a kind of "bacterial hibernation."

Endothelial Cells These are the flat, sort of plate-shaped cells that line the surface of all blood vessels, heart, and lymphatics within the body. Endothelial cells possess transmembrane (i.e., through the cell membrane) molecules known as adhesion molecules, which selectively allow the passage (from bloodstream to tissues) of some molecules (e.g., leukocytes, monocytes, hormones, etc.). Endothelial cells are packed much tighter together in the capillaries that provide blood to the brain. This tighter packing limits the size and kind of molecules that can pass into the brain. This blood-brain barrier serves to protect the sensitive brain tissue from pathogens or harmful molecules. See ENDOTHELIUM, ADHESION MOLECULES, MONOCYTES, MITOGEN, SELECTINS, BLOOD-BRAIN BARRIER (BBB), LECTINS, and ELAM-1.

Endothelin A peptide that causes arteries to contract (which consequently causes blood pressure to increase). See also PEPTIDE and ATRIAL PEPTIDES.

Endothelium The layer of epithelial cells that line blood vessels throughout the body. The layer selectively allows the passage (from bloodstream to tissues) of nutrients, hormones, and other molecules that are essential for tissue growth and function. The endothelium is involved in the recovery and recycling of old red blood cells. It also produces two compounds that prevent blood clotting: prostacyclin and Von Willebrand factor. See also ENDOTHELIAL CELLS, SELECTINS, LECTINS, and ADHESION MOLECULES.

Endotoxin A lipopolysaccharide (fat/sugar complex; poison, also known as LPS) which forms an integral part of the cell wall of gram negative bacteria. It is only released when the cell is ruptured. It can cause, among other things, septic shock and tissue damage. Pharmaceutical preparations are routinely tested for the presence of endotoxins. This is one reason why pharmaceuticals must be prepared in a sterile environment. See also SEPSIS, BACTERIA, LIPIDS, POLYSACCHARIDES, TOXIN, GRAM-NEGATIVE (G−), and GOOD MANUFACTURING PRACTICES (GMP).

Engineered Antibodies Chimeric monoclonal antibodies, produced via genetic engineering of human antibody-producing cells (clones). For example, the genes coding for antilymphoma binding sites from a rat have been inserted into human antibody-producing cells to yield rat (antigen) binding sites mounted on human antibody "stems." See also CHIMERIC PROTEINS, MONOCLONAL ANTIBODIES (MAb), ANTIBODY, GENETIC ENGINEERING, COMBINING SITE, LYMPHOCYTE, and SEMISYNTHETIC CATALYTIC ANTIBODY.

Enhanced Nutrition Crops See "NUTRIENT ENHANCED™".

Enkephalins A class of hormones produced in the brain that act as natural painkillers. Discovered by John Hughes and Hans Kosterlitz in 1975, they are some of the endorphins. See also ENDORPHINS.

Ensiling The fermentation of (usually chopped-up) agricultural vegetation in order to preserve it. It is carried out for 1–2 weeks, using either indigenous microorganisms (e.g., *Lactobacillus* spp.) or introduced microorganisms (to speed up the process, yield product containing more nutrients for livestock, etc.), in the absence of oxygen (to prevent the growth of aerobic mold fungi). When indigenous microorganisms are used, *Lactobacillus* spp. become the dominant microorganisms present, and heat is generated by the microorganisms within the vegetative mass (optimum temperature is 25–30°C, which is 77–86°F). Lactic acid is produced by the microorganisms, which inhibits the growth of bacteria that would normally putrefy the vegetation. See also FERMENTATION, MICROORGANISM, AEROBIC, FUNGUS, and OPTIMUM TEMPERATURE.

Enterohemorrhagic *E. coli* The several dozen (approximately 60 known, as of 1996) serotypes (strains) of *E. coli* bacteria that cause internal hemorrhaging in humans that ingest those bacteria. The toxin produced by these particular *E. coli* bacteria attacks the human kidney, which often leads to kidney failure and/or death of infected humans. See also *ESCHERICHIA COLIFORM* 0157:H7, TOXIN, and SEROTYPES.

Enzyme An organic, protein-based catalyst that is not itself used up in the reaction. It is naturally produced by living cells to catalyze biochemical reactions. Each enzyme is highly specific with regard to the type of chemical reaction that it catalyzes, and to the substances (called substrates) upon which it acts. This specific catalytic activity and its control by other biochemical constituents are of primary importance in the physiological functions of all organisms. Although all enzymes are proteins, they may, and usually do, contain additional nonprotein components called coenzymes that are essential for catalytic activity. See also APOENZYME, CATALYST, COENZYME, HOLOENZYME, SUBSTRATE (CHEMICAL), PROTEIN, HORMONE, and EXTREMOZYMES.

Enzyme Denaturation The loss of enzyme (catalytic) activity due to loss of the correct functional structure of the protein. Denaturation may be caused by factors such as exposure to heat and organic solvents, degradation of the enzyme molecule by proteases, oxygen, and acid or alkaline pH. See also ENZYME, CONFORMATION, DENATURATION, and EXTREMOZYMES.

Enzyme Derepression Commonly known as induction (of an enzyme). Initially a repressor protein is bound to a specific region of DNA. This binding inhibits transcription to mRNA, thus blocking the synthesis of the protein (enzyme) specified by the mRNA. When present, the inducer molecule binds to the repressor protein and inactivates it. Thus the inhibition caused by the repressor protein is overcome and mRNA can be synthesized, which consequently leads to synthesis of the mRNA-specified protein (enzyme). The word derepression is sometimes used because the repressor protein is, by itself, active in repressing protein (enzyme) synthesis. Its repressive action is mitigated (derepressed) by the inducer molecule. Hence, derepression (or unrepression) of repression equals induction. See also CONTINUOUS PERFUSION, ENZYME REPRESSION, ENZYME, and REPRESSION (OF AN ENZYME).

Enzyme Immunoassay (EIA) See ELISA.

Enzyme Repression Inhibition of enzyme synthesis caused by the availability of the product of that enzyme. On a molecular level a repressor molecule (which could be, for example, the amino acid arginine) combines with a specific repressor protein that is present in the cell. This repressor molecule/repressor protein complex is then able to bind to a specific region of DNA at the initial end of the gene which is called the operator region. It is in this region where the synthesis of mRNA is initiated. The repressor "roadblock" thus stops the synthesis of mRNA, and therefore the synthesis of the protein is also blocked. See also ENZYME, REPRESSION (OF AN ENZYME), and ENZYME DEPRESSION.

Enzyme-Linked Immunosorbent Assay See ELISA.

Eosinophils Polymorphonuclear leukocytes made in the bone marrow. They circulate in the blood for a number of hours (three to eight) and then migrate into the tissue where they reside. They kill parasites too large to be phagocytized by secreting substances that kill the parasites (hookworms, trichinosis, etc.). They also inhibit histamine release from mast cells and secrete chemicals that neutralize histamine. Allergy causes an increase in eosinophils. GM-CSF stimulates eosinophil production. See also POLYMORPHONUCLEAR LEUKOCYTES (PMN), BASOPHILS, ANTIGEN, and CELLULAR IMMUNE RESPONSE.

EPD See EXPECTED PROGENY DIFFERENCES.

Epidermal Growth Factor (EGF) A protein of 53 amino acids that greatly increases growth/reproduction of epidermal (skin) cells. This protein also increases growth of wool in sheep. High concentrations of epidermal growth factor are found in human tears. EGF was discovered by Stanley Cohen. See also EGF RECEPTOR, NERVE GROWTH FACTOR (NGF), AMINO ACID, and FILLER EPITHELIAL CELLS.

Epimerase An enzyme capable of the reversible interconversion of two epimers. See also ENZYME and EPIMERS.

Epimers Two stereoisomers differing in configuration. See also CONFIGURATION and STEREOISOMERS.

Episome (of a bacterium) An independent genetic element (DNA) that occurs inside bacterium in addition to the normal bacterial cell genome. The episome can replicate either as an autonomous unit or as one integrated into the host genome. The F (fertility) factor is an episome. See also GENOME, PLASMID, BACTERIA, and DEOXYRIBONCULEIC ACID (DNA).

Epistasis Interaction between nonallelic genes in which the presence of a certain allele at one locus prevents expression of an allele at a different locus. See also ALLELE, GENE, EXPRESS, and LOCUS.

Epithelial Projections Projections that anchor the epidermis (surface skin) to the dermis (subsurface tissue). Growth of these projections is increased by epidermal growth factor during the wound healing process. See also EPIDERMAL GROWTH FACTOR (EGF).

Epithelium The prefix "epi-" means on, above, or upon. The membranous cellular tissue that covers a free surface or lines a tube or cavity of an animal body. It serves to enclose and protect the other tissues, to produce secretions and excretions, and to function in assimilation. See also ASSIMILATION.

Epitope Also called antigenic determinant. The specific group of atoms (on an antigen molecule) that is recognized by (that antigen's) antibodies. See also ANTIBODY, ANTIGEN, and IDIOTYPE.

EPO See ERYTHROPOIETIN and EUROPEAN PATENT OFFICE.

EPPO See EUROPEAN PLANT PROTECTION ORGANIZATION.

EPSP Synthase Enolpyruvyl-shikimate phosphate synthase. An enzyme produced by virtually all plants, it is essential in a plant's metabolism biochemical pathway, and for the biosynthesis (i.e., creation) of aromatic (ring-shaped) amino acids. Some (glyphosate-containing) herbicides kill unwanted plants (e.g., weeds) by inhibiting EPSP synthase. By incorporating a gene that causes (over-) production of CP4 EPSP synthase into several crops (e.g., soybeans, cotton, etc.), scientists have been able to help those crops to survive post-emergence application(s) of glyphosate-containing herbicide. Additional resistance to glyphosate-containing herbicide can be conferred to plants via incorporation into plants of a gene (GO) which causes those plants to produce glyphosate oxidase. See also ENZYME, METABOLISM, GENE, PAT GENE, BAR GENE, GENETIC ENGINEERING, SOYBEAN PLANT, CORN, GLYPHOSATE, GLYPHOSATE OXIDASE, CP4 EPSPS, HERBICIDE-TOLERANT CROP, and CHLOROPLAST TRANSIT PEPTIDE (CTP).

EPSPS See EPSP SYNTHASE and CP4 EPSPS.

Erythrocytes (red blood cells) Hemoglobin-containing cells (manufactured in the bone marrow) that transport the oxygen from the lungs to the body tissues where it is needed.

Erythropoiesis The formation of red blood cells from certain stem cells. Stimulated by the protein erythropoietin. See also STEM CELLS and ERYTHROPOIETIN (EPO).

Erythropoietin (EPO) A glycoprotein hormone produced in the kidneys that stimulates stem cells in the bone marrow to increase the number of red blood cells. Erythropoietin can be used to help correct a variety of anemias. See also GLYCOPROTEIN, HORMONE, ERYTHROCYTES, and STEM CELLS.

***Escherichia coli* 0157:H7** See *ESCHERICHIA COLIFORM* 0157:H7.

Escherichia coli See *ESCHERICHIA COLIFORM (E. COLI).*

Escherichia coliform *(E. coli)* A bacterium that commonly inhabits the human intestine as well as the intestine of other vertebrates (i.e., animals possessing a skeleton). The most thoroughly studied of all bacteria, *Escherichia coli* is used in many microbiological experiments. It has historically been considered the workhorse of genetic engineering research, and genetically engineered versions have been used to produce human proteins (e.g., insulin). One of the more exotic uses of genetically engineered *Escherichia coli* was to make indigo dye (originally discovered in 1983, using indole or tryptophan as starting materials). In 1993, Burt D. Ensley and coworkers at Amgen discovered a way to genetically engineer *Escherichia coli* to produce indigo from glucose starting material. *Escherichia coli* has 4,288 genes. See also TRYPTOPHAN (trp), BACTERIA,

GENETIC ENGINEERING, GENE, RECOMBINANT DNA (rDNA), and *ESCHERICHIA COLIFORM* 0157:H7 *(E. COLI* 0157:H7*)*.

***Escherichia coliform* 0157:H7** *(E. coli* 0157:H7*)* The particular strain (serotype) of *Escherichia coliform (E. coli)* bacteria that causes often-fatal diarrhea, internal bleeding, and kidney damage in humans. Although cattle were susceptible to *E. coli*'s toxins prior to the 1980s, they eventually developed resistance. That meant that the cattle could carry these bacteria without getting sick, and transmit *E. coli* 0157:H7 to humans whenever conditions allow (e.g., when *E. coli* 0157:H7-infected cattle are slaughtered and people consume the meat without first heating it to a high enough temperature to kill the *E. coli* 0157:H7). Some varieties of *E. coli* 0157:H7 are resistant to the antibiotics tetracycline and streptomycin.

In 1996, researchers at Cornell University in New York State, U.S.A. discovered that nonambulatory cows (that could not walk) were approximately four times as likely as other cows to test positive for *E. coli* 0157:H7. Other research in Canada indicates that fasting of cattle (common occurrence of nonambulatory cows) tends to alter the pH inside the cow's rumen (stomach) in a way that encourages the proliferation of *E. coli* 0157:H7 instead of the bacteria that normally populate the rumen. See also *ESCHERICHIA COLIFORM (E. COLI),* BACTERIA, SEROTYPE, TOXIN, BIOLUMINESCENCE, and STRAIN.

Essential Amino Acids Those amino acids that cannot be synthesized by humans and most other vertebrates, and therefore must be obtained from the diet. They are phenylalanine, valine, threonine, tryptophan, isoleucine, methionine, histidine, arginine, leucine, and lysine (glycine and proline for poultry). See AMINO ACID, LYSINE (lys), METHIONINE (met), and OPAGUE-2.

Essential Fatty Acids The group of polyunsaturated fatty acids of plants that are required in the human diet, because the human body cannot manufacture them, yet must have them for proper functioning. See also FATTY ACID, SOYBEAN OIL, LECITHIN, and FATS.

Essential Nutrients Chemical compounds in foods that are required for (consuming organism's) life, growth, or tissue repair, and cannot be synthesized by that organism. See ESSENTIAL AMINO ACIDS, ESSENTIAL FATTY ACIDS, and ESSENTIAL POLYUNSATURATED FATTY ACIDS.

Essential Polyunsaturated Fatty Acids See ESSENTIAL FATTY ACIDS.

EST Abbreviation for EXPRESSED SEQUENCE TAGS.

Estrogen A female sex hormone, secreted by the ovaries, that promotes estrus and helps to regulate the pituitary gland's production of luteinizing hormone (LH) and follicle-stimulating hormone (FSH). Estrogen is also responsible for the development of female secondary sex characteristics (e.g., smaller body size, lack of facial hair, higher pitch voice in humans). See also

HORMONE, PITUITARY GLAND, FOLLICLE-STIMULATING HORMONE (FSH), TESTOSTERONE, LUTEINIZING HORMONE (LH), and HYPOTHALAMUS.

Etiological Agent (of a disease) The microorganism (or other agent) that causes the disease. See also PATHOGEN and ETIOLOGY.

Etiology The science (study) of the cause (source) of a disease. See also PATHOGEN and ETIOLOGICAL AGENT (OF A DISEASE).

Eucaryote Also spelled eukaryote. A cell characterized by compartmentalization (by membranes) of its extensive internal structures; or an organism made up of such cells. For example, eucaryotes possess a distinct membrane-surrounded nucleus containing the DNA. Eucaryotic cells (e.g., human cells) are much larger and more complex than procaryotic cells (e.g., bacteria). The cells of all higher organisms, both plant and animal, are eucaryotic. See also PROCARYOTES, CELL, and DEOXYRIBONUCLEIC ACID (DNA).

Eugenics First formulated by Francis Galton, who was a contemporary of Gregor Mendel in the 19th century, eugenics is the concept that a species can be "improved" by encouraging reproduction of only those organisms in that species that possess "desired" traits.

This belief became popular in a number of countries during the early 20th century. Margaret Sanger, founder of America's Planned Parenthood organization, referred to African Americans as "human weeds" and called for "more children from the fit, less from the unfit." Based upon Charles Darwin's written assertion that "the civilized races of man will almost certainly exterminate and replace the savage races," a number of large genocides were committed by national governments. See also GENETICS, GENE, TRAIT, GENOTYPE, HEREDITY, HERITABILITY, and GENOME.

Euploid A cell carrying an exact multiple of the haploid chromosome number. For example, a diploid possesses twice the haploid number of chromosomes. See also HAPLOID, DIPLOID, and CHROMOSOMES.

European Corn Borer (ECB) Also known as pyralis. Latin name *Ostrinia nubialis*, it is an insect whose larvae eat and bore into the corn/maize plant (*Zea mays L.*). In doing so, they can act as vectors (i.e., carriers) of the fungi known as *Aspergillus flavus* (source of aflatoxin) or *Fusarium moniliforme* (source of fumonisin). ECB control can be effected via some of the following methods:

- spraying of conventional synthetic chemical pesticides
- spraying of pesticides produced via promulgation of *Bacillus thuringiensis (B.t.)* bacteria
- incorporating a (protoxin) gene from *Bacillus thuringiensis (B.t.)* into the DNA of the corn plant, so that the plant itself produces *B.t.* protoxin.

As part of Integrated Pest Management (IPM), farmers can utilize:

(a) Corn possessing *Bacillus thuringiensis (B.t.)* gene(s) to control populations of ECB without applying insecticides
(b) The parasitic *Euplectrus comstockki* wasp to help control the ECB. When that wasp's venom is injected into ECB larva, it stops the larva from molting (and thus maturing).
(c) Other additional methods, alone or in concert with above

See CORN, AFLATOXIN, INTEGRATED PEST MANAGEMENT (IPM), *BACILLUS THURINGIENSIS (B.t.)*, *B.t. KURSTAKI*, FUSARIUM, FUSARIUM MONILIFORME, and PROTOXIN.

European Medicines Evaluation Agency (EMEA) A London-based agency of the European Union (EU) that began operation in 1995. It coordinates drug licensing and safety matters throughout the nations of the EU. Its licensing/approval process is compulsory throughout the EU. See also COMMITTEE FOR PROPRIETARY MEDICINAL PRODUCTS (CPMP), MEDICINES CONTROL AGENCY (MCA), FOOD AND DRUG ADMINISTRATION (FDA), KOSEISHO, BUNDESGESUNDHEITSAMT (BGA), COMMITTEE ON SAFETY IN MEDICINES, and COMMITTEE FOR VETERINARY MEDICINAL PRODUCTS (CVMP).

European Patent Convention An international patent treaty signed in 1973, by which the countries of Europe agreed to recognize and honor the patents granted by each country, plus those patents granted by the European Patent Office (EPO). Plant varieties or animal breeds were initially excluded from patentability by the European Patent Convention. In 1997, the European Parliament removed that exclusion. See also EUROPEAN PATENT OFFICE (EPO), U.S. PATENT AND TRADEMARK OFFICE (USPTO), PLANT'S NOVEL TRAIT (PNT), PLANT BREEDER'S RIGHTS (PPR), and UNION FOR PROTECTION OF NEW VARIETIES OF PLANTS (UPOV).

European Patent Office (EPO) The Munich, Germany based agency of the European Union (EU)—established in 1977—that is responsible for common patent protection matters for all of the (EU) member countries, plus the non-EU countries of Switzerland and Liechtenstein. The European Patent Office originally did not allow a "plant or animal breed" to be patented, whereas its American counterpart—the U.S. Patent and Trademark Office (USPTO), does allow patenting of microbes, plants, and animals (e.g., those which have been genetically engineered by man). In 1997, the European Parliament removed that exclusion, thus making the two patent systems compatible. See also EUROPEAN PATENT CONVENTION, MICROBE, GENETIC ENGINEERING, BIOTECHNOLOGY, AMERICAN TYPE CULTURE COLLECTION (ATCC), U.S. PATENT AND TRADEMARK OFFICE (USPTO), PLANT'S NOVEL TRAIT (PNT), PLANT BREEDER'S RIGHTS (PBR), and UNION FOR PROTECTION OF NEW VARIETIES OF PLANTS (UPOV).

European Plant Protection Organization (EPPO) One of the international SPS standard-setting organizations that develops plant health standards, guidelines and recommendations (e.g., to prevent transfer of a plant disease or plant pest from one country to another). Its secretariat is in Paris, France.

EPPO is one of the organizations within the International Plant Protection Convention (IPPC), and it covers the countries of Europe. See also INTERNATIONAL PLANT PROTECTION CONVENTION (IPPC), NORTH AMERICAN PLANT PROTECTION ORGANIZATION (NAPPO), SPS, PLANT'S NOVEL TRAIT (PNT), and PLANT BREEDER'S RIGHTS (PBR).

Event Refers to each instance of a genetically engineered organism. For example, the same gene inserted by man into a given plant genome at two different locations (i.e., loci) along that plant's DNA would be considered two different "events." Alternatively, two different genes inserted into the same locus of two same-species plants would also be considered two different "events."

Generally speaking, the world's regulatory agencies confer new biotech-derived product approvals in terms of events. See GENETIC ENGINEERING, GENETICALLY ENGINEERED ORGANISM (GEO), GENE, DEOXYRIBONUCLEIC ACID (DNA), LOCUS, LOCI, GENOME, and MUTUAL RECOGNITION AGREEMENTS (MRAs).

Ex vivo (testing) The testing of a substance by exposing it to (excised) living cells (but not to the whole, multicelled organism) in order to ascertain the effect of the substance (e.g., pharmaceutical) on the biochemistry of the cell. See also *IN VITRO* and *IN VIVO*.

Ex vivo (therapy) Removal of cells (e.g., certain blood cells) from a patient's body, alteration of those cells in one or more therapeutic ways, followed by re-insertion of the altered cells into the patient's body. See also *IN VITRO* and *IN VIVO*.

Excision The cutting out of a piece of damaged or defective DNA by enzymes. DNA damage might be constituted by the presence of a thymine dimer which inactivates that part of the DNA. The region of the dimer is cut out and it is then repaired. See also RECOMBINATION, GENOME, and INFORMATIONAL MOLECULES.

Excitatory Amino Acids (EAAs) Amino acids present in the brain, which can kill brain cells when in excess (results from strokes, which cause the release of too many EAAs in the brain). Some spiders paralyze their prey with venom that contains a substance that blocks the action of EAAs (thus may be used to prevent brain damage in stroke victims). See also AMINO ACID.

Exclusion Chromatography See GEL FILTRATION.

Exergonic Reaction A chemical reaction with a negative standard free energy change (i.e., a "downhill" reaction). A reaction which releases energy (exothermic; in the form of heat). See also ENDERGONIC REACTION and FREE ENERGY.

Exobiology Extraterrestrial biology.

Exocytosis The releasing of an entity that was bound inside an "endosome" (e.g., inside a cell). See ENDOCYTOSIS.

Exoglycosidase An enzyme that hydrolyzes (cuts) only a terminal (i.e., end) bond in the oligosaccharide (molecular) branch(es) of a glycoprotein. See also ENDOGLYCOSIDASE, GLYCOPROTEIN, and RESTRICTION ENDOGLYCOSIDASES.

Exon The segment of a eucaryotic gene that is transcribed into an mRNA (messenger RNA) molecule; it codes for a specific domain of a protein. See also PROTEIN, EUCARYOTE, MESSENGER RNA (mRNA), GENE, and HOMEOBOX.

Exonuclease An enzyme that hydrolyzes (cuts) only a terminal phosphodiester bond of a nucleic acid. See also HYDROLYZE.

Exotoxin Proteins (toxins) produced by certain bacteria that are released by the bacteria into their surroundings (growth medium). Produced by primarily Gram-positive bacteria. Diphtheria toxin was the first one discovered. Other exotoxins cause botulism, tetanus, gas gangrene, and scarlet fever. Exotoxins are generally more potent and specific in their actions than endotoxins. See also ENDOTOXIN and TOXIN.

Expected Progeny Differences (EPD) Numerical rankings of (livestock) parental genetics, in terms of an animal's genetic impact on progeny's four following commercial traits:

(1) Number of progeny born alive
(2) Weight of progeny at weaning age
(3) Number of days required to reach slaughter weight, when fed adequately
(4) Carcass lean meat versus fat percentages

EPDs allow a farmer to estimate differences in performance of future offspring (of a given parent) versus offspring produced by parents of average genetic value. For example, a boar (male pig) possessing an EPD of −4 for "number of days required to reach slaughter weight" produces offspring that reach slaughter weight in four fewer days (of feeding time) than offspring that are sired by a boar possessing an EPD of 0. See also GENETICS, TRAIT, PHENOTYPE, GENOTYPE, and BEST LINEAR UNBIASED PREDICTION (BLUP).

"Explosion" Method [to introduce foreign (new) genes into plant cells] A technique for gene-into-cell introduction in which the gene (genetic material) is driven into plant cells by the force of an explosion (vaporization) of a drop of water (to which the gene and gold particles have been added). The explosion is caused by application of high-voltage electricity to the drop of gene-laden water; the water then vaporized explosively, driving the "shot" (gold particles) and genetic material through the cell membrane. The plant cell then heals itself (reseals the hole where the gene entered), incorporates the new gene into its genetic complement, and produces whatever product (e.g., a protein) that the newly introduced gene codes for. See also *AGROBACTERIUM TUMEFACIENS,* CODING SEQUENCE, GENETIC ENGINEERING, VECTOR, "SHOTGUN" METHOD [TO INTRODUCE FOREIGN (NEW) GENES INTO PLANT CELLS], GENE, GENOME, and RIBOSOMES.

Express To translate the cell's genetic information stored in the DNA (gene) into a specific protein (synthesized by the cell's ribosome system). See also RIBOSOMES, GENE, DEOXYRIBONUCLEIC ACID (DNA), CELL, TRANSCRIPTION, TRANSLATION, MESSENGER RNA (mRNA), and TRANSCRIPTION UNIT.

Expressivity The intensity with which the effect of a gene is realized in the phenotype. The degree to which a particular effect is expressed by individuals. See also PHENOTYPE, EXPRESS, and RIBOSOMES.

Extension (in nucleic acids) The nucleic acid strand elongation (lengthening) that occurs in a polymerization reaction. See also NUCLEIC ACIDS and POLYMER.

Extranuclear Genes Genes that reside within the cell, but *outside* the nucleus. Generally, extranuclear genes reside inside organelles such as mitochondria and chloroplasts. See also GENE, CELL, NUCLEUS, COPY NUMBER, ORGANELLES, CHLOROPLASTS, and MITOCHONDRIA.

Extremophilic Bacteria Bacteria that live and reproduce outside (either colder or hotter) the typical temperature range of 40°F (4°C) to 140°F (60°C) that bacteria tend to be found in, on Earth. Other extremes are high pressure (e.g., at the ocean bottom), salt saturation, (e.g., the Dead Sea), pH lower than 2 (e.g., coal deposits), pH higher than 11 (e.g., sewage sludge). See also THERMOPHILIC BACTERIA, THERMOPHILE, and THERMODURIC.

Extremozymes Enzymes within the microorganisms (e.g., extremophilic bacteria) that populate extreme environments. Because extremozymes can catalyze reactions under high pressure, high temperatures, etc., they are increasingly being used as catalysts for industrial processes. See also EXTREMOPHILIC BACTERIA, ENZYME, and ARCHAEA.

F

F1 Hybrids The first-generation offspring of crossbreeding; also known as first filial hybrids. They tend to be more healthy, productive, and uniform than their parents. See also GENETICS and HYBRIDIZATION (PLANT GENETICS).

FACS See FLUORESCENCE ACTIVATED CELL SORTER.

Factor VIII Also known as antihemophilic globulin (AHG) or antihemophilic factor VIII. A protein factor in the blood serum that is instrumental in the "cascade" of chemical reactions (involving 17 blood components) that leads to clot formation following a cut or other wound to tissue. Also, a deficiency of AHG is the cause of the classical type of hemophilia sometimes known as hemophilia A. See also FIBRIN and FIBRONECTIN.

Facultative Anaerobe An organism that will grow under either aerobic or anaerobic conditions. See also AEROBE and ANAEROBE.

Facultative Cells Cells that can live either in the presence or absence of oxygen. See also AEROBE and ANAEROBE.

FAD See FLAVIN ADENINE DINUCLEOTIDE.

FAO Food and Agriculture Organization of the United Nations. See CONSULTATIVE GROUP ON INTERNATIONAL AGRICULTURAL RESEARCH and CODEX ALIMENTARIUS COMMISSION.

Fats Energy storage substances produced by animals and some plants, which consist of a combination of fatty acids.

The content levels of individual fatty acids vary somewhat with the diet of the animal (i.e., for animal fat) and vary somewhat with the plant's growing conditions (i.e., for plant fat also known as vegetable oil). No natural fat is either totally saturated or unsaturated.

When eaten, fats are generally not absorbed directly through the intestinal wall. They are first emulsified, then hydrolyzed by the lipase enzyme. The components (i.e., fatty acids, cholesterol, monoacylglycerol, phospholipids, etc.) form micelles that pass through the intestinal wall and are absorbed by the body. When fats are oxidized in cells, they provide energy for the body. Some of the energy is released as heat and some is stored in the form of adenosine triphosphate (ATP), which "fuels" metabolic processes. See FATTY ACID, HYDROLYSIS, HYDROLYTIC CLEAVAGE, HYDROLYZE, LIPASE, MONOUNSATURATED FATS, SATURATED FATTY ACIDS, TRIGLYCERIDES, MICELLE, CELL, METABOLISM, and DIGESTION (WITHIN ORGANISMS).

Fatty Acid A long-chain aliphatic acid found in natural fats and oils. Fatty acids are abundant in cell membranes and (after extraction/purification) are widely used as industrial emulsifiers, for example, phosphatidylcholine (lecithin). In general, fats possessing the highest levels of saturated fatty acids tend to be solid at room temperature; and those fats possessing the highest levels of unsaturated fatty acids tend to be liquid at room temperature. That rule of thumb was the original "dividing line" between compounds called *fats* and *oils*, respectively. In general, saturated fatty acids tend to be more stable (resistant to oxidation and thermal breakdown) than unsaturated fatty acids.

Fatty acids in biological systems (e.g., produced by plants in oilseeds, etc.) tend to contain an even number of carbon atoms in their molecular "backbone," typically between 14 and 24 carbon atoms. The molecular backbone (alkyl chain) may be saturated (no double bonds) or it may contain one or more double bonds. The configuration of the double bonds in most unsaturated fatty acids is *CIS*. See also ESSENTIAL FATTY ACIDS, LAURATE, PHYTOCHEMICALS, SATURATED FATTY ACIDS, LECITHIN, UNSATURATED FATTY ACID, MONOUNSATURATED FATS, POLYUNSATURATED FATTY ACIDS (PUFA), LPAAT PROTEIN, STEAROYL-ACP DESATURASE, SOYBEAN OIL, CANOLA, FATS, OLEIC ACID, and *TRANS* FATTY ACIDS.

Federal Insecticide, Fungicide, and Rodenticide Act (FIFRA) A law enacted by the United States Congress in 1972. During 1994, the U.S. Environmental Protection Agency (EPA) proposed that the substances produced by plants for their defense against pests and diseases would be regulated by EPA under FIFRA. See also TOXIC SUBSTANCES CONTROL ACT (TSCA), GENETICALLY ENGINEERED MICROBIAL PESTICIDES (GEMP), WHEAT TAKE-ALL DISEASE, and *BACILLUS THURINGIENSIS (B.t.)*.

Feedback Inhibition Inhibition of the first enzyme in a metabolic pathway by the end product of that pathway. This is a method of shutting down a metabolic pathway that is producing a product that is no longer needed. See also METABOLISM, ENZYME, and EFFECTOR.

Feedstock Raw material(s) used for the production of chemicals; or growth substrates of microbes (e.g., yeasts or bacteria that require a solid phase to attach themselves to).

Fermentation A term first used with regard to the foaming that occurs during the manufacture of wine and beer. The process dates back to at least 6,000 B.C. when the Egyptians made wine and beer by fermentation. From the Latin word *fermentare*, "to cause to rise." The term "fermentation" is now used to refer to so many different processes that fermentation is no longer accepted for use in most scientific publications. Three typical definitions are given below:

(1) A process in which chemical changes are brought about in an organic substrate through the actions of enzymes elaborated (produced) by microrganisms.

(2) The enzyme-catalyzed, energy-yielded pathway in cells by which "fuel" molecules such as glucose are broken down anaerobically (in the absence of oxygen). One product of the pathway is always the energy-rich compound adenosine triphosphate (ATP). The other products are of many types: alcohol, glycerol, and carbon dioxide from yeast fermentation of various sugars; butyl alcohol, acetone, lactic acid, and acetic acid from various bacteria; citric acid, gluconic acid, antibiotics, vitamin B_{12} and B_2 from mold fermentation. The

Japanese utilize a bacterial fermentation process to make the amino acid, L-glutamic acid, a derivative of which is widely used as a flavoring agent.

(3) An enzymatic transformation of organic substrates (feedstocks), especially carbohydrates, generally accompanied by the evolution of gas. A physiological counterpart of oxidation, permitting certain organisms to live and grow in the absence of air; used in various industrial processes for the manufacture of products such as alcohols, acids, and cheese by the action of yeasts, molds, and bacteria. Alcoholic fermentation is the best known example. Also known as zymosis. See also ZYMOGENS. The leavening of bread depends on the alcoholic fermentation of sugars. The dough rises due to production of carbon dioxide gas that remains trapped within the viscous dough.

See also SUBSTRATE (CHEMICAL), ADENOSINE TRIPHOSPHATE (ATP), MICROORGANISM, ENZYME, FEEDSTOCK, and CARBOHYDRATES (SACCHARIDES).

Ferritin An iron-protein complex (a metalloprotein) that occurs in living tissues. Functions in iron storage in spleen. See also HEMOGLOBIN.

Ferrobacteria Also called iron bacteria. Any of a group of bacteria that oxidize iron as a source of energy. The oxidized iron in the form $Fe(OH)_3$ is then deposited in the environment by secretion from the bacterium. The energy obtained from these reactions is used to carry on processes in which the basic substances needed by the bacterium are manufactured. These bacteria are commonly found in seepage waters of coal and iron mining areas where iron compounds abound. Ferrobacteria are not disease producers (i.e., pathogenic), but they are important as scavengers. Sometimes they create a nuisance by multiplying so profusely in iron water pipes that they stop the flow of water. Ferrobacteria have been active through long periods of geologic time. For example, the great Mesabi iron (ore) seam of America's Lake Superior region is thought to be a product of ferrobacteria activity. See also PATHOGEN.

Ferrochelatase A mitochondrial enzyme that catalyzes the incorporation of iron into the protoporphyria molecule. See also MITOCHONDRIA, ENZYME, CATALYST, and PORPHYRINS.

Ferrodoxin An iron- and sulfur-containing protein important in the electron-transfer processes of photosynthesis in plants. It also plays a role in the metabolism of some bacteria and was first found in an anaerobic bacterium. See also PHOTOSYNTHESIS and METABOLISM.

Fertility Factor (F) A type of transmissible (i.e., can enter other cells) plasmid that is often found in *Escherichia coli (E. coli)*. See also PLASMID, VECTOR, and *ESCHERICHIA COLI (E. COLI)*.

Fertilization The union of the (haploid) male and (haploid) female germ cells (sex cells or gametes) to produce a diploid zygote. Fertilization marks the start of development of a new individual (organism), the beginning of cell differentiation. See also GERM CELL.

FGF See FIBROBLAST GROWTH FACTOR (FGF).

FGMP See FOOD GOOD MANUFACTURING PRACTICE (FGMP).

FIA Refers to immunodiagnostic tests that are based on fluorescence tracers (labels). See also IMMUNOASSAY, FLUORESCENCE, and RADIOIMMUNOASSAY.

Fibrin The ordered fibrous array of fibrin monomers, called a fibrin-platelet clot (blood clot), which spontaneously assembles from fibrin monomers (which themselves are formed by the thrombin-catalyzed conversion of fibrinogen into fibrin). Fibrinogen itself is the product of a controlled series of zymogen activation steps (enzymatic cascade) triggered initially by substances that are released from body tissues as a consequence of trauma (harm) to them. See also FIBRONECTIN, ZYMOGENS, and LIPOPROTEIN-ASSOCIATED COAGULATION (CLOT) INHIBITOR (LACI).

Fibrinogen See FIBRIN and LIPOPROTEIN-ASSOCIATED COAGULATION (CLOT) INHIBITOR (LACI).

Fibrinolytic Agents Blood-borne compounds that activate fibrin in order to dissolve blood clots. See also TISSUE PLASMINOGEN ACTIVATOR (tPA), THROMBOLYTIC AGENTS, and FIBRIN.

Fibroblast Growth Factor (FGF) First described in the mid-1970s by Dr. Gospodarowicz and fellow researchers at the University of California, San Francisco. It is a protein that stimulates the formation/development of blood vessels and fibroblasts (precursors to collagen, the connective tissue "glue" that holds cells together). FGF also is mitogenic (causes cells to divide and multiply) for both fibroblasts and endothelial cells, and attracts those two cell types (i.e., is chemotactic). Dr. Gospodarowicz named the FGF originally derived from bovine (cow) brain tissue to be Acidic FGF. Dr. Gospodarowicz named the FGF originally derived from bovine pituitary tissue to be Basic FGF. This was due to their identical *biological* activity, but differing isoelectric points (i.e., the former being acidic, and the latter being basic). Basic FGF is, however, ten times more "potent" than acidic FGF in most bioassays. See also ANGIOGENIC GROWTH FACTORS, PROTEIN, FIBROBLASTS, PITUITARY GLAND, COLLAGEN, MITOGEN, ENDOTHELIAL CELLS, CHEMOTAXIS, BIOLOGICAL ACTIVITY, BIOASSAY, ACID, and BASE.

Fibroblasts Cells that are precursors to the connective tissue cells found in the skin. They make structural proteins like collagen, which gives skin its strength. Because fibroblasts do not express antigens on their cell surfaces (free standing, separated), fibroblasts possess potential for use in making artificial organs (e.g., artificial pancreas for diabetics), since recipent immune system cannot recognize the fibroblast cells as foreign. See also CELLULAR IMMUNE RESPONSE, HUMORAL IMMUNITY, GRAFT-VERSUS-HOST DISEASE (GVHD), XENOGENEIC ORGANS, CELL, FIBROBLAST GROWTH FACTOR (FGF), and COLLAGEN.

Fibronectin An adhesive glycoprotein that forms a link between the epithelial cells and the connective tissue matrix (essential for blood clotting). Research has indicated that fibronectin may solve the problem of getting new cells to stick to existing tissue, once a growth factor has caused them to grow (e.g., when growth factor is administered after a serious wound to tissue). See also FIBRIN, GLYCOPROTEIN, GROWTH FACTOR, and ORGANOGENESIS.

Field Inversion Gel Electrophoresis (FIGE) A chromatographic procedure for the separation of a mixture of molecules by means of a two-dimensional electrical field, applied across a gel matrix containing those molecules. For example, FIGE is commonly used to separate mixtures of large DNA molecules by their size and (electrical) charge. FIGE can be used to separate (resolve) DNA molecules up to 2000 Kbp in length. See also CHROMATOGRAPHY, ELECTROPHORESIS, KILOBASE PAIRS (Kbp), POLYACRYLAMIDE GEL ELECTROPHORESIS (PAGE), and DEOXYRIBONUCLEIC ACID (DNA).

FIFRA See FEDERAL INSECTICIDE, FUNGICIDE, and RODENTICIDE ACT (FIFRA).

Filler Epithelial Cells Skin cells that initially form under a scab in the wound healing process, in response to stimulation by epidermal growth factor (EGF). See also EPIDERMAL GROWTH FACTOR (EGF).

Fingerprinting See PEPTIDE MAPPING ("FINGERPRINTING") and COMBINATORIAL CHEMISTRY.

First Filial Hybrids See F1 HYBRIDS.

Flagella A protein-based, flexible, whip-like organ of locomotion found on some microorganisms. With these, microorganisms are able to swim. Flagella are usually very long and there are usually only one or two per cell. The tails of sperm cells are examples of flagella. Flagella are used in the swimming motion of bacteria towards sources of nutrients in a process called chemotaxis. Singular: flagellum. See also CILIA, CHEMOTAXIS, BACTERIA, and PROTEIN.

Flanking Sequence A segment of DNA molecule that either precedes or follows the region of interest on the molecule. See also DEOXYRIBONUCLEIC ACID (DNA).

Flavin Also known as lyochrome. One of a group of pale yellow, greenly fluorescing biological pigments widely distributed in small quantities in plant and animal tissues. Flavins are synthesized only by bacteria, yeast, and green plants; for this reason, animals are dependent on plant sources for riboflavin (vitamin B_2), the most prevalent member of the group.

Flavin Adenine Dinucleotide (FAD) The coenzyme of some

oxidation-reduction enzymes; it contains riboflavin. See also FLAVIN and COENZYME.

Flavin Mononucleotide (FMN) Riboflavin phosphate, a coenzyme of certain oxido-reduction enzymes. See also COENZYME.

Flavin Nucleotides Nucelotide coenzymes (FMN and FAD) containing riboflavin. See also FLAVIN MONONUCLEOTIDE (FMN) and FLAVIN ADENINE DINUCLEOTIDE (FAD).

Flavin-Linked Dehydrogenases Dehydrogenases are enzymes (involved in removing hydrogen atoms from their substrate) which require one of the riboflavin coenzymes, FMN or FAD, in order to function. See also DEHYDROGENASES, FLAVIN MONONUCLEOTIDE (FMN), FLAVIN ADENINE DINUCLEOTIDE (FAD), and SUBSTRATE (CHEMICAL).

Flavinoids See FLAVONOIDS.

Flavonoids A category of phytochemicals, that are beneficial to the health of humans that consume them. Hundreds of flavonoids are naturally produced (by plants) in common human foods. One flavonoid is quercetin, a non-nutritive antioxidant produced in almonds. See PHYTOCHEMICALS.

Flavoprotein An enzyme containing a flavin nucleotide as a prosthetic group. See also PROSTHETIC GROUP.

FLK-2 Receptors See TOTIPOTENT STEM CELLS.

Flora The microorganisms found in a given situation, e.g., reservoir flora (the microorganisms present in a given municipal water reservoir) or intestinal flora (the microorganisms found in the intestines).

Floury-2 A gene in corn/maize (*Zea mays L.*) that (when present in the DNA of a given plant) causes that plant to produce seed that contains higher-than-traditional levels of the amino acids methionine and tryptophan. See also GENE, CORN, METHIONINE (met), HIGH-METHIONINE CORN, ESSENTIAL AMINO ACIDS, VALUE-ENHANCED GRAINS, and DEOXYRIBONUCLEIC ACID (DNA).

Fluorescence The reaction of certain molecules upon absorption of specific amount/wavelength of light; in which those molecules emit (reradiate) light energy possessing a longer wavelength than the original light absorbed. All cells will naturally fluoresce, at least a bit.

Human colon cancer cells, and precursor cells, fluoresce much more (and emit much more red light when they fluoresce) than noncancerous cells; which may lead to a new and better means of early detection. See also CELL, CANCER, FIA, and BRIGHT GREENISH-YELLOW FLUORESCENCE (BGYF).

Fluorescence Activated Cell Sorter (FACs) A machine that is used to sort cells from a mixed group of cells (e.g., to remove only the cells into which a new gene has been inserted via genetic engineering techniques). The desired cells are first labeled with a specific fluorescent dye, then passed through a flow chamber that is illuminated by a laser beam, which causes the labeled cells to fluoresce (i.e., glow). The molecules of the fluorescent dye, which "stick" to only one type of cell in the mixture, contain chromophores that can be elevated to an excited, unstable state via irradiation with specific wavelength(s) of light. Those chromophores remain in that excited state for a maximum of 10^{-9} seconds before releasing their energy by emitting light, and returning to their unexcited "ground" state. This fluorescence (glow) is a measurable property and the FACS machine utilizes it to separate the desired cells from the rest of the mixture. See also BASOPHILIC, GENETIC ENGINEERING, CELL, and FLUORESCENCE.

Follicle Stimulating Hormone (FSH) A protein hormone used in conventional medical therapy in an attempt to increase production of sperm in men (inside the follicles of the testes). See also THYROID STIMULATING HORMONE (TSH), GRAVE'S DISEASE, PROTEIN, HORMONE, and PITUITARY GLAND.

Food and Drug Administration (FDA) The federal agency charged with approving all pharmaceutical and food ingredient products sold within the United States. See also KOSEISHO, COMMITTEE FOR PROPRIETARY MEDICINAL PRODUCTS (CPMP), COMMITTEE FOR VETERINARY MEDICINAL PRODUCTS (CVMP), COMMITTEE ON SAFETY IN MEDICINES, "TREATMENT" IND REGULATIONS, KEFAUVER RULE, IND, IND EXEMPTION, PHASE I CLINICAL TESTING, EUROPEAN MEDICINES EVALUATION AGENCY (EMEA), MEDICINES CONTROL AGENCY (MCA), and BUNDESGESUNDHEITSAMT (BGA).

Food Good Manufacturing Practice (FGMP) The Food and Drug Administration's (FDA's) approval mechanism for a process to manufacture a given food or food additive. It is implemented instead of specific regulations (such as those used to dictate processes in simple food manufacture, as in beef packing), due to the newness of the technology, and may later be superceded (due to further advances in the technology). See also FOOD AND DRUG ADMINISTRATION (FDA).

Footprinting A technique used by researchers to determine precisely *where* (on DNA molecule) certain DNA-binding proteins make specific contact with that DNA molecule. For example, certain types of drugs act by binding tightly to certain DNA molecules in specific locations (e.g., in order to halt cancerous growth of cells, etc.). See also DEOXYRIBONUCLEIC ACID (DNA), PROTEIN, and GENOTOXIC.

For Treatment IND See "TREATMENT" IND REGULATIONS.

Formaldehyde Dehydrogenase An enzyme which catalyzes the oxidation of formaldehyde to formic acid (formate at intracellular pH). It requires NAD (i.e., nicotinamide-adenine dinucleotide) as an electron acceptor. It is important in the metabolism of methanol. See also METABOLISM, ENZYME, NAD (NADH, NADP, NADPH), and CATALYST.

Forward Mutation A mutation from the wild (natural) type to the mutant (type). See also MUTATION and WILD TYPE.

FOS See FRUCTOSE OLIGOSACCHARIDES.

FOSHU A Japanese government designation meaning "Foods of Specified Health Use." See NUTRACEUTICALS and PHYTOCHEMICALS.

Foundation on Economic Trends A small group that lobbies against biotechnology. See also BIOTECHNOLOGY.

Frameshift A shift (displacement) of the reading frame in a DNA or RNA molecule. Frameshifts generally result from the addition or deletion of one or more nucleotides to/from the DNA or RNA molecule. See also READING FRAME, CODON, GENETIC CODE, MUTATION, DEOXYRIBONUCLEIC ACID (DNA), NUCLEOTIDE, and RIBONUCLEIC ACID (RNA).

Free Energy The component of the total energy of a system that can do work at a constant temperature and pressure. Also known as Gibbs free energy.

Fructose Oligosaccharides A family of oligosaccharides, some of which help to foster the growth of bifidobacteria in the lower colon of monogastric animals (e.g., humans, pigs, etc.). Those bifidobacteria generate certain short-chain fatty acids, which are absorbed by the colon and result in a reduction of cholesterol levels in the bloodstream. See OLIGOSACCHARIDES, CHOLESTEROL, HIGH-DENSITY LIPOPROTEINS (HDLPs), LOW-DENSITY LIPOPROTEINS (LDLPs), BACTERIA, FATTY ACID, and MANNANOLIGOSACCHARIDES (MOS).

Fumarase (fum) An enzyme that catalyzes the hydration (addition of hydrogen atoms) of fumaric acid to maleic acid, as well as the reverse dehydration reaction (removal of hydrogen atoms). See also ENZYME and CATALYST.

Fumaric Acid ($C_4H_4O_4$) A dicarboxylic organic acid produced commercially by chemical synthesis and fermentation; the *trans* isomer of maleic acid; colorless crystals, melting point 87°C (191°F); used to make resins, paints, varnishes and inks, in food, as a mordant (dye fixer/stabilizer), and as a chemical intermediate. Also known as BOLETIC ACID. See ACID and ISOMER.

Fumonisins Mycotoxins that are primarily produced by the fungus *Fusarium moniliforme*. See MYCOTOXINS, FUNGUS, *FUSARIUM*, *FUSARIUM MONILIFORME*, and EUROPEAN CORN BORER (ECB).

Functional Foods See NUTRACEUTICALS and PHYTOCHEMICALS.

Functional Genomics Study of, or discovery of, what traits/functions are conferred to an organism by given (gene) sequences. Typically, functional genomic study follows after discovery of gene sequences found via structural genomics study.

Some methods utilized to determine which traits/functions result from which gene(s) are:

(a) Site-directed mutagenesis (SDM), to compare two same-species organisms possessing two different genes at the same site on the genome
(b) Antisense DNA sequence, to compare two same-species organisms (one of which has gene at same site "turned off" via antisense DNA)
(c) Reporter gene, to compare two same-species organisms (with two different genes at same site on genome) via a "reporter" gene adjacent to gene/site, to detect presence of desired trait/function

See GENOMICS, TRAIT, GENE, GENOTYPE, PHENOTYPE, POLYGENIC, STRUCTURAL GENE, STRUCTURAL GENOMICS, DEOXYRIBONUCLEIC ACID (DNA), SEQUENCE (OF A DNA MOLECULE), PLEIOTROPIC, GENETIC CODE, INFORMATIONAL MOLECULES, POINT MUTATION, SITE-DIRECTED MUTAGENESIS (SDM), ANTISENSE (DNA SEQUENCE), REPORTER GENE, and POSITIONAL CLONING.

Fungicide Any chemical compound that is toxic to fungi. See also BIOCIDE and FUNGUS.

Fungus (plural: fungi) Any of a major group of saprophytic and parasitic plants that lack chlorophyll and flowers, including molds, toadstools, rusts, mildews, smuts, mushrooms, and yeasts.

Furanose A sugar molecule containing the five-membered furan ring. See SUGAR MOLECULES.

Fusarium A genus of fungus, also known as "scab," that infests certain grains (e.g., wheat *Triticum aestivum*, corn or maize *Zea mays L.*, etc.) during growing seasons in which climate (e.g., high humidity, cool weather) and other conditions combine to enable rapid growth/proliferation of the fungus. In wheat, (*fusarium* head blight) fungus infestation causes the wheat plant to weaken and to produce empty seed heads, which reduces yield.

As a by-product of its metabolism, *fusarium* produces fumonisins (one of which is known as DON or "vomitoxin"), a group of mycotoxins. Fumonisin B_1 is the most prevalent *fusarium*-produced mycotoxin in corn (maize). Its presence can cause livestock to refuse to eat infested feed, decrease reproductive efficiency in swine, and even kill horses (via equine leukoencephalomalacia).

When consumed by humans, fumonisin B_1 induces cell death via apoptosis; and the tissues that are adjacent to killed cells respond with cell replication/proliferation to replace the lost cells.

Fumonisin B_1 inhibits the enzyme ceramide synthetase (which is crucial to the biosynthetic pathway for the creation of sphingolipids in cells), resulting in accumulation of sphinganine in cells, and decreases ceramides and complex sphingolipids. These internal changes signal the cells to die via apoptosis ("programmed cell death"), especially liver and kidney cells.

Maximum fumonisin content allowed in flour (for U.S. bread) is one part per million. Maximum fumonisin content allowed in U.S. malting barley (*Hordeum vulgare*) is zero.

In 1997, Iowa State University research showed that *B.t.* corn varieties (which express the *B.t.* protoxin in the corn ears) have significantly less ear mold caused by *Fusarium* fungi. That is because the European corn borer (ECB) is a vector (carrier) of *Fusarium*. See FUNGUS, MYCOTOXINS, METABOLISM, APOPTOSIS, ENZYME INHIBITION, LIPIDS, VOMITOXIN, DON, *BACILLUS THURINGIENSIS (B.t.)*, EUROPEAN CORN BORER (ECB), and CD95 PROTEIN.

Fusarium moniliforme One of the *Fusarium* fungi. See *FUSARIUM*, FUMONISINS, and FUNGUS.

Fusion Protein A protein consisting of all or part of the amino acid sequences (known as the "domain") of two or more proteins. Formed by fusing the two protein-encoding genes (which causes the ribosome to subsequently produce the fusion protein). This fusion is often done deliberately, either to put the expression of one of the (fused) genes under the control of the strong promoter for the first gene, or to allow the gene of interest (which is difficult to assay) to be more easily studied via substituting some of the (gene) protein with a more easily measured (assayed) function. For example, fusing a difficult-to-study gene with the β-galactosidase gene, the (protein) product of which can easily be measured (assayed) using chromatography. See also PROTEIN, AMINO ACID, SEQUENCE (OF A PROTEIN MOLECULE), GENE, RIBOSOMES, PROMOTER, ASSAY, CODING SEQUENCE, and DOMAIN (OF A PROTEIN).

Fusion Toxin A fusion protein that consists of a toxic protein (domain) plus a cell receptor binding region (protein domain). The cell receptor portion (of the total fusion toxin molecule) delivers the toxin directly to the (diseased) cell, thus sparing other healthy tissues from the effect of the toxin. See also FUSION PROTEIN, TOXIN, RICIN, PROTEIN, PROTEIN ENGINEERING, DOMAIN (OF A PROTEIN), RECEPTORS, and ENDOCYTOSIS.

Fusogenic Agent Any compound, virus, etc., that causes cells to fuse together. For example, one of the effects of the HIV (i.e., AIDS-causing) viruses is to cause the T cells of the human immune system to fuse (causing collapse of the immune system). See also ACQUIRED IMMUNE DEFICIENCY SYNDROME (AIDS), HUMAN IMMUNODEFICIENCY VIRUS (TYPE 1 and TYPE 2), HELPER T CELLS (T4 CELLS), and ADHESION MOLECULE.

Futile Cycle An enzyme-catalyzed set of cyclic reactions that results in release of thermal energy (heat) through the hydrolysis of ATP (adenosine triphosphate). The hydrolysis of ATP is normally coupled to other cycles and reactions in which the energy released is metabolically used. However, futile cycles would appear to waste the energy of ATP as heat—except when one is shivering to keep warm. The production of heat by shivering is an example of the futile cycle. See also ADENOSINE TRIPHOSPHATE (ATP), ENZYME, and HYDROLYSIS.

G

G+ See GRAM-POSITIVE.

G− See GRAM-NEGATIVE.

G-Proteins (Guanine Nucleotide Binding Proteins) Discovered by Rodbell and co-workers at America's NIH, and Gilman and co-workers at the American University of Virginia–Charlottesville, during the 1970s–80s. These are proteins embedded in the surface membrane of cells. G-proteins "receive chemical signals" from outside the cell (e.g., hormones) and "pass the signal" into the cell, so that cell can "respond to the signal." For example, a hormone, drug, neurotransmitter, or other "signal" binds to a receptor molecule on the surface of the cell's exterior membrane. That receptor then activates the G-protein, which causes an effector inside cell to produce a second "signal" chemical inside cell, which causes cell to react to the original external chemical signal. Dysfunction of G-proteins causes the salt and water losses inherent in cholera (the body's compromised immune defense inherent in pertussis), and is believed responsible for some symptoms of diabetes and alcoholism. See also PROTEIN, SIGNALLING, SIGNAL TRANSDUCTION, HORMONE, CELL, BETA CELLS, and GTPases.

GalNAc *N*-acetyl-D-galactosamine.

Galactose (gal) A monosaccharide occurring in both levo (L) and dextro (D) forms as a constituent of plant and animal oligosaccharides (lactose and raffinose) and polysaccharides (agar and pectin). Galactose is also known as cerebrose. See also STEREOISOMERS, DEXTROROTARY (D) ISOMER, and LEVOROTARY (L) ISOMER.

GALT See GUT-ASSOCIATED LYMPHOID TISSUE.

Gamete A germ or reproductive cell. In animals (and humans) the functional, mature, male gamete is called a spermatozoon; in plants it is called a spermatozoid. In both animals and plants the female gamete is called the ovum, or egg. See also OOCYTES.

Gamma Globulin A type of blood protein that plays a major role in the process of immunity (immune system response). Sometimes the term "gamma globulin" refers to a whole group of blood proteins that are known as antibodies or immunoglobulins (Ig). Most often, however, it applies to a particular immunoglobulin, designated as IgG, believed to be the most abundant type of antibody in the body. See also ANTIBODY,

GUT-ASSOCIATED LYMPHOID TISSUE (GALT), PROTEIN, and IMMUNOGLOBULIN.

Gamma Interferon Produced by T lymphocytes. See INTERFERONS and T LYMPHOCYTES.

GAP A double-stranded DNA is said to be "gapped" when one strand is missing over a short region of the molecule. See also DEOXYRIBONUCLEIC ACID (DNA).

Gated Transport (of a protein) One of three means for a protein molecule to pass between compartments within eucaryotic cells. The compartment "wall" (membrane) possesses a "sensor" (receptor) that detects the presence of a correct protein (e.g., after that protein has been synthesized in the cell's ribosomes), then opens a "gate" (pore) in the membrane to allow that protein to pass from the first compartment to the second compartment. See also PROTEIN, EUCARYOTE, CELL, RIBOSOMES, SIGNALLING, VESICULAR TRANSPORT (OF A PROTEIN).

GDH Gene See GLUTAMATE DEHYDROGENASE.

GDNF See GLIAL DERIVED NEUROTROPHIC FACTOR.

Gel A colloid, where the dispersed phase is liquid and the dispersion medium is solid.

Gel Electrophoresis See POLYACRYLAMIDE GEL ELECTROPHORESIS (PAGE), and ELECTROPHORESIS.

Gel Filtration Also known as exclusion chromatography. An effective technique for separating molecules (such as peptide mixtures) on the basis of size. This is accomplished by passing a solution of the molecules to be separated over a column of Sephadex®, for example; which is a polymerized carbohydrate derivative that contains tiny holes. The holes are of such a size that some of the smaller molecules diffuse into them and are in this way retained (held back) while the larger molecules are not able to get into the holes and pass on by the solid phase (Sephadex®, in this example). This, simplistically, is how separation is effected. See also ELECTROPHORESIS, CHROMATOGRAPHY and FIELD INVERSION GEL ELECTROPHORESIS.

GEM A project conducted under the auspices of the United States Department of Agriculture, in concert with 16 American Universities and 20 corn (maize) seed companies. This acronym stands for Germ plasm Enhancement for Maize. GEM's intent is to cross exotic (not in current use) germ plasm with commercial maize lines in order to increase corn yield. See CORN, GERM PLASM, HYBRIDIZATION (PLANT GENETICS), and PLEIOTROPIC.

GEMP (Genetically Engineered Microbial Pesticide) See GENETICALLY ENGINEERED MICROBIAL PESTICIDE and INTEGRATED PEST MANAGEMENT (IPM).

Gene A natural unit of the hereditary material, which is the physical basis for the transmission of the characteristics of living organisms from one generation to another. The basic genetic material is fundamentally the same in all living organisms: it consists of chain-like molecules of nucleic acids—deoxyribonucleic acid (DNA) in most organisms and ribonucleic acid (RNA) in certain viruses—and is usually associated in a linear arrangement that (in part) constitutes a chromosome.

The segment of DNA that is involved in producing a polypeptide chain. It includes regions preceding and following the coding region (leader and trailer) as well as intervening sequences (introns) between individual coding segments (exons). See also INFORMATIONAL MOLECULES, DEOXYRIBONUCLEIC ACID (DNA), RIBONUCLEIC ACID (RNA), CHROMOSOMES, MESSENGER RNA (mRNA), CODON, INTRON, EXON, and CODING SEQUENCE.

Gene Amplification The copying of segments (e.g., genes) within the DNA or RNA molecule. This can be done by man (e.g., polymerase chain reaction), can be caused by certain chemical carcinogens (e.g., phorbol ester), or occur naturally (e.g., in procaryotes and certain lower eucaryotes). The five primary techniques that are used by man to perform gene amplification are: 1) Polymerase Chain Reaction (PCR), 2) Ligase Chain Reaction (LCR), 3) Self-sustained Sequence Replication (SSR), 4) Q-beta Replicase Technique, and 5) Strand Displacement Amplification (SDA). See also GENE, Q-BETA REPLICASE TECHNIQUE, POLYMERASE CHAIN REACTION (PCR), CARCINOGEN, PROCARYOTES, and EUCARYOTE.

Gene Delivery (gene therapy) The insertion of genes (e.g., via retroviral vectors) into selected cells in the body in order to:

(a) cause those cells to produce specific therapeutic agents (e.g., growth hormone in livestock, factor VIII in hemophiliacs, insulin in diabetics, etc.). A potential way of curing some genetic diseases, in that the inserted gene will produce the protein and/or enzyme that is missing in the body due to a defective gene (thus causing the genetic disease). Approximately 3,000 genetic diseases are known to man. Examples of genetic diseases include cystic fibrosis, sickle cell anemia, Huntington's disease, phenylketonuria (PKU), Tay-Sach's disease, ADA deficiency (adenosine deaminase enzyme deficiency) and thalassemia.

(b) cause those cells to become (more) susceptible to a conventional therapeutic agent that previously was ineffective against that particular condition/disease (e.g., insertion of Hs-tk gene into brain tumor cells to make those tumor cells susceptible to the Syntex drug Ganciclovir)

(c) cause those cells to become less susceptible to a conventional therapeutic agent (e.g., insert genes into healthy tissue in order to enable that healthy tissue to resist the harmful effects of such conventional chemotherapy agents as vincristine)

(d) counter the effects of abnormal (damaged) tumor suppressor genes via insertion of normal tumor suppressor genes
(e) cause expression of ribozymes that cleave oncogenes (cancer-causing genes)
(f) be used for other therapeutic uses of genes in cells.

See also TUMOR SUPPRESSOR GENES, ONCOGENES, CANCER, p53 GENE, TUMOR, PROTO-ONCOGENES, RETROVIRAL VECTORS, RETROVIRUSES, HUNTINGTON'S DISEASE, GENETIC CODE, INFORMATIONAL MOLECULES, DEOXYRIBONUCLEIC ACID (DNA), CHROMOSOMES, HORMONE, ENZYME, PROTEIN, and GENETIC TARGETING.

Gene Machine An instrument which, when fed information on the amino acid sequence of a protein (usually via a protein sequencer), will automatically produce polynucleotide gene segments to code for that protein. See also SEQUENCING (OF DNA MOLECULES), SYNTHESIZING (OF DNA MOLECULES), GENE, AMINO ACID, and PROTEIN.

Gene Manipulation See GENETIC ENGINEERING.

Gene Map See LINKAGE MAP, GENETIC MAP, and PHYSICAL MAP (OF GENOME).

Gene Mapping See SEQUENCING (OF DNA MOLECULES), GENETIC MAP, LINKAGE MAP, and PHYSICAL MAP (OF GENOME).

Gene Probe See DNA PROBE.

Gene Replacement Therapy See GENE DELIVERY (GENE THERAPY).

Gene Silencing The suppression of gene expression (e.g., gene for polygalacturonase which causes fruit to ripen) via sense or antisense genes. See ANTISENSE (DNA SEQUENCE), TRANSWITCH, SENSE, and POLYGALACTURONASE (PG).

Gene Splicing The enzymatic attachment (joining) of one gene (or part of a gene) to another; also removal of introns and splicing of exons during mRNA synthesis. See also SPLICING and MESSENGER RNA (mRNA).

Gene "Stacking" See "STACKED" GENES.

Gene Targeting See GENETIC TARGETING, GENE SPLICING, GENE DELIVERY, and GENETIC ENGINEERING.

Gene Technology Office An agency of the Australian government, established in 1997, to oversee and regulate all genetic engineering activities conducted in the country of Australia. See GENETIC ENGINEERING, RECOMBINANT DNA ADVISORY COMMITTEE (RAC), ZKBS (CENTRAL COMMISSION ON BIOLOGICAL SAFETY), INDIAN DEPARTMENT OF BIOTECHNOLOGY, and COMMISSION OF BIOMOLECULAR ENGINEERING.

Gene Therapy See GENE DELIVERY (GENE THERAPY).

Generation Time The time required for a population of cells to double. The average time required for a round of cell division. See also CELL and MITOSIS.

Genetic Code The set of triplet code words in DNA coding for all of the amino acids. There are more than 20 different amino acids and only four bases (adenine, thymine, cytosine, and guanine). The mRNA code is a triplet code, that is, each successive "frame" of three nucleotides (sometimes called a codon) of the mRNA corresponds to one amino acid of the protein. This rule of correspondence is the genetic code. The genetic code consists of 64 entries—the 64 triplets possible when there are four possible nucleotides, each of which can be at any of three places ($4 \times 4 \times 4 = 64$). A triplet code was required because a doublet code would have only been able to code for ($4 \times 4 = 16$) sixteen amino acids. A triplet code allows for the coding of 64 theoretical amino acids. Since only a little over 20 exist, there is some redundancy in the system. Hence some certain amino acids are coded for by two or three different triplets. See also MESSENGER RNA (mRNA), DEOXYRIBONUCLEIC ACID (DNA), and INFORMATIONAL MOLECULES.

Genetic Engineering The selective, deliberate alteration of genes (genetic material) by man. This term has come to have a very broad meaning including the manipulation and alteration of the genetic material (constitution) of an organism in such a way as to allow it to produce endogenous proteins with properties different from those of the normal, or to produce entirely different (foreign) proteins altogether. Other words applicable to the same process are gene splicing, gene manipulation, or recombinant DNA technology (techniques). See also GENE, INFORMATIONAL MOLECULES, CHROMOSOMES, GENE AMPLIFICATION, VECTOR, PLASMID, GENE SPLICING, DEOXYRIBONUCLEIC ACID (DNA), TRANSGENIC (ORGANISM), GMO, RECOMBINANT DNA (rDNA), RECOMBINATION, HETEROKARYON, HEREDITY, MESSENGER RNA (mRNA), HETERODUPLEX, POSITIVE AND NEGATIVE SELECTION (PNS), POLYMERASE CHAIN REACTION (PCR) TECHNIQUE, and BIOTECHNOLOGY.

Genetic Linkage See LINKAGE and LINKAGE GROUP.

Genetic Map A diagram showing the relative sequence and position of specific genes along a chromosome molecule. See also POSITION EFFECT, GENE, GENOME, CHROMOSOMES, and PHYSICAL MAP (OF GENOME).

Genetic Marker See MARKER (GENETIC MARKER).

Genetic Probe See DNA PROBE.

Genetic Targeting The insertion of antisense DNA molecules *in vivo* into selected cells of the body in order to block the activity of undesirable genes. These genes might include oncogenes, or genes crucial to the life cycle of parasites such as trypanosomes (cause sleeping sickness). See also ANTISENSE (DNA SEQUENCE), GENE, GENE DELIVERY (GENE THERAPY), ONCOGENES, and DENDRIMERS.

Genetically Engineered Microbial Pesticides (GEMP) One or more microbes that have been genetically engineered in such a way as to cause them to be effective in combatting pest(s) that attack crops or livestock. For example, a microbe that naturally attacks a crop pest could be genetically engineered to make the microbe more potent, or more durable in field environment when applied to field via selected method of microbe application. See also MICROBE, GENETIC ENGINEERING, WHEAT TAKE-ALL DISEASE, BACULOVIRUS, *BACILLUS THURINGIENSIS (B.t.)*, FEDERAL INSECTICIDE, FUNGICIDE, AND RODENTICIDE ACT (FIFRA), and TOXIC SUBSTANCES CONTROL ACT (TSCA).

Genetically Engineered Organism (GEO) See GEO.

Genetically Manipulated Organism (GMO) See GMO.

Genetically Modified Organism (GMO) See GMO.

Genetically Modified Microorganism (GMM) See GMM.

Genetics The branch of biology concerned with heredity, it was literally invented by Gregor Mendel in the 19th century. It is a study of the manner in which genes operate and are transmitted from parents to offspring. It involves the study of the mechanism of gene action—the manner in which the genetic material (DNA) affects physiological reactions within the cell. See also GENE, DEOXYRIBONUCLEIC ACID (DNA), HYBRIDIZATION (PLANT GENETICS), and HEREDITY.

Genistein (Gen) One of several phytochemicals produced by the soybean plant as a defense against certain plant diseases; and to signal *rhizobia* bacteria (to produce nitrogen for the soybean plant via colonization of its roots, followed by nitrogen fixation from the air). Genistein is also produced as a by-product of mycobacterium fermentation (process used to produce commercial amounts of certain antibiotics). Genistein is an isoflavone, a steroid-like compound that can be lethal to certain animal cells via its kinase-inhibiting properties. Genistein fights cancer (tumor cells) by inhibiting protein tyrosine kinase and topoisomerase II.

Attached to a pharmaceutical "guided missile" such as a monoclonal antibody or the CD4 protein, genistein is potentially useful for treatment against some tumors and has been investigated as a possible treatment against B-cell precursor leukemia. See also IMMUNOTOXIN, MONOCLONAL ANTIBODIES (MAb), CD4 PROTEIN, GENETIC ENGINEERING, NITROGEN FIXATION, PHYTOCHEMICAL, FUSION PROTEIN, FUSION TOXIN, SOLUBLE CD4, ISOFLAVONES, SOYBEAN PLANT, RICIN, TYROSINE (tyr), STEROID, and INHIBITION.

Genistin See GENISTEIN (Gen).

Genome The entire hereditary material (which was proven by Oswald Avery in 1944 to be DNA) in a cell. It is composed of one or more chromosomes, depending on the complexity of the organism. See also DEOXYRIBONUCLEIC ACID (DNA) and CHROMOSOMES.

Genomics The scientific study of genes and their role in an organism's structure, growth, health, disease (and/or resistance to disease, etc.). See also GENOTYPE, GENE, GENETIC MAP, GENETIC TARGETING, GENETICS, GENETIC CODE, SEQUENCING (OF DNA MOLECULES), INFORMATIONAL MOLECULES, DEOXYRIBONUCLEIC ACID (DNA), FUNCTIONAL GENOMICS, GENE AMPLIFICATION, CODING SEQUENCE, and STRUCTURAL GENOMICS.

Genosensors Biosensors (electronic) that can detect the individual nucleotides that comprise a genome (DNA) molecule. Automated genosensors enable rapid, nondestructive sequencing of DNA molecules. See also GENOME, NUCLEOTIDE, DEOXYRIBONUCLEIC ACID (DNA), SEQUENCING (OF DNA MOLECULES), TEMPLATE, BIOSENSORS (ELECTRONIC), FOOTPRINTING and NANOTECHNOLOGY.

Genotoxic Refers to compounds that interfere with normal functioning of genetic material (i.e., DNA). For example, the antitumor antibiotic family of duocarmycin drugs. See DEOXYRIBONUCLEIC ACID (DNA), GENOTOXIC CARCINOGENS, and FOOTPRINTING.

Genotoxic Carcinogens Compounds that act directly on the genetic material (i.e., DNA) of an organism, thus causing cancer in that organism. Of the numerous chemicals that have been documented to be human carcinogens, the majority of them are genotoxic. See also CARCINOGEN, CANCER, GENE, and DEOXYRIBONUCLEIC ACID (DNA).

Genotype The total genetic, or hereditary, constitution that an individual receives from its parents. An individual organism's genotype is distinguished from its phenotype, which is its appearance or observable character. See also TRAIT, PHENOTYPE, and WILD TYPE.

Gentechnik Gesetz (Gene Technology Law) The 1990 law that governs recombinant DNA research and development in the country of Germany. It was amended January 1, 1994 to make it somewhat less restrictive. See also ZKBS (CENTRAL COMMISSION ON BIOLOGICAL SAFETY), RECOMBINANT DNA ADVISORY COMMITTEE (RAC), GENETIC ENGINEERING, RECOMBINANT DNA (rDNA), RECOMBINATION, BIOTECHNOLOGY, BUNDESGESUNDHEITSAMT (BGA), and INDIAN DEPARTMENT OF BIOTECHNOLOGY.

Genus A group of closely related species. See also SPECIES and CLADES.

GEO Genetically engineered organism. See also GENETIC ENGINEERING, GMO, GENE, GENE SPLICING, and GMM.

Geomicrobiology Applications of microbiological knowledge to an understanding of geological phenomena. See also FERROBACTERIA.

German Gene Law See GENTECHNIK GESETZ (GENE TECHNOLOGY LAW).

Germ Cell The sex cell (sperm or egg). It differs from other cells in that it contains only half (haploid) the usual number of chromosomes. See also GAMETE and HAPLOID.

Germ Plasm The total genetic variability to an organism, represented by the pool of germ cells or seed. See also GERM CELL and GEM.

GH See GROWTH HORMONE.

Gibberellins Plant hormones that, among other functions, regulate the growth of grass (after that gibberellin is activated by an enzyme).

In 1996, Lew Mander and Richard Pharis discovered an analogue (i.e., a chemical that is similar) to grass gibberellin that does *not* cause grass to grow. When this analogue is sprayed onto grass, it mixes into the naturally occurring grass gibberellin and significantly slows grass growth (thus reducing the amount of mowing required for lawns, golf courses, etc.). See also PLANT HORMONE, ENZYME, and ANALOGUE.

Glial Derived Neurotrophic Factor (GDNF) A neurotrophic factor that assists the survival and functional activity of the brain's dopaminergic neurons. Because dopaminergic neurons typically deteriorate and die in brains of the victims of Parkinson's disease, it is possible that GDNF may someday be used in treatment of Parkinson's disease. See also NEUROTRANSMITTER and PARKINSON'S DISEASE.

Globular Protein A soluble protein in which the polypeptide chain is tightly folded in three dimensions to yield a globular (roughly oval, circular) shape. See also PROTEIN FOLDING, POLYPEPTIDE (PROTEIN), CONFORMATION, and TERTIARY STRUCTURE.

GLP See GOOD LABORATORY PRACTICES (GLP).

GLQ223 See TRICHOSANTHIN.

Glucagon A hormone produced by the pancreas that causes the breakdown of glycogen in the liver. Glycogen is a form of storage sugar and its breakdown releases glucose for energy production. See also GLYCOGEN.

Glucocerebrosidase (trade name Ceredase) An enzyme used in treatment of inherited Gaucher's disease in which there is abnormal deposition of glucocerebrosides (hydrophobic lipid molecules that contain a hydrophilic sugar head group). Gaucher's disease is an enzyme deficiency disease that may be amenable to cure by incorporation of the gene coding for glucocerebrosidase into the patient's genome via gene delivery techniques. See also ENZYME and GENE DELIVERY (GENE THERAPY).

Glucogenic Amino Acid Amino acids whose carbon chains can be metabolically converted into glucose or glycogen. See also GLUCONEOGENESIS.

Gluconeogenesis The net biosynthesis (formation) of new glucose from noncarbohydrate precursors such as pyruvate, lactate, certain amino acids and intermediates of the citric acid cycle. See also CARBOHYDRATES (SACCHARIDES), GLUCOSE (GLc), and CITRIC ACID CYCLE.

Glucose (GLc) A prime fuel for the generation of energy by organisms. It is broken down (to obtain energy) via a metabolic process called glycolysis. Glucose is a hexose, a sugar possessing six carbon atoms in its molecule. The six carbon atoms are connected to each other to form a closed ring structure known as a hexose (6) ring.

Animal cells store glucose in the form of glycogen (sometimes called animal starch), a large branched polymer of glucose units. Plant cells store glucose in the form of starch, a large polymer of glucose units.

Yeasts and bacteria store glucose in the form of dextran, a polymer of glucose units. The difference between the forms of storage glucose is (1) in the size (molecular weight) of the final polymer formed, (2) in the type of linkages that connect the single glucose units together in the branched molecule, and (3) in the degree of branching which occurs in the polymer. Note that a glucose polymer does not consist of just a single long straight chain. The backbone chain has other polymer chains branching off of it. The whole molecule may be visualized as looking somewhat like a tree without the trunk. The other very abundant polymer formed by glucose units is structural in nature and is called cellulose. It is the most abundant cell wall and structural polysaccharide in the plant world. Hence, glucose is used not only as an energy source, but also as a structural material. See also AMYLOSE, AMYLOPECTIN, GLYCOLYSIS, GLYCOGEN, STARCH, DEXTRAN, and CELLULOSE.

Glucose Oxidase An enzyme that breaks down sugar molecules (causing oxygen consumption in an organism). See also GLUCOSE, GLYCOLYSIS, and SUGAR MOLECULES.

Glufosinate See PAT GENE, HERBICIDE-TOLERANT CROP, GENE, and GLUTAMINE SYNTHETASE.

Gluphosinate See PAT GENE, HERBICIDE-TOLERANT CROP, GENE, and GLUTAMINE SYNTHETASE.

Glutamate Dehydrogenase An enzyme found naturally in certain soil bacteria, which helps those bacteria to utilize soilborne nitrogen. When its gene (GDH gene) is inserted into corn plant via genetic engineering, the resultant plant production of glutamate dehydrogenase enables that corn plant to better utilize soilborne nitrogen. As a result, that genetically engineered corn (*Zea mays L.*) has protein yield increase of approximately seven percent, according to research begun in 1981 by David Lightfoot. See ENZYME, BACTERIA, GENE, CORN, NITROGEN CYCLE, DEHYDROGENASES, PROTEIN, and GENETIC ENGINEERING.

Glutamic Acid A dicarboxylic amino acid of the α-ketoglutaric acid family.

Glutamine An amino acid; the monamide of glutamic acid. Glutamine is of fundamental importance for amino acid biosynthesis in all forms of life. See GLUTAMINE SYNTHETASE, AMINO ACID, and PAT GENE.

Glutamine Synthetase An enzyme that catalyzes the synthesis of glutamine (which is crucial for amino acid biosynthesis). See GLUTAMINE, ENZYME, PAT GENE, and AMINO ACID.

Glutathione A tripeptide that is found in all cells of higher animals. Composed of the amino acids glutamic acid, cysteine, and glycine. The cysteine possesses a sulfhydryl group that makes glutathione a weak reducing agent. See also REDUCTION (IN A CHEMICAL REACTION).

Glyceraldehyde (D- and L-) One of the smallest monosaccharides, it is called an aldose because it contains an aldehyde group. Glyceraldehyde has a single asymmetric carbon atom, thus there are two stereoisomers (D-glyceraldehyde and L-glyceraldehyde). See also MONOSACCHARIDES and STEREOISOMERS.

Glycetein See ISOFLAVONES.

Glycine (gly) The simplest (and smallest) of the amino acids found in proteins. It is the only amino acid that does not have an asymmetric carbon atom within its molecule. Thus, it is not optically active. See also AMINO A0CID, PROTEIN, STEREOISOMERS, and OPTICAL ACTIVITY.

Glycine max See SOYBEAN PLANT.

Glycobiology The study of the involvement (function) of sugars in biological processes. See also GLUCOSE (GLc), GLUCOSE OXIDASE, GLYCOGEN, GLYCOLIPID, GLYCOLYSIS, GLYCOPROTEIN, GLYCOSIDASES, GLYCOSIDE, and GLYCOSYLATION (TO GLYCOSYLATE).

Glycocalyx A polysaccharide matrix that is involved (in some microorganisms) in firm attachment of the organism to a solid surface.

Glycoform One of several molecular arrangements that a given glycoprotein can possess [varieties are determined by the attachment of various oligosaccharide(s)]. Some glycoforms of a given glycoprotein may exhibit greater or lesser biological activity (e.g., pharmaceutical effectiveness for biotherapeutic glycoproteins) because the oligosaccharide units of the glycoprotein molecule mediate interactions of the glycoprotein with the cells of the body. See also GLYCOPROTEIN and OLIGOSACCHARIDES.

Glycogen A polymer of glucose with a branching, tree-like molecular structure. It is the chief storage form of carbohydrates in animals. In mammals, glycogen is stored mainly in the liver and muscles. Its molecular weight may be several million. See also GLUCOSE (GLc), GLUCAGON, and MOLECULAR WEIGHT.

Glycolipid A lipid containing at least one carbohydrate group within its molecule. See also LIPIDS, GLYCOPROTEIN, GLYCOSYLATION, and GLYCOLYSIS.

Glycolysis A metabolic process in which sugars are broken down into smaller compounds with the release of energy. This series of chemical reactions is found in plant and animal cells as well as in many microorganisms. Except for the final reaction in the series, the chemical reaction pathway of glycolysis is the same as that for fermentation. See also GLUCOSE (GLc), METABOLISM, and FERMENTATION.

Glycoprotein A conjugated protein containing at least one carbohydrate (oligosaccharide) group within its molecule. A commonly occurring category of glycoproteins found in nature is called mucoproteins. These are protein-polysaccharide compounds that occur in the tissues, particularly in mucous secretions. Other glycoproteins include lymphokines (e.g., interleukins), hormones (e.g., somatotropins), receptors (e.g., GP120), enzymes (e.g., tissue plasminogen activator), and some therapeutics (e.g., CD4PE40). See also GLYCOFORM, CONJUGATED PROTEIN, GP120 PROTEIN, CONJUGATE, PROTEIN, OLIGOSACCHARIDES, and POLYSACCHARIDES.

Glycoprotein C A blood clot–regulating glycoprotein. See PROTEIN C and GLYCOPROTEIN.

Glycoprotein Remodeling The use of restriction endoglycosidases to (enzymatically) remove sugar (i.e., oligosaccharide) "branches" from glycoprotein (i.e., part protein, part oligosaccharide) molecules. One reason to perform such glycoprotein remodeling would be to remove one or more oligosaccharide branches so that the glycoprotein is less or no longer antigenic (i.e., triggers an immune response). This allows the glycoprotein to be injected into the body (e.g., for pharmaceutical purposes) without incurring an unwanted immune response. See also GLYCOPROTEIN, RESTRICTION ENDOGLYCOSIDASES, ENZYME, OLIGOSACCHARIDES, ANTIGEN, CELLULAR IMMUNE RESPONSE, HUMORAL IMMUNITY, ANTIBODY, EPITOPE, and HAPTEN.

Glycosidases Enzymes that catalyze the cleavage (hydrolysis) of glycosidic molecular bonds. For example, lysozyme (an enzyme found in human tears) lyses (cuts up) certain bacteria by cleaving the (β configuration) glycosidic linkages (bonds) between the monosaccharide units that (when linked) comprise the polysaccharide component of the bacterial cell walls. A bacterial cell devoid of a cell wall usually bursts. See also ENDOGLYCOSIDASE, EXOGLYCOSIDASE, and RESTRICTION ENDOGLYCOSIDASES.

Glycoside Any of a group of compounds that yield sugar molecules on hydrolysis. All parts of a glycoside compound may be sugar molecules, so that sucrose, raffinose, starch, and cellulose—all of which hydrolyze into sugar molecules—may all be considered to be glycosides. However, the name (glycoside) is usually applied to a compound in which part of the molecule is not a sugar. This nonsugar component is called the aglycon.

Glycosylation (to glycosylate) Addition of oligosaccharide units (e.g., to protein molecules). The oligosaccharide units are linked to either asparagine side chains by *N*-glycosidic bonds or to serine and threonine side chains by *O*-glycosidic bonds. See also OLIGOSACCHARIDES, PROTEIN, and GOLGI BODIES.

Glycosyltransferases A class of enzymes (transferases) that catalyze the addition (chemical reaction) of specific sugars (molecular groups) to oligosaccharides, glycoproteins or glycosides. See also OLIGOSACCHARIDES, MONOSACCHARIDES, ENZYME, GLYCOPROTEIN, GLYCOSIDE, and TRANSFERASES.

Glyphosate An active ingredient in some herbicides, it kills plants (e.g., weeds) by inhibiting the crucial plant enzyme EPSP synthase. See ENZYME, EPSP SYNTHASE, CP4 EPSPS, GLYPHOSATE OXIDASE, GLYPHOSATE-TRIMESIUM, and GLYPHOSATE ISOPROPYLAMINE SALT.

Glyphosate Isopropylamine Salt One of several forms of active ingredient utilized in some glyphosate-based herbicides. See GLYPHOSATE, EPSP SYNTHASE, CP4 EPSPS, GLYPHOSATE OXIDASE, and GLYPHOSATE-TRIMESIUM.

Glyphosate Oxidase An enzyme that (via catalysis) chemically breaks down glyphosate (i.e., the active ingredient in some herbicides). Glyphosate oxidase is produced in nature by acclimated microorganisms.

In 1988, Michael Heitkamp discovered a strain of *Pseudomonas* bacteria which possessed a gene (GO) that caused those particular *Pseudomonas* bacteria to produce unusually large amounts of glyphosate oxidase. That GO gene can be incorporated into a variety of crop plants (e.g., soybean, cotton, etc.) in order to enable those plants to survive post-emergence applications of glyphosate-containing herbicide.

Additionally, a plant can be genetically engineered to survive post-emergence applications of glyphosate-containing herbicide via insertion of gene (cassette) for plant production of the enzyme CP4 EPSPS. See also ENZYME, ACCLIMATIZATION, STRAIN, *PSEUDOMONAS FLUORESCENS*, GENE, GENETIC ENGINEERING, BACTERIA, MICROORGANISM, SOYBEAN PLANT, EPSP SYNTHASE, CP4 EPSPS, CASSETTE, and GLYPHOSATE.

Glyphosate-Trimesium One of several forms of active ingredient utilized in some glyphosate-based herbicides. See GLYPHOSATE, EPSP SYNTHASE, CP4 EPSPS, GLYPHOSATE OXIDASE, and GLYPHOSATE ISOPROPYLAMINE SALT.

GMO Genetically manipulated organism, or genetically modified organism. See GENE, GENE SPLICING, and GENETIC ENGINEERING.

GMP See GOOD MANUFACTURING PRACTICES (GMP).

GMS Genetically modified soya. See GMO and SOYBEAN PLANT.

GNE Group of National Experts on Safety in Biotechnology. The group of people within the OECD that developed OECD's guidelines for nations to utilize in their safety evaluations of foods derived from biotechnology. See ORGANIZATION FOR ECONOMIC COOPERATION AND DEVELOPMENT (OECD), BIOTECHNOLOGY, and GENETIC ENGINEERING.

GO Gene See GLYPHOSATE OXIDASE.

Golgi Bodies (also known as Golgi complexes) The primary "sorting centers" of cells, and the mechanism for glycosylation of (i.e., adding oligosaccharide and polysaccharide branches onto) proteins; before those proteins are then transported by transfer vesicles to lysosomes, secretory vesicles, or the plasma membrane. Visually, a Golgi complex is a stack of flattened membranous sacs (usually six sacs in mammal cells and twenty sacs in plant cells). See also GLYCOSYLATION (TO GLYCOSYLATE), CELL, OLIGOSACCHARIDES, POLYSACCHARIDES, PROTEIN, and LYSOSOME.

Good Laboratory Practice for Nonclinical Studies (GLPNC) The Good Laboratory Practice (GLP) that is required by the U.S. Food and Drug Administration (FDA) for studies of the safety and toxicological effects of new drugs for livestock. See also GOOD LABORATORY PRACTICES (GLP) and NADA.

Good Laboratory Practices (GLP) A set of rules and regulations issued by the Food and Drug Administration (FDA) that establishes broad methodological guidelines for procedures and record keeping. They are to be followed in laboratories involved in the testing and/or preparation of pharmaceuticals. GLPs also apply to the Environmental Protection Agency (EPA) (e.g., toxicity testing of new herbicides).

Good Manufacturing Practices (GMP) The set of general methodologies, practices, and procedures mandated by the Food and Drug Administration (FDA) which is to be followed in the testing and manufacture of pharmaceuticals. The purpose of GMPs is essentially to provide for record keeping and in a wider context to protect the public. GMP guidelines exist instead of specific regulations due to the newness of the technology, and may later be superceded (modified) due to further advances in technology and understanding. See also cGMP.

GP120 Protein An adhesion molecule (glycoprotein) on the envelope (surface membrane) of HIV (i.e., AIDS-causing) viruses that directly interacts with the CD4 protein on helper T cells; enabling the HIV viruses to bind to and infect helper T cells. In 1994, a group at America's Scripps Research Institute led by Dennis Burton and Carlos Barbas III announced that they had generated a recombinant human antibody to the GP120 protein; which neutralized more than 75% of HIV isolates that it was tested against. This advance holds the potential to someday lead to a vaccine against AIDS. See also MONOCLONAL ANTIBODIES (MAb), HUMAN IMMUNODEFICIENCY VIRUS (HIV), ACQUIRED IMMUNE DEFICIENCY SYNDROME (AIDS), SOLUBLE CD4, CD4 PROTEIN, HELPER T CELLS (T4 CELLS), CD44 PROTEIN, ADHESION MOLECULE, CONSERVED, GLYCOPROTEIN, SELECTINS, LECTINS, and PROTEIN.

Graft-versus-Host Disease (GVHD) The rejection of transplanted organs by the recipient's immune system. Also known as hyperacute rejection. It is caused by the attack of the recipient's T lymphocytes (i.e., T cells, a certain class of white blood cells) on the transplanted organ. The recipient's T cells are able to distinguish between self and foreign cells, and are hence able to recognize the foreign (nonself) cells of the transplanted organ. They then, naturally, try to destroy the "foreign invaders" in the body. This then constitutes rejection of the transplanted organ. From this it should be understood that there is nothing wrong with the body, but that it is behaving exactly as it should. See also CELLULAR IMMUNE RESPONSE, HUMORAL IMMUNITY, XENOGENEIC ORGANS, FIBROBLASTS, and CYCLOSPORIN.

Gram Molecular Weight The weight in grams of a compound that is numerically equal to its molecular weight; the weight of one mole (6.02×10^{23} molecules). See also MOLECULAR WEIGHT and MOLE.

Gram Stain Devised by Hans Christian Joachim Gram in 1884, this is a test that illuminates the composition/makeup of the physical structure of the cell wall of bacteria being tested. It is utilized to judge the effectiveness of a given chemical compound (e.g., an antibiotic) against bacteria types.

The test consists of a differential staining procedure, which allows most bacteria to be visually separated into two groups, known as Gram-Positive (G+) and Gram-Negative (G−). An antibiotic is defined in terms of the group of (pathogenic) bacteria that it is effective against, which is known as that antibiotic's "spectrum of activity." An antibiotic is said to have a spectrum of activity against gram-positive bacteria, gram-negative bacteria, or the bacteria of *both* groups. An antibiotic that is effective against both groups of bacteria is termed "broad spectrum" or "wide spectrum." See also BACTERIA, GRAM-POSITIVE, GRAM-NEGATIVE, PATHOGENIC, CELL, and ANTIBIOTIC.

Gram-Negative (G−) Pertaining to one of the most important ways of classifying bacteria by means of the differences in the way they stain. The set of bacteria that are not able to be stained (blue) when treated with the gram staining procedure. Gram negativity (and gram positivity) is conferred not by the chemical constituents of the bacteria, but rather by the physical structure of the bacteria cell wall. The staining procedure involves the staining of all cells in a sample with a blue dye. Gram-negative bacteria have a very thin peptidoglycan cell wall (capsule). Hence, the washing procedure, which is an integral part of the overall staining procedure, washes out the blue dye (known as crystal violet). This leaves the gram-negative bacteria colorless. The cells are then stained with a red acidic counterstain (dye) such as acid fuchsin or safranine. After treatment with counterstain the gram-negative cells are red and the gram-positive cells are blue. See also GRAM-POSITIVE, BACTERIA, CELL, and GRAM STAIN.

Gram-Positive (G+) Pertaining to bacteria, holding the color of the primary stain (blue) when treated with Gram's stain (a commercial staining agent), or Gentian violet solution.

In contrast to the gram-negative bacteria, the gram-positive bacteria possess a much thicker peptidoglycan cell wall (capsule). Because of this, the blue crystal violet dye (with which the bacteria were stained) does not wash out of the cell and the bacteria appear blue under the microscope. See also GRAM-NEGATIVE, BACTERIA, CELL, GRAM STAIN, and CAPSULE.

Granulation Tissue A mixture of proteins and cells produced by the fibroblast growth that results from a wound. See also FIBROBLASTS and PROTEIN.

Granulocidin A protein produced by white blood cells, which has demonstrated (in the laboratory) an ability to kill a broad spectrum of pathogens. See also PATHOGEN and PROTEIN.

Granulocyte Colony Stimulating Factor (G-CSF) A colony stimulating factor (CSF; a protein) that stimulates production of granulocytes, particularly neutrophils. See also COLONY STIMULATING FACTORS, GRANULOCYTES, and NEUTROPHILS.

Granulocyte-Macrophage Colony Stimulating Factor (GM-CSF) or **Granulocyte-Monocyte Colony Stimulating Factor** (GM-CSF) A colony stimulating factor (CSF; a protein) that stimulates production of granulocytes/macrophages/monocytes. See also COLONY STIMULATING FACTORS (CSFs), MACROPHAGE, and MONOCYTES.

Granulocytes (polymorphonuclear granulocytes) Phagocytic (scavenging, ingesting) cells that are part of the immune system. When their cell nucleus is segmented into lobes and they have granule-like inclusions within their cytoplasm (the neutrophils, eosinophils, and basophils) they are collectively known as polymorphonuclear granulocytes. See also PHAGOCYTE.

GRAS List A list of food additives/ingredients considered to be Generally Recognized as Safe, by the American Government's Food and Drug Administration (FDA). This list of additives is judged to be safe by a panel of FDA pharmacologists and toxicologists, who base their judgment upon data that is available for each ingredient. In practice, those additives for which extensive experience of common use in foods (without known ill effects) has been accumulated over time (e.g., common table salt) are often approved by the FDA due more to the "common use factor" than to any toxicology data, per se. See also FOOD AND DRUG ADMINISTRATION (FDA), DELANEY CLAUSE, PHARMACOLOGY, and CANOLA.

Grave's Disease An autoimmune disease in which the body's immune system attacks the body's own thyroid gland by manufacturing antibodies which resemble molecules of thyroid stimulating hormone (TSH). TSH is a hormone that "stimulates" the thyroid gland to increase the metabolic rate of the body. Thus, when the TSH-like antibodies attach themselves to the thyroid gland, the gland mistakes the antibodies for TSH molecules. The thyroid gland causes the body's metabolism to "go into overdrive"; causing fast heartbeat, agitation, sleeplessness, etc. Grave's disease is also thought to sometimes cause erratic or violent behavior. See also AUTOIMMUNE DISEASE, ANTIBODY, THYROID GLAND, THYMUS, METABOLISM, and THYROID STIMULATING HORMONE (TSH).

GRF See GROWTH HORMONE RELEASING FACTOR.

GRH See GROWTH HORMONE RELEASING FACTOR.

Group of National Experts on Safety in Biotechnology See GNE.

Growth (microbial) An increase in the number of cells. See also GENERATION TIME.

Growth Curve The change in the number of cells in a growing culture as a function of time. See also GENERATION TIME.

Growth Factor A specific substance that must be present in the tissues or growth medium (when *in vitro*) in order for the cells to multiply. See also FIBROBLAST GROWTH FACTOR (FGF), NERVE GROWTH FACTOR (NGF), EPIDERMAL GROWTH FACTOR (EGF), ANGIOGENIC GROWTH FACTORS, ANGIOGENIN, and BONE MORPHOGENETIC PROTEIN (BMP).

Growth Hormone (GH) A hormone produced by the anterior pituitary gland. This hormone is a protein (somatotropin) and can be obtained from the bodies of animals, or produced by genetically engineered microorganisms. Its major action in humans (human growth hormone) is a generalized stimulation of skeletal growth. However, human growth hormone (HGH) is also known to affect the growth of other tissues, to be important in fat, protein, and carbohydrate metabolism, and to enhance the effects of various other hormones. See also BOVINE SOMATOTROPIN (BST), PORCINE SOMATOTROPIN (PST), and PITUITARY GLAND.

Growth Hormone–Releasing Factor (GRF or GHRF) Also termed growth hormone–releasing hormone (GRH). A factor that causes the release of growth hormone. It is 44 amino acids in length. See also GROWTH HORMONE (GH), GROWTH FACTOR, AMINO ACID, and HORMONE.

GT-AG Rule Describes the presence of these constant dinucleotides at the first two and last two positions of introns of nuclear genes. See also INTRON and GENE.

GTO Abbreviation for Gene Technology Office. See GENE TECHNOLOGY OFFICE.

GTP See GMP.

GTPases Guanosine triphosphatases. These are G-proteins (enzymes) which are crucial for growth, movement, and maintenance of the cell's shape. When active, GTPases are bound to cell membranes (surfaces) by an isoprene molecule (receptor). See G-PROTEINS, ENZYME, CELL, PHOSPHORYLATION, RECEPTORS, and PROTEIN.

GTS Glyphosate tolerant soybean. See HERBICIDE-TOLERANT CROP, SOYBEAN PLANT, CP4 EPSPS, and GLYPHOSATE.

GTS Glufosinate-ammonium tolerant soybean. See HERBICIDE-TOLERANT CROP, SOYBEAN PLANT, PAT GENE, and GLUFOSINATE.

Guanine A purine base. It occurs naturally as a fundamental component of nucleic acids. See PURINE and NUCLEIC ACIDS.

Gut-Associated Lymphoid Tissues (GALT) A variety of specialized lymph-reticular tissues that line the inside of an animal's digestive system. GALT include Peyer's Patches, the appendix, and small solitary lymphoid tissues in the gut. They constitute the intestinal immune system (response to antigens). See also LYMPHOCYTE, PEYER'S PATCHES, ANTIGEN, HUMORAL IMMUNITY, CELLULAR IMMUNE RESPONSE, and EDIBLE VACCINES.

H

H. pylori A bacteria that has been linked (e.g., cause) to gastric ulcers and other gastric problems in humans. That link was first announced by Barry Marshall in the early 1990's. See also BACTERIA and *HELICOBACTER PYLORI* BACTERIA.

H. virescens See *HELIOTHIS VIRESCENS*.

H. zea See *HELICOVERPA ZEA*.

Habitat The natural environment of an organism within an ecosystem. The place, in an ecosystem, where an organism lives. See also ECOLOGY.

HAC See HUMAN ARTIFICIAL CHROMOSOMES (HAC).

HACCP See HAZARD ANALYSIS AND CRITICAL CONTROL POINTS (HACCP).

Hairpin Loop A section of highly curving, single-stranded DNA or RNA formed when a long piece (string) of the DNA or RNA bends back on itself and hydrogen-bonds (is able to base pair) in some regions to form double-stranded regions. The structure can be visualized by taking a human hair, bending it back on itself and holding it in such a way as to half its original length. The section where the two ends of hair lie next to each other represents the section of double-stranded DNA or RNA. At one end the hair will have to make a sharp turn and will form a loop. This loop represents the single-stranded hairpin loop. See also RIBONUCLEIC ACID (RNA) and DEOXYRIBONUCLEIC ACID (DNA).

Halophile Microorganisms that require NaCl (salt) for growth (they are called obligate halophiles). Those that do not require it, but can grow in the presence of high NaCl concentrations, are called facultative halophiles. Natural habitats containing high salt concentrations are, for example, the Great Salt Lake in Utah, the Dead Sea in Israel and the Caspian Sea in Russia. See also HABITAT.

Haploid A cell with one set of chromosomes; half as many chromosomes as the normal somatic body cells contain. A characteristic of sex cells. See also GAMETE.

Haplophase A phase in the life cycle of an organism in which it has only one copy of each gene. The organism is then said to be haploid. Yeast can exist as true haploids. Humans are haploid for only a few genes and cannot exist as true haploids. See also HAPLOID.

Hapten A small foreign molecule that will stimulate an immune system response (e.g., antibody production) if the small molecule (now called a haptenic determinant) is attached to a macromolecule (carrier) to make it large enough to be recognized by the immune system. See also EPITOPE, CELLULAR IMMUNE RESPONSE, and HUMORAL IMMUNITY.

Hardening See COLD HARDENING and HYDROGENATION.

Harvesting A term used to describe the recovery of microorganisms from a liquid culture (in which they have been grown by man). This is usually accomplished by means of filtration or centrifugation. See also MICROORGANISM, CULTURE MEDIUM, ULTRACENTRIFUGE, and DIALYSIS.

Harvesting Enzymes Enzymes that are used to gently dissociate (i.e., break apart) cells in living tissues in order to produce single, separate cells that can then be established and propagated in a cell culture reactor. Harvesting enzymes are also used to dissociate cells that have been grown for some time in a cell culture reactor. See also CELL CULTURE, MAMMALIAN CELL CULTURE, ENZYME, and CULTURE MEDIUM.

Hazard Analysis and Critical Control Points (HACCP) A quality control program (for food processing) to systematically prevent hazards (e.g., pathogens) from entering the production process. HACCP was initially developed in the 1950's by the Pillsbury Company to supply food products for astronauts in America's space program.

Under HACCP, food processors/handlers must analyze and identify in advance the points where hazards are most likely to occur, and eliminate them. For example, because melons lie in pathogen-contaminated dirt while growing, a "critical control point" for restaurants serving sliced melon is cleansing of the knife after each melon is cut (to prevent the knife carrying pathogens from one infected melon to other melons). See PATHOGEN and RAPID MICROBIAL DETECTION (RMD).

Heat-Shock Proteins See STRESS PROTEINS.

Heavy-Chain Variable (VH) Domains The regions (domains) of the antibody (molecule's) "heavy chain" that vary in their amino acid sequence. The "chains" (of atoms) comprising the antibody (immunoglobulin) molecule consist of a region of variable (V) amino acid sequence and a region in which the amino acid sequence remains constant (C).

An antibody molecule possesses two antigen binding sites, and it is the variable domains of the light (VL) and heavy (VH) chains which contribute to this (antigen binding ability). See also ANTIBODY, PROTEIN, IMMUNOGLOBULIN, SEQUENCE (OF A PROTEIN MOLECULE), ANTIGEN, AMINO

ACID, COMBINING SITE, DOMAIN (OF A PROTEIN), and LIGHT-CHAIN VARIABLE (VL) DOMAINS.

Helicobacter pylori Bacteria. See *H. PYLORI.*

***Helicoverpa zea** (H. zea)* Known as the corn earworm (when it is on corn plants), and known as the tomato fruitworm (when it is on tomato plants), this is one of three insect species that is called "bollworms" (when on cotton plants). *H. zea* is one of the insects that can act as a vector (carrier) of *Aspergillus flavus* fungus. See *B.t. KURSTAKI*, *HELIOTHIS VIRESCENS*, FUNGUS, *PECTINOPHORA GOSSYPIELLA*, *ASPERGILLUS FLAVUS*, and CORN.

Heliothis virescens (*H. virescens*) Known as the tobacco budworm (when it is on tobacco plants), this is one of three insect species that is called "bollworms" (when they are on cotton plants).

As part of Integrated Pest Management (IPM), farmers can utilize the parasitic *Euplectrus comstockki* wasp to help control the tobacco budworm/cotton bollworm. When that wasp's venom is injected into *Heliothis* larva, it stops the larva from molting (and thus maturing). See *B.T. KURSTAKI*, *HELICOVERPA ZEA (H. ZEA)*, *PECTINOPHORA GOSSYPIELLA*, and INTEGRATED PEST MANAGEMENT (IPM).

Helix A spiral, staircase-like structure with a repeating pattern described by two simultaneous operations (rotation and translation). It is one of the natural conformations exhibited by biological polymers. See also BIOMIMETIC MATERIALS, and ANALOGUE.

Helper T Cells (T4 cells) T cells (lymphocytes) which bind B cells (upon recognizing a foreign epitope on B cell surface). The binding stimulates B cell proliferation by secreting B cell growth factor. See also B CELLS, CYTOKINES, T CELL, T CELL RECEPTORS, and SUPPRESSOR T CELLS.

Hematologic Growth Factors (HGF) A class of colony stimulating factors (proteins) that stimulates bone marrow cells to produce certain types of red and white blood cells. Some colony stimulating factors are:

(1) Granulocyte-macrophage colony stimulating factor (GM-CSF)
(2) Granulocyte-monocyte colony stimulating factor
(3) Granulocyte colony stimulating factor (GM-CSF)
(4) Erythropoietin (EPO)
(5) Interleukin-3 (IL-3)
(6) Macrophage colony stimulating factor (M-CSF).

Hematopoietic Growth Factors Growth factors that stimulate the body to produce blood cells. See also GROWTH FACTOR and INTERLEUKIN-6 (IL-6).

Heme The iron-porphyrin prosthetic group of a class of proteins called "heme proteins." See also PROSTHETIC GROUP, CHELATING AGENT, PROTEIN, and TRANSFERRIN.

Hemoglobin An oxygen-transporting respiratory pigment. It is carried in the red blood cells (erythrocytes), and is responsible for the red color of the blood. It is composed of two pairs of identical polypeptide chains and iron-containing heme groups, comprising the (total) hemoglobin molecule. The molecular structure of hemoglobin was determined by Max Perutz in 1959. A disease known as sickle-cell anemia is caused by (genetically induced) small change in the hemoglobin molecule's structure (in victims of that disease). See also HEME, POLYPEPTIDE, GENETICS, HEREDITY, ERYTHROCYTES, and PROTEIN STRUCTURE.

Hemostasis See FIBRIN.

Heparin A polysaccharide sulfuric acid ester found in liver, lung, and other tissues that prolongs the clotting time of blood by preventing the formation of fibrin. Used in vascular surgery and in treatment of postoperative thrombosis and embolism. See also FIBRIN and THROMBOSIS.

Herbicide Resistance See HERBICIDE-TOLERANT CROP.

Herbicide-Resistant Crop See HERBICIDE-TOLERANT CROP.

Herbicide-Tolerant Crop Crop plants, cultivated by man, which have been altered to be able to survive application(s) of one or more herbicides by the incorporation of certain gene(s), via either genetic engineering or traditional breeding techniques. For example, crops (e.g., soybean, canola, cotton, corn/maize, etc.) are made tolerant to glyphosate-containing herbicides by insertion (via genetic engineering techniques) of the transgene for CP4 EPSPS. Corn (maize) is made tolerant to imidazolinone-containing herbicides by adding (via traditional breeding techniques) the imidazolinone-tolerant trait. That trait is imparted by the T-Gene, IT-Gene, or the IR-Gene. See also GENE, GENETIC ENGINEERING, EPSP SYNTHASE, GLYPHOSATE OXIDASE, PAT GENE, BAR GENE, GENETICS, ALS GENE, EPSP SYNTHASE, CP4 EPSPS, CHLOROPLAST TRANSIT PEPTIDE (CTP), TRANSGENE, TRAIT, CANOLA, SOYBEAN PLANT, and CORN.

Heredity Transfer of genetic information from parent cells to progeny. See also INFORMATIONAL MOLECULES, GENE, GENETIC CODE, GENOME, GENETICS, GENOTYPE, DEOXYRIBONUCLEIC ACID (DNA), HERITABILITY, and QUANTITATIVE TRAIT LOCI (QTL).

Heritability The fraction of variation (of an individual's given trait) that is due to genetics. For example, if a pig's trait (e.g., weight at birth) is 30% heritable, that means that 30% of the (birthweight) difference between that individual pig and its (statistically representative) group of contemporaries (pigs) is due to genetics. The other 70% would be due to factors such as nutrition of the mother during pregnancy, etc. See also HEREDITY, TRAIT, GENETICS, INFORMATIONAL MOLECULES, GENE, GENETIC CODE, GENOME, GENOTYPE, DEOXYRIBONUCLEIC ACID (DNA), and QUANTITATIVE TRAIT LOCI (QTL).

Hetero- A chemical nomenclature prefix meaning "different." For example, a heterocyclic compound is one with a (ring) structure made up of more than one kind of atom. See the following eleven words.

Heterocyclic See HETERO-.

Heteroduplex A DNA molecule, the two strands of which come from different individuals so that there may be some base pairs or blocks of base pairs that do not match. Can arise from mutation, recombination, or by annealing DNA single strands *in vitro.* See also DEOXYRIBONUCLEIC ACID (DNA).

Heterogeneous (catalysis) Catalysis occurring at a phase boundary, usually a solid-fluid interface. See also HETERO-, HETEROGENEOUS (MIXTURE), and CATALYST.

Heterogeneous (chemical reaction) A chemical reaction in which the reactants are of different phases; for example, gas with liquid, liquid with solid, or a solid catalyst with liquid or gaseous reactants. See also HETERO-, HETEROGENEOUS (CATALYSIS), and CATALYST.

Heterogeneous (mixture) One that consists of two or more phases such as liquid-vapor, or liquid-vapor-solid. See also HETERO-.

Heterokaryon A fused cell containing nuclei of different species. See also NUCLEOID.

Heterologous Proteins Those proteins produced by an organism that is not the wild type source of those proteins. For example, bacteria have been genetically engineered to produce human growth hormone and bovine (i.e., cow) somatotropin. See also PROTEIN, WILD TYPE, GROWTH HORMONE (GH), BOVINE SOMATOTROPIN (BST), and HOMOLOGOUS PROTEIN.

Heterology A sequence of amino acids in two or more proteins that are not identical to each other. See also AMINO ACID, PROTEIN, and HOMOLOGY.

Heterosis Also known as "hybrid vigor." See F1 HYBRIDS.

Heterotroph An organism that obtains nourishment from the ingestion and breakdown of organic matter.

Heterozygote An individual organism with different alleles at one or more particular loci. See also ALLELE.

Hexadecyltrimethylammonium Bromide (CTAB) A solvent that is widely utilized to dissolve plant DNA samples (e.g., when a scientist wants to sequence that sample of plant DNA). CTAB solvent helps the scientist to separate out contaminants that are commonly present in samples from plant tissues (i.e., polysaccharides, quinones, etc.) because DNA molecules are much more soluble in CTAB than are the contaminant molecules. See DEOXYRIBONUCLEIC ACID (DNA), POLYSACCHARIDES, SEQUENCING (OF DNA MOLECULES), and SDS.

Hexose See GLUCOSE (GLc).

HF Cleavage A research process in which hydrofluoric acid is used to sequentially remove side-chain protective groups from peptide chains. Also used to remove the resin support from peptides that have been prepared via solid phase peptide synthesis. The HF cleavage reaction is a temperature-dependent process. See also PROSTHETIC GROUP and SYNTHESIZING (OF PROTEINS).

High-Density Lipoproteins (HDLPs) Lipoproteins that can help move excess low-density lipoproteins (i.e., "bad" cholesterol, which can clog arteries) out of the human body by binding to the low-density lipoproteins (also known as LDL cholesterol) in the blood and then attaching to special LDLP receptor molecules in the liver. The liver then clears those (bound) low-density lipoproteins out of the body as a part of regular liver functions. See LOW-DENSITY LIPOPROTEINS (LDLP), RECEPTORS, and WATER SOLUBLE FIBER.

High-Lysine Corn Developed in America in the mid-1960's, these are corn varieties possessing the opague-2 gene. The opague-2 gene causes such corn to contain 0.30–0.55% lysine (i.e., 50–80% more than traditional No. 2 yellow corn). High-lysine corn is particularly useful for feeding of swine, since traditional No. 2 yellow corn does not contain enough lysine for optimal pig growth. See CORN, LYSINE (lys), OPAGUE-2, GENE, VALUE-ENHANCED GRAINS, "IDEAL PROTEIN" CONCEPT, and MAL (MULTIPLE ALEURONE LAYER) GENE.

High-Methionine Corn Developed in America in the mid-1960's, these are corn varieties possessing the floury-2 gene. The floury-2 gene causes such corn to contain slightly higher levels of methionine than traditional No. 2 yellow corn. High-methionine corn is particularly useful for feeding of poultry, since traditional No. 2 yellow corn does not contain enough methionine for optimal poultry growth. See also METHIONINE (met), FLOURY-2, GENE, VALUE-ENHANCED GRAINS, OPAGUE-2, "IDEAL PROTEIN" CONCEPT, and MAL (MULTIPLE ALEURONE LAYER) GENE.

High-Oil Corn Conceived in 1896 at the University of Illinois, high-oil corn (HOC) is defined to be corn possessing a kernel oil content of 5.8% or greater. Traditional No. 2 yellow corn tends to contain 3.5% or less oil content. See also VALUE-ENHANCED GRAINS, CORN, and CHEMOMETRICS.

High-Oleic Oil Soybeans Soybeans from soybean plants which have been genetically engineered to produce soybeans bearing oil that contains more than 80% oleic acid, instead of the typical 24% oleic acid content of soybean oil produced from traditional varieties of soybeans. See SOYBEAN PLANT, SOYBEAN OIL, FATTY ACID, MONOUNSATURATED FATS, GENETIC ENGINEERING, and OLEIC ACID.

High-Phytase Corn (or high-phytase soybeans, etc.) Crop plants that have been genetically engineered to contain in their grain/seed high(er) levels of the enzyme phytase (which aids digestion and absorption of phosphate in that grain/seed). High-phytase grains or oilseeds are particularly useful for the feeding of swine and poultry, since traditional No. 2 yellow corn (maize) or traditional soybean varieties, do not contain phytase in amounts needed for complete digestion/absorption of phosphate naturally contained in those traditional soybeans and corn (maize) in the form of phytate. See PHYTASE, ENZYME, PHYTATE, and VALUE-ENHANCED GRAINS.

High-Stearate Soybeans Soybean plant varieties which have been genetically engineered so their beans contain more than the (historic average) typical 3% stearic acid content in the soybean oil. See STEARATE, VALUE-ENHANCED GRAINS, SOYBEAN PLANT, and SOYBEAN OIL.

High-Sucrose Soybeans Another name for low-stachyose soybeans because the soybeans replace the (reduced) stachyose with (additional) sucrose. See LOW-STACHYOSE SOYBEANS, STACHYOSE, VALUE-ENHANCED GRAINS, SOYBEAN PLANT, and SUGAR MOLECULES.

Histamine A base that is naturally present in ergot (a fungus) and plants; it is also naturally produced by basophils (basophilic leukocytes) in the human body. It is formed from histidine by decarboxylation, and is held to be responsible for the dilation and increased permeability of blood vessels which play a major role in allergic reactions. See also HISTIDINE (HIS) and BASOPHILS.

Histidine (his) A basic amino acid that is essential in the nutrition of the rat. It is formed by the decomposition of most proteins (as globin). See also PROTEIN.

Histiocyte See MACROPHAGE.

Histoblasts See B LYMPHOCYTES.

Histones Proteins rich in basic amino acids (e.g., lysine) found complexed with chromosomes of all eucaryotic cells except sperm where the DNA is specifically complexed with another group of basic proteins, the protamines. See also CHROMOSOMES, CHROMATIDS, and CHROMATIN.

Histopathologic Refers to changes in tissue caused by a disease. For example, certain diseases (e.g., jaundice) cause the skin to turn yellow. See also PATHOGENIC, VIRUS, CANCER, and ADHESION MOLECULE.

HIV-1 and HIV-2 See HUMAN IMMUNODEFICIENCY VIRUS (TYPE 1 and TYPE 2).

HLA See HUMAN LEUKOCYTE ANTIGENS.

HNGF Human nerve growth factor. See NERVE GROWTH FACTOR (NGF).

HOC See HIGH-OIL CORN.

Hollow Fiber Separation (of proteins) The separation of proteins from a mixture by means of "straining" the mixture through hollow, semipermeable fibers (e.g., polysulfone fibers) under pressure. The hollow fibers are constructed in such a way that they have very tiny (molecular size) holes in them. In this way large molecules are retained in the original liquid while smaller molecules, which are able to pass through the holes, are filtered out. See also DIALYSIS, PROTEIN, and ULTRAFILTRATION.

Holoenzyme The entire, functionally complete enzyme. The term is used to designate an enzyme that requires a coenzyme in order for it to function (possess catalytic abilities). The holoenzyme consists of the protein part (apoenzyme) plus a dialyzable, nonprotein coenzyme part that is bound to the apoenzyme protein. See also COENZYME, APOENZYME, and DIALYSIS.

Homeobox A short sequence of DNA that is 180 base pairs long and located in the 3′ exon of certain genes of the *Drosophila* fly (where they were discovered by Walter Gehring during the 1970s).

In the 1980s, Jani Christian Nusslein-Volhard discovered that one homeobox was attached (in adjacent exon) to each of the genes that are responsible for embryonic development (i.e., "switched on" only in an embryo that is developing into an adult), in a wide variety of species including invertebrates, birds, and mammals. Thus, it is now possible to locate many embryonic-development genes in many species by using a DNA probe (made via a *Drosophila* homeobox DNA sequence) to find homeobox sequences attached to those embryonic-development genes. In such a role, the respective homeobox sequences attached to each gene are known as DNA markers. See also GENE, DEOXYRIBONUCLEIC ACID (DNA), DNA PROBE, DNA MARKER, SEQUENCE (OF A DNA MOLECULE), BASE PAIR (bp), *DROSOPHILA*, EXON, and SPECIES.

Homeostasis A tendency toward maintenance of a relatively stable internal environment in the bodies of higher animals through a series of interacting physiological processes. An example is the mammal's maintenance of a constant body temperature despite extremes in weather temperature.

Homing Receptor Also known as L-selectin. See SELECTINS, LECTINS, and ADHESION MOLECULES.

Homologous (chemically) See HOMOLOGY.

Homologous (chromosomes) Chromosomes or chromosome segments that are identical with respect to their constituent genetic loci and their visible structure. See also CHROMOSOMES and LOCUS.

Homologous Protein A protein having identical functions and similar properties in different species. For example, the hemoglobins that perform identical functions in the blood of different species.

Homology A sequence of amino acids in two or more proteins that are identical to each other. Nucleic-acids homology refers to complementary strands that can hybridize with each other.

Homotropic Enzyme An allosteric enzyme whose own substrate functions as an activity modulator. See also ENZYME.

Homozygote An organism in which the corresponding genes (alleles) on the two genomes are identical. An organism which possesses an identical pair of alleles in regard to a given (genetic) characteristic. See also GENE, ALLELE, GENOME, GENOTYPE, PHENOTYPE, HOMOZYGOUS, and HETEROZYGOTE.

Homozygous In a diploid organism, a state where both alleles of a given gene are the same. See also HETEROZYGOTE, ALLELE, DIPLOID, DIPLOPHASE, and HOMOZYGOTE.

Hormone A type of chemical messenger (peptide), occurring both in plants and animals, that acts to inhibit or excite metabolic activities (in that plant or animal) by binding to receptors on specific cells to deliver its "message." A hormone's site of production is distant from the site of biological activity (i.e., where the message is delivered). See also PEPTIDE, MINIMIZED PROTEINS, and SIGNALLING.

Host Cell A cell whose metabolism is used for growth and reproduction by a virus. Also the cell into which a plasmid is introduced (in recombinant DNA experiments).

Host Vector (HV) System The host is the organism into which a gene from another organism is transplanted. The guest gene is carried by a vector (i.e., a larger DNA molecule, such as a plasmid, or a virus into which that gene is inserted) which then propagates in the host.

Hot Spots Sites in genes at which events, such as mutations, occur with unusually high frequency. See also GENE, JUMPING GENES, MUTATION, and TRANSLOCATION.

HPLC High performance liquid chromatography. See CHROMATOGRAPHY.

HSOD See HUMAN SUPEROXIDE DISMUTASE (hSOD).

HTC See herbicide-tolerant crop. See also STS, PAT GENE, EPSP SYNTHASE, ALS GENE, BAR GENE, CP4 EPSPS, and GLYPHOSATE OXIDASE.

Human Artificial Chromosomes (HAC) Chromosomes that have been synthesized (made) from chemicals that are identical to chromosomes within human cells. See also YEAST ARTIFICIAL CHROMOSOMES (YAC), CHROMOSOMES, *ARABIDOPSIS THALIANA*, and SYNTHESIZING (OF DNA MOLECULES).

Human Chorionic Gonadotropin A human hormone. In 1986, Mark Bogart discovered that elevated levels of human chorionic gonadotropin in pregnant women are correlated with babies (later) born with Down Syndrome. See HORMONE.

Human Colon Fibroblast Tissue Plasminogen Activator A second generation tissue plasminogen activator (tPA), which has the clot-sensitive activation of plasminogen with potentially greater selectivity and (clot) specificity. See also TISSUE PLASMINOGEN ACTIVATOR (tPA).

Human EGF-Receptor-Related Receptor (HER-2) A gene that appears to be directly related to human breast cancer mortality. The more copies of the HER-2 gene (in a patient's breast tumor cells) the more dismal that patient's prospects for survival.

Human Gamma-Glutamyl Transpeptidase A glycoprotein that is thought to possess a different oligosaccharide when it is produced by a (liver) tumor cell instead of a healthy cell. Thus, it is a possible early warning marker for liver cancer. See also GLYCOPROTEIN and OLIGOSACCHARIDES.

Human Growth Hormone (HGH) See GROWTH HORMONE (GH).

Human Immunodeficiency Virus Type I (HIV-1) and **Human Immunodeficiency Virus Type 2** (HIV-2) Two viruses identified (so far) which cause acquired immune deficiency syndrome (AIDS). HIV-1 and HIV-2 show a preferential tropism (affinity) toward the helper T cells, although other immune system (and nervous system) cells are also infected. The GP120 envelope (surface) protein of HIV-1 and HIV-2 directly interacts (binds) with the CD4 proteins (receptors) on the surface of helper T cells, enabling the viruses to bind (attach to) and infect the helper T cells. In order to successfully enter and infect cells, the HIV must also bind with CKR-5 proteins (receptors) located on the surface of cells of most humans. In 1996, Nathaniel Landau and Richard Koup discovered that approximately one percent of humans carry a gene for a version of CKR-5 receptor that resists entry to cells by HIV. As of 1996, a total of nine separate strains (serotypes) of Human Immunodeficiency Virus were known; identified by the letters A, B, C, D, E, F, G, H, I. See also CD4 PROTEIN, ADHESION MOLECULE, GP120 PROTEIN, ACQUIRED IMMUNE DEFICIENCY SYNDROME (AIDS), RECEPTORS, TROPISM, HELPER T CELLS (T4 CELLS), T CELL RECEPTORS, VIRUS, and SEROTYPES.

Human Leukocyte Antigens (HLA) A very complex array of six proteins that cover the surface of leukocytes (and the bone marrow cells that produce leukocytes). These HLA are usually different (i.e., a non-match) for individuals that are not genetically related to each other (e.g., a father-son or a father-daughter), so have been used in the past to prove paternity. HLA must also be matched (as nearly as possible) for successful bone marrow transplants, to prevent the donated bone marrow (and the marrow recipient) from "rejecting" each other. See also LEUKOCYTES, ANTIGEN, MAJOR HISTOCOMPATIBILITY COMPLEX (MHC), PROTEIN, and GRAFT-VERSUS-HOST DISEASE (GVHD).

Human Protein Kinase C An enzyme that is involved in the control of blood coagulation and fibrinolysis. See also FIBRIN.

Human Superoxide Dismutase (hSOD) An enzyme that "captures" oxygen free radicals (oxygen atoms bearing an extra electron) e.g., which are generated within an occluded (closed off) blood vessel. Oxygen free radicals are generated within occluded blood vessels when a blood clot blocks arteries in the heart, causing a heart attack. These oxygen free radicals are highly energized and can cause damage to blood vessel walls after the clot is dissolved (e.g., with tissue plasminogen activator), so hSOD may profitably be administered in conjunction with clot-dissolving pharmaceuticals to minimize damage when occluded arteries are reopened.

Research indicates that hSOD may help protect elderly patients from the lethal effects of influenza (i.e., the flu), because influenza often causes overproduction of free radicals in the victim's body. See also XANTHINE OXIDASE and TISSUE PLASMINOGEN ACTIVATOR (tPA).

Recent research indicates that hSOD may be made more effective when administered in combination with certain copper/zinc compounds to bolster its efficacy. See also PEG-SOD (POLYETHYLENE GLYCOL SUPEROXIDE DISMUTASE), CATALASE, XANTHINE OXIDASE, and TISSUE PLASMINOGEN ACTIVATOR (tPA).

Human Thyroid-Stimulating Hormone (hTSH) A naturally occurring hormone that causes the thyroid gland to develop. See also HORMONE.

Humoral Immune Response See HUMORAL IMMUNITY.

Humoral Immunity The immune system response consisting of the soluble blood serum components that fight an infection (e.g., antibodies, complement proteins, cecrophins, etc.). See also ANTIBODY, COMPLEMENT (COMPONENT OF IMMUNE SYSTEM), COMPLEMENT CASCADE, CECROPHINS (LYTIC PROTEINS), CELLULAR IMMUNE RESPONSE, and IMMUNOGLOBULIN (IgA, IgE, IgG, and IgM).

Huntington's Disease A disease of humans, in which a defective gene causes the production of a (mutant) protein that kills brain cells. That gene was discovered in 1994 by an international collaboration of scientists that included Francis Collins (University of Michigan), David Housman (Massachusetts Institute of Technology), John Wasmuth (University of California), Hans Lehrach (Imperial Research Cancer Fund in London), Peter Harper (University of Wales) and Gusella and colleagues (Massachusetts General Hospital). Normally, the "Huntington's gene" contains the three base molecules cytosine-adenine-guanine (abbreviated: CAG) in a linear repetition of approximately 10–20 times within the DNA of that gene. In a mutant (defective) Huntington's gene, that cytosine-adenine-guanine sequence is instead found in a linear repetition of far greater length than the normal 10–20 times. Thus, the mutant Huntington's gene is far larger than the normal gene; and this enlargement apparently causes the mutant protein to be produced that kills brain cells. See also GENE, PROTEIN, MUTATION, MUTANT, GENETIC CODE, INFORMATIONAL MOLECULES, HEREDITY GENETICS, BASE PAIR (bp), DEOXYRIBONUCLEIC ACID (DNA), and CODING SEQUENCE.

Hybridization (molecular genetics) The pairing (tight physical bonding) of two complementary single strands of RNA and/or DNA to give a double-stranded molecule. See also ANNEAL, STICKY ENDS,

RIBONUCLEIC ACID (RNA), BIOSENSORS (ELECTRONIC), BIOSENSORS (CHEMICAL), HYBRIDIZATION SURFACES, DNA PROBE, and DEOXYRIBONUCLEIC ACID (DNA).

Hybridization (plant genetics) The mating of two plants from different species or *genetically very different* members of the same species to yield hybrids (first filial hybrids) possessing some of the characteristics of each parent. Those (hybrid) offspring tend to be more healthy, productive, and uniform than their parents—a phenomenon known as "hybrid vigor." See F1 HYBRIDS, SPECIES, TRANSGRESSIVE SEGREGATION, GENETICS, and GEM.

Hybridization Surfaces Various physical substrates (surfaces) onto which have been "attached" genetic materials (DNA, RNA, oligonucleotides, etc.). Relevant complementary genetic materials (e.g., DNA, RNA, oligonucleotides, etc.) then are hybridized onto those attached-to-surface genetic materials for various specific purposes (e.g., detection of the presence of those unattached genetic materials, in the case of biosensor's hybridization surface).

One of the technologies that can be utilized to assay (evaluate) DNA from hybridization surfaces is Matrix-Assisted Laser Desorption Ionization Time of Flight Mass Spectrometry (MALDI-TOF-MS). See SUBSTRATE (STRUCTURAL), HYBRIDIZATION (MOLECULAR GENETICS), COMPLEMENTARY DNA (c-DNA), DEOXYRIBONUCLEIC ACID (DNA), RIBONUCLEIC ACID (RNA), NANOCRYSTAL MOLECULES, DOUBLE HELIX, BIOSENSORS (ELECTRONIC), BIOSENSORS (CHEMICAL), OLIGONUCLEOTIDE, OLIGONUCLEOTIDE PROBES, MALDI-TOF-MS, and ASSAY.

Hybridoma The cell line produced by fusing a myeloma (tumor cell) with a lymphocyte (which makes antibodies); it continues indefinitely to express the immunoglobulins (antibodies) of both parent cells. See also MONOCLONAL ANTIBODY (MAb) and AGING.

Hybrid Vigor See F1 HYBRIDS and HYBRIDIZATION (PLANT GENETICS).

Hydrazine A chemical with formula N_2H_4. Used as a rocket fuel, and in the hydrazinolysis of glycoproteins. See also HYDRAZINOLYSIS (OF GLYCOPROTEINS, TO ISOLATE UNREDUCED OLIGOSACCHARIDE SIDE CHAINS), GLYCOPROTEIN, and REDUCTION (IN A CHEMICAL REACTION).

Hydrazinolysis (of glycoproteins to isolate unreduced oligosaccharide side chains) A technique that used the chemical hydrazine to separate and isolate the oligosaccharide portion from the protein portion of a glycoprotein. The hydrazine chemically "chews up" the polypeptide (i.e., protein) portion of a glycoprotein molecule, leaving the intact oligosaccharides behind. It can subsequently be analyzed (after chromatographic separation from the peptide pieces and other chemical components). See also REDUCTION (IN A CHEMICAL REACTION) HF CLEAVAGE, POLYPEPTIDE

(PROTEIN), GLYCOPROTEIN, SEQUENCING (OF OLIGOSACCHARIDES), HYDRAZINE, and CHROMATOGRAPHY.

Hydrofluoric Acid Cleavage See HF CLEAVAGE.

Hydrogenation A chemical reaction/process in which hydrogen atoms are added to molecules (e.g., of unsaturated fatty acids) in edible oils. In the case of fatty acids, the fraction of each isomeric form (*trans* vs. *cis* fatty acids) and the molecular chain length (of the fatty acids present) have a large impact on the melting characteristics of each (fat or oil), with shorter-chain fats melting at lower temperature.

Hydrogenation is the most common chemical reaction utilized in the edible oils (processing) industry. Hydrogenation increases the solids (i.e., crystalline fat) content of edible fats/oils, and improves their resistance to thermal and atmospheric oxidation (e.g., for frying of foods). Those increases in solids and resistance to oxidation result from the reduction in the fat/oil relative unsaturation, plus increased geometric and positional isomerization of the fat/oil molecules.

The edible oil/fat hydrogenation reaction is accomplished by treating fats/oils with pressurized hydrogen gas in the presence of a catalyst. As a result, the (usually) liquid oils are converted to more-saturated fats, which are semisolids at an ambient temperature of 72°F (22°C).

The presence of *trans* fatty acids in hydrogenated edible oils can be reduced significantly via changes in catalyst, temperature, pressure, etc. utilized in the hydrogenation reaction.

In general, natural oils and fats possessing melting points lower than 121°F (50°C) are nearly completely absorbed in the digestive system of typical humans. See also FATTY ACID, MONOUNSATURATED FATS, SATURATED FATTY ACIDS, DEHYDROGENATION, ESSENTIAL FATTY ACIDS, LAURATE, LECITHIN, TRIGLYCERIDES, UNSATURATED FATTY ACID, SOYBEAN OIL, OXIDATION, ISOMER, STEREOISOMERS, CATALYST, SUBSTRATE (CHEMICAL), and *trans* FATTY ACIDS.

Hydrolysis Literally, means "cleaved by water." It is used for a chemical reaction in which the chemical bond attaching an atom, or group of atoms to the (rest of the) molecule is cleaved, followed by attachment of a hydrogen atom at the same chemical bond.

Hydrolytic Cleavage A chemical reaction in which a portion (e.g., an atom or a group of atoms) of a molecule is "cut" off the molecule via hydrolysis. See also HYDROLYSIS.

Hydrolyze To "cut" a chemical bond (i.e., with a molecule) via hydrolysis. See also HYDROLYSIS.

Hydrophilic This term means water loving or having a great affinity for water. It is used to describe molecules or portions of molecules that have an affinity for water. The property of having an affinity for water at an oil-water interface. For example, ordinary sugar that dissolves readily in water is said to be hydrophilic (i.e., a molecule that is "water loving"). See also AMPHIPHILIC MOLECULES.

Hydrophobic This term means water hating or having a great dislike for water. It is used to describe molecules or portions of molecules that have very little or no affinity for water. The property of having an affinity for oil (nonpolar environments) at an oil-water interface. For example, a nonpolar hydrocarbon such as butane (as used in lighters) that will not dissolve in water, but which will dissolve (be miscible) in oil is said to be hydrophobic (i.e., a molecule that is "water hating"). See also AMPHIPHILIC MOLECULES.

Hydroxylation Reaction A chemical reaction in which one or more hydroxyl groups (i.e., the —OH group) is introduced (i.e., is chemically attached) to a molecule.

Hyperacute Rejection See GRAFT-VERSUS-HOST DISEASE (GVHD).

Hyperchromicity The increase in optical density that occurs when DNA is denatured. See also DEOXYRIBONUCLEIC ACID (DNA), DENATURED DNA, and OPTICAL DENSITY (OD).

Hyperthermophilic (organisms) See THERMOPHILE and THERMOPHILIC BACTERIA.

Hypostasis Interaction between nonallelic genes in which one gene will not be expressed in the presence of a second. See also EPISTASIS, GENE, EXPRESS, and ALLELE.

Hypothalamus A part of the brain structure, lying near base of brain, it regulates a number of hormones. As a part of the brain, it constantly receives (neurochemical) signals from nerve cells (neurons). The hypothalamus monitors those signals, and converts them into hormonal "signals" [e.g., it generates a "burst" of hormones in response to certain visual stimuli, certain physical (e.g., sexual) stimuli, etc.].

Also, the hypothalamus is able to monitor and detect changes in the blood levels of hormones coming from endocrine glands. For example, the metabolic hormone insulin (from the pancreas) and the reproductive hormone estrogen (from the ovaries) both trigger changes in function in the hypothalamus. The hypothalamus regulates biological processes (e.g., metabolic rate, appetite, etc.). A major function of the hypothalamus is to control reproduction, via secretion of gonadotropin-releasing hormone (GnRH) from the tips of hypothalamic nerve fibers that extend downward toward (into) the pituitary gland. Similarly, the hypothalamus also helps to control the body's growth (from birth until the end of puberty) via secretion of growth hormone–releasing factor (GHRF) to the pituitary gland. See also HORMONE, ENDOCRINE HORMONES, ENDOCRINE GLANDS, ENDOCRINOLOGY, PITUITARY GLAND, GROWTH HORMONE (GH), NEUROTRANSMITTER, and GROWTH HORMONE–RELEASING FACTOR (GHRF).

I

IBA See INDUSTRIAL BIOTECHNOLOGY ASSOCIATION.

IBG See INTERNATIONAL BIOTECHNOLOGY GROUP.

ICAM Intercellular adhesion molecule. See ADHESION MOLECULE.

IDE "Investigational Device Exemption" application to the Food and Drug Administration (FDA) seeking approval to begin clinical studies of a new medical device.

"Ideal Protein" Concept This refers to the protein content in the feed ration (food) eaten by livestock or poultry. Feed that contains "ideal protein" contains protein(s) that (when digested by the animal) yields all the essential amino acids, in proper proportions, for the growth and/or maintenance needs of that animal.

"Ideal protein" varies for different species (e.g., pigs require different amino acids/rations than chickens do). "Ideal protein" varies for different stages in the life of a given animal (e.g., poultry requires more sulfur-containing amino acids, such as methionine, during stages when feather growth is at a comparatively high rate).

The animal's requirement for one essential amino acid is proportionally linked to the animal's requirement for another. Increasing the supply of one essential amino acid in the animal's diet would improve that animal's (growth) performance if no other amino acid(s) were limiting. See PROTEIN, ESSENTIAL AMINO ACIDS, ESSENTIAL NUTRIENTS, METHIONINE (met), DIGESTION (WITHIN ORGANISMS), and HIGH-METHIONINE CORN.

Idiotype The region of the antibody molecule that enables each antibody to recognize a specific foreign structure (i.e., epitope or hapten) is said to have an idiotype (for that epitope or hapten). An identifying characteristic (or property) of the thing we are talking about. See also EPITOPE, HAPTEN, ANTIGEN, and ANTIBODY.

IDM See INTEGRATED DISEASE MANAGEMENT.

IFBC See INTERNATIONAL FOOD BIOTECHNOLOGY COUNCIL.

IFN-Alpha Alpha interferon. See INTERFERONS.

IFN-Beta Beta interferon. See INTERFERONS.

IGF-1 See INSULIN-LIKE GROWTH FACTOR-1.

IGF-I See INSULIN-LIKE GROWTH FACTOR-1.

IGF-2 See INSULIN-LIKE GROWTH FACTOR-2.

IGF-II See INSULIN-LIKE GROWTH FACTOR-2.

IL-1 See INTERLEUKIN-1.

IL-Ira See INTERLEUKIN-1 RECEPTOR ANTAGONIST.

Immune Response See CELLULAR IMMUNE RESPONSE, ANTIBODY, and HUMORAL IMMUNITY.

Immunoassay The use of antibodies to identify and quantify (measure) substances by a variety of methods. The binding of antibodies to antigen (substance being measured) is often followed by tracers, such as fluorescence or (radioactive) radioisotopes, to enable measurement of the substance. See also ANTIBODY, TRACER (RADIOACTIVE ISOTOPIC METHOD), ANTIGEN, ELISA (TEST FOR PROTEINS), RADIOIMMUNOASSAY, and EIA.

Immunoconjugate A molecule that has been formed by attachment to each of two originally different molecules. One of these is an antibody and, hence, the word "immunoconjugate." Classic organic drug molecules such as methotrexate, adriamycin chlorambucil, etc.; radionuclides; enzymes; toxins; and ribosome-inhibiting proteins may be conjugated to antibodies. The salient point is that the antibody portion of the conjugate is there to "steer" the biologically active molecule to its target. See also CONJUGATE.

Immunocontraception Any process or procedure in which an organism's immune system is utilized to attack or inactivate the reproductive cells (e.g., sperm) within the organism. See also CELLULAR IMMUNE RESPONSE, ANTIBODY, HUMORAL IMMUNITY, and GERM CELL.

Immunogen See also ANTIGEN.

Immunoglobulin (IgA, IgE, IgG, and IgM) A class of (blood) serum proteins representing antibodies. Often used, along with the more specific monoclonal antibodies, in health diagnostic reagents. In certain people who are genetically predisposed to foodborne allergies, immunoglobulin-E (IgE) initiates an immune system response to antigen(s) present on protein molecule(s) in the particular food that person is allergic to.

Severe allergic reactions to foods may lead to death. See also PROTEIN, ANTIGEN, ALLERGIES (FOODBORNE), ANTIBODY, and IMMUNOASSAY.

Immunosuppressive That which suppresses the immune system response (e.g., certain chemicals). See also CELLULAR IMMUNE RESPONSE and HUMORAL IMMUNITY.

Immunotoxin A conjugate formed by attaching a toxic molecule (e.g., ricin) to an agent of the immune system (e.g., a monoclonal antibody), that is specific for the pathogen or tumor to be killed. The immune system–agent portion (of the conjugate) delivers the toxic chemical directly to the specified (disease) site, thus sparing other healthy tissues from the effect of the toxin. See also RICIN and MONOCLONAL ANTIBODIES (MAb).

In situ In the natural or original position (e.g., inside the body).

In vitro In an unnatural position (e.g., outside the body, in the test tube). "*In vitro*" is Latin for "in glass." For example, the testing of a substance, or the experimentation in (using) a "dead" cell-free system. See also *IN VITRO* SELECTION.

***In vitro* Evolution** See *IN VITRO* SELECTION, COMBINATORIAL CHEMISTRY, TARGET, and MOLECULAR DIVERSITY.

***In vitro* Selection** A search process (e.g., for a new pharmaceutical) that first involves the construciton of a large "pool" of polynucleotide sequences (at least some of which are likely to possess the desired pharmaceutical properties), synthesized by a totally random process. This is followed by repeated cycles of screening (for those sequences possessing desired properties) and/or enriching, and amplification (of the screened/enriched sequences). Common amplification techniques include Polymerase Chain Reaction (PCR), Ligase Chain Reaction (LCR), Self-sustained Sequence Replication (SSR), Q-beta Replicase Technique, and Strand Displacement Amplification (SDA). See also *IN VITRO*, AMPLIFICATION, GENE AMPLIFICATION, POLYMERASE CHAIN REACTION (PCR), Q-BETA REPLICASE TECHNIQUE, NUCLEOTIDE, DEOXYRIBONUCLEIC ACID (DNA), SYNTHESIZING (OF DNA MOLECULES), OLIGONUCLEOTIDE, DNA PROBE, and GENE MACHINE.

***In-vitro* Evolution** See *IN VITRO* SELECTION.

***In-vitro* Selection** See *IN VITRO* SELECTION.

In vivo The testing of a substance or experimentation in (using) a living, whole organism. An *in vivo* test is one in which an experimental substance is injected into an animal such as a rat in order to ascertain its effect on the organism.

Inclusion Bodies See REFRACTILE BODIES (RB).

IND "Investigational New Drug" application to the Food and Drug Administration (FDA) seeking approval to begin clinical studies of a new pharmaceutical. See also "TREATMENT" IND, IND (REGULATIONS) EXEMPTION, PHASE I CLINICAL TESTING, and FOOD AND DRUG ADMINISTRATION (FDA).

IND Exemption A permit by the Food and Drug Administration (FDA) to begin clinical trials on humans (of a new pharmaceutical) after toxicity data has been reviewed and approved by the FDA. See also KEFAUVER RULE, IND, and PHASE I CLINICAL TESTING.

Indian Department of Biotechnology The governmental body in India that regulates all recombinant DNA research. It is the Indian counterpart of the American Government's Recombinant DNA Advisory Committee (RAC). See also RECOMBINANT DNA ADVISORY COMMITTEE (RAC), ZKBS (CENTRAL COMMISSION ON BIOLOGICAL SAFETY), GENETIC ENGINEERING, RECOMBINANT DNA (rDNA), RECOMBINATION, BIOTECHNOLOGY, GENE TECHNOLOGY OFFICE, and COMMISSION OF BIOMOLECULAR ENGINEERING.

Induced Fit A substrate-induced change in the shape of an enzyme molecule that causes the catalytically functional groups of the enzyme to assume positions that are optimal for catalytic activity to occur. See also ENZYME.

Inducers Molecules that cause the production of larger amounts of the enzymes that are involved in the uptake and metabolism of the inducer (such as galactose). Inducers may be enzyme substrates. See also INDUCIBLE ENZYMES.

Inducible Enzymes Enzymes whose rate of production can be increased by the presence of certain chemical molecules.

Industrial Biotechnology Association (IBA) An American trade association of companies involved in biotechnology. Formed in 1981, the IBA tended to consist of the larger firms involved in biotechnology. In 1993, the Industrial Biotechnology Association (IBA) was merged with the Association of Biotechnology Companies (ABC) to form the Biotechnology Industry Organization (BIO). See also ASSOCIATION OF BIOTECHNOLOGY COMPANIES (ABC), BIOTECHNOLOGY INDUSTRY ORGANIZATION (BIO), and BIOTECHNOLOGY.

Informational Molecules Molecules containing information in the form of specific sequences of different building blocks. They include proteins and nucleic acids. See also HEREDITY, GENE, GENETIC CODE, GENOME, GENOTYPE, MESSENGER RNA (mRNA), DEOXYRIBONUCLEIC ACID (DNA), and RIBONUCLEIC ACID (RNA).

Ingestion Taking a substance into the body. For example, the amoeba surrounds a food particle, then ingests the particle.

Inhibition The suppression of the biological function of an enzyme or system by chemical or physical means. See also APTAMERS and ENZYME.

Initiation Factors Specific proteins required to initiate synthesis of a polypeptide on ribosomes. See RIBOSOMES, PROTEIN, and POLYPEPTIDE (PROTEIN).

Insitu See IN SITU.

Insulin A protein hormone normally secreted by the β cells of the pancreas (when stimulated by glucose, and the parasympathetic nervous system). Insulin and glucagon are the most important regulators of fuel (food) metabolism. In essence, insulin signals the "fed" state to the body's cells, which stimulates the storage of fuels and the synthesis of proteins in a variety of ways.

The disease known as diabetes results from a body's inability to produce insulin. In 1922, Canadian scientists Frederick Banting, Charles Best, J. J. R. MacLeod and J. B. Collip succeeded in extracting insulin from the pancreas of slaughtered livestock (cows, pigs) in a form that could be injected into diabetes patients as a substitute for human insulin.

The English biochemist, Fred Sanger, was first to determine the complete amino acid sequence of the insulin molecule. In 1977, the American scientist Howard Goodman, collaborating with William Rutter, announced the first cloning of insulin genes. This led to human insulin production by genetically engineered microorganisms (approved by FDA in 1982). See also ISLETS OF LANGERHANS, HORMONE, PROTEIN, GLUCOSE (GLc), AMINO ACID, POLYPEPTIDE (PROTEIN), SEQUENCE (OF PROTEIN MOLECULE), GENETIC ENGINEERING, and GLUCAGON.

Insulin-Like Growth Factor-1 (IGF-1) A protein hormone that is produced by the body's bone cells (when those bone cells have been stimulated by parathyroid hormone and/or estrogen), which is a promoter of bone formation and follicle development (in ovaries).

Another function of IGF-1 is to facilitate the transport of amino acids into cells, and further inhibit protein breakdown in cells.

If the body is injured, IGF-1 works with platelet-derived growth factor (PDGF) to stimulate fibroblast and collagen cell division/metabolism to cause healing of wounds and bones. IGF-1 also occurs naturally in cow's milk. See also FIBROBLASTS, AMINO ACID, COLLAGEN, ESSENTIAL AMINO ACIDS, DIGESTION (WITHIN ORGANISMS), METABOLISM, PROTEIN, MESSENGER RNA (mRNA), and UBIQUITIN.

Insulin-Like Growth Factor-2 (IGF-2) See INSULIN-LIKE GROWTH FACTOR-1 (IGF-1).

Integrated Crop Management See INTEGRATED PEST MANAGEMENT.

Integrated Disease Management See INTEGRATED PEST MANAGEMENT.

Integrated Pest Management (IPM) A holistic (system) approach utilized by farmers to try to control agricultural pests (e.g., tobacco budworm, European corn borer, soybean cyst nematode, etc.) which was initially developed as a methodology by Ray Smith and Perry Adkisson. IPM also helps to control plant diseases. For example, farmers can plant buckwheat near their cornfields in order to help control European corn borer (ECB), a serious pest of corn (maize) *Zea mays L.* Lacewing beetles, which prey on European corn borers, are attracted by the buckwheat and consume ECB in the corn while they live in the buckwheat areas. Because European corn borer is a vector (carrier) of disease microorganisms like *Aspergillus flavus* and *Fusarium*, this lacewing beetle (IPM) control of ECB also helps reduce those diseases. See *HELIOTHIS VIRESCENS* (*H. VIRESCENS*), EUROPEAN CORN BORER (ECB), LOW-TILLAGE CROP PRODUCTION, NO-TILLAGE CROP PRODUCTION, SOYBEAN CYST NEMATODES (SCN), CORN, SOYBEAN PLANT, and *BACILLUS THURINGIENSIS (B.t.)*.

Integrins A class of proteins that is found on the surface (membranes) of cells, and that function as cellular adhesion receptors. For example, integrin $\alpha_v B_3$ is a receptor on the surface of endothelial cells in tumors. It binds angiogenic endothelial cells, enabling them to form new blood vessels. See also ADHESION MOLECULES, PROTEIN, GLYCOPROTEINS, CELL, RECEPTORS, LECTINS, SELECTINS, SIGNAL TRANSDUCTION, ANGIOGENESIS, and TUMOR.

Intercellular Adhesion Molecule (ICAM) See ADHESION MOLECULE.

Interferons A family of small (cytokines) proteins (produced by vertebrate cells following a virus infection) possessing potent antiviral effects. Secreted interferons bind to the plasma membrane of other cells in the organism and induce an antiviral state in them (conferring resistance to a broad spectrum of viruses). Three classes of interferons have been isolated and purified, so far: α-interferon (originally called leukocyte interferon), β-interferon (beta interferon or fibroblast interferon), and γ-interferon (gamma interferon or immune interferon, a lymphokine). These proteins have been cloned and expressed in *Escherichia coli (E. coli),* which has enabled large quantities to be produced for evaluation of the interferons as possible antiviral and anticancer agents. To date, interferons have been used to treat Kaposi's sarcoma, hairy cell leukemia, venereal warts, multiple sclerosis, and hepatitis. See also ALPHA INTERFERON, BETA INTERFERON, CYTOKINES, PROTEIN, LYMPHOKINES, and *ESCHERICHIA COLI (E. COLI)*.

Interleukin-1 (IL-1) A cytokine (glycoprotein) release by activated macrophages, during the inflammatory stage of immune system response to an infection, which promotes the growth of epithelial (skin) cells and white blood cells. Recent research has indicated that too much IL-1 is linked to the development of rheumatoid arthritis, diabetes, inflammatory bowel disease, and other autoimmune diseases. See also MACROPHAGE, AUTOIMMUNE DISEASE, ADHESION MOLECULE, TUMOR NECROSIS FACTOR (TNF), CYTOKINES, GLYCOPROTEIN, WHITE BLOOD CELLS, ISLETS OF LANGERHANS, EPITHELIUM, and INTERLEUKIN-1 RECEPTOR ANTAGONIST (IL-Ira).

Interleukin-1 Receptor Antagonist (IL-1ra) A glycoprotein (produced by macrophages in response to presence of Interleukin-1, and endotoxin in tissues) that preferentially binds to those cell receptors in the body that typically bind the lymphokine, Interleukin-1 (IL-1). When manufactured by man (e.g., via genetic engineering) and injected into the body in large quantities. IL-Ira can block the deleterious effects of (too much) Interleukin-1. See also INTERLEUKIN-1 (IL-1), RECEPTORS, RECEPTOR FITTING, GLYCOPROTEIN, MACROPHAGE, ENDOTOXIN, ADHESION MOLECULE, CELLULAR IMMUNE RESPONSE, PROTEIN, LYMPHOKINES, and ANTAGONISTS.

Interleukin-2 (IL-2) Known as T cell growth factor. A cytokine (glycoprotein) secreted by (immune system response) stimulated helper T cells which promotes the proliferation/differentiation of more helper T cells, and promotes the growth of lymphocytes to combat an infection. Interleukin-2 also stimulates the lymphocytes to produce gamma interferon. It is gamma interferon that prompts the cytotoxic T cells to attack virus-infected cells and kill the virus within them. The structure of the gene that codes for synthesis of IL-2 (by immune system cells) was determined by Tadatsugu Taniguchi in 1983. See also IMMUNE RESPONSE, HUMORAL IMMUNITY, CYTOKINES GLYCOPROTEIN, CYTOTOXIC T CELLS, T CELLS, HELPER T CELLS, T CELL RECEPTORS, and INTERFERONS.

Interleukin-3 (IL-3) A hematologic growth factor (glycoprotein) cytokine that stimulates the proliferation of a wide range of white blood cells (to combat an infection). See also HEMATOLOGIC GROWTH FACTORS (HGF), CYTOKINES, and WHITE BLOOD CELLS.

Interleukin-4 (IL-4) A cytokine (glycoprotein) that stimulates production of antibody-producing B cells and promotes cytotoxic T cell (i.e., killer T cells) growth. See also ANTIBODY, CYTOTOXIC T CELLS, B CELLS, GLYCOPROTEIN, and CYTOKINES.

Interleukin-5 (IL-5) A cytokine (glycoprotein) that stimulates eosinophil growth. See EOSINOPHILS, PROTEIN, GLYCOPROTEIN, CYTOKINES, and CELLULAR IMMUNE RESPONSE.

Interleukin-6 (IL-6) A cytokine (glycoprotein) that is pleiotropic (i.e., stimulates several different types of immune system cells); and is a hematopoietic growth factor. See also HEMATOPOIETIC GROWTH FACTORS, GROWTH FACTOR, GLYCOPROTEIN, PLEIOTROPIC, MACROPHAGE, and CYTOKINES.

Interleukin-7 (IL-7) A cytokine (glycoprotein) synthesized in the bone marrow that stimulates early (fetal) proliferation and differentiation of B cells and T cells. May be useful in regenerating lymphoid cells in patients whose immune systems have been devastated by cancer chemotherapy. See also CYTOKINES, GLYCOPROTEIN, STEM CELL ONE, T CELLS, and CANCER.

Interleukin-8 (IL-8) A basic polypeptide (glycoprotein) with heparin-binding activity. Endogenous endothelial IL-8 appears to regulate transvenular traffic during acute inflammatory responses. See also POLYPEPTIDE, GLYCOPROTEIN, HEPARIN, ENDOTHELIAL CELLS, ENDOTHELIUM, POLYMORPHONUCLEAR LEUKOCYTES (PMN), and CELLULAR IMMUNE RESPONSE.

Interleukin-9 (IL-9) A cytokine (glycoprotein) that is released at sites in the body where inflammation has occurred. See CYTOKINES, GLYCOPROTEIN, and CELLULAR IMMUNE RESPONSE.

Interleukin-12 (IL-12) A cytokine (glycoprotein) produced by the body, which serves to activate the immune system against certain tumors and pathogens. See also CYTOKINES, GLYCOPROTEIN, TUMOR, TUMOR-ASSOCIATED ANTIGENS, MAJOR HISTOCOMPATIBILITY COMPLEX (MHC), T CELL RECEPTORS, CYTOTOXIC T CELLS, and PATHOGEN.

Intermediary Metabolism The chemical reactions that take place in the cell that transform the complex molecules derived from food into the small molecules needed for the growth and maintenance of the cell. See METABOLISM, CELL, and DIGESTION (WITHIN ORGANISMS).

International Food Biotechnology Council (IFBC) An organization that was established in 1988 by the Industrial Biotechnology Association (IBA) and the International Life Sciences Institute (ILSI), in order to "produce a (recommended) set of guidelines that could be used to assess the safety of genetically altered foods." See also GNE, INDUSTRIAL BIOTECHNOLOGY ASSOCIATION (IBA), INTERNATIONAL LIFE SCIENCES INSTITUTE (ILSI), SENIOR ADVISORY GROUP ON BIOTECHNOLOGY, BIOTECHNOLOGY INDUSTRY ORGANIZATION (BIO), GENETIC ENGINEERING, POLYGALACTURONASE, ANTISENSE (DNA SEQUENCE), BIOTECHNOLOGY, and BACTERIOCINS.

International Life Sciences Institute (ILSI) A research institute.

International Office of Epizootics (OIE) One of the three international SPS standard-setting organizations that is recognized by the World Trade Organization (WTO), the OIE is an international veterinary organization headquartered in Paris. The OIE was established in 1924, originally as part of the League of Nations, and is the worldwide authority for development of animal health and zoonoses standards, guidelines, and recommendations. See also SPS, INTERNATIONAL PLANT PROTECTION CONVENTION (IPPC), ZOONOSES, and WORLD TRADE ORGANIZATION (WTO).

International Plant Protection Convention (IPPC) One of the three international SPS standard-setting organizations that is recognized by the World Trade Organization (WTO), the IPPC is the worldwide authority for development of plant health standards, guidelines, and recommendations (e.g., to prevent transfer of a plant disease from one country to another). The treaty establishing the IPPC was signed in 1952. IPPC standards are set (and enforced) via regional SPS institutions such as the North American Plant Protection Organization (NAPPO), European Plant Protection Organization (EPPO), etc. See SPS, EUROPEAN PLANT PROTECTION ORGANIZATION (EPPO), INTERNATIONAL OFFICE OF EPIZOOTICS (OIE), WORLD TRADE ORGANIZATION (WTO), and NORTH AMERICAN PLANT PROTECTION ORGANIZATION (NAPPO).

International Society for the Advancement of Biotechnology (ISAB) A nonprofit organization of individuals that was started in 1994 "to advance and promote the general welfare of the science and commercialization of genetic engineering and industrial biotechnology." See also GENETIC ENGINEERING, BIOTECHNOLOGY, AMERICAN SOCIETY FOR BIOTECHNOLOGY (ASB), and BIOTECHNOLOGY INDUSTRY ORGANIZATION (BIO).

International Union for Protection of New Varieties of Plants (UPOV) See UNION FOR PROTECTION OF NEW VARIETIES OF PLANTS (UPOV).

Introgression The incorporation of transgenes (genes from transgenic organisms) into a wild type's genome. See also TRANSGENIC (ORGANISM), WILD TYPE, GENOME, GENE, and TRANSLOCATION.

Intron A (intervening sequence) segment of deoxyribonucleic acid (DNA) that is transcribed, but is removed from within the transcript by splicing together the sequences (exons) on either side of it (in the molecule). It is generally considered a nonfunctioning portion of the molecule. See also TRANSCRIPTION, DEOXYRIBONUCLEIC ACID (DNA), and EXON.

Inulin A fructose oligosaccharide (FOS) that is naturally produced in more than 30,000 plants. See FRUCTOSE OLIGOSACCHARIDES.

Invasin A transmembrane (i.e., through the membrane of the cell) protein that enables bacterial cells to invade normal (body) cells. See also CD4 PROTEIN, RECEPTORS, T CELL RECEPTORS, and ENDOCYTOSIS.

Inverted Micelle See REVERSE MICELLE (RM). Also see MICELLE for comparison.

Investigational New Drug See IND.

Invitro See IN VITRO.

Invivo See IN VIVO.

Ion An atom or molecule possessing a positive or a negative electrical charge. Ions are produced by the dissociation (coming-apart) of (electrolyte) molecule resulting from electrolyte dissolving in solution. One example is the dissociation of common table salt (i.e., sodium chloride) in water, which results in positively charged sodium ions (called cations) and negatively charged chloride ions (called anions). Ions play critically important roles in may biological processes such as nerve activity. See also CHELATION, CHELATING AGENT, and CITRIC ACID.

Ion-Exchange Chromatography Separation of ionic compounds (which include nucleic acids and proteins) in a chromatographic column containing a polymeric resin (i.e., the stationary phase) having fixed charge groups. The process works in that the charges of the column (stationary phase) interact with the opposite charges of the material dissolved in the solution that is flowing through the column (mobile phase). The charge interaction between the column material and, say, the protein has the effect of slowing down the rate of movement of the protein through the column. The other molecules, meanwhile, which do not interact with the column, flow right on through. This then constitutes the separation process. See also CHROMATOGRAPHY.

IPM See INTEGRATED PEST MANAGEMENT (IPM).

IPPC See INTERNATIONAL PLANT PROTECTION CONVENTION.

Iron Bacteria See FERROBACTERIA.

Islets of Langerhans (also called beta cells) Cells in the pancreas that produce insulin in response to the presence of glucose (sugar) in the bloodstream. The failure of insulin production results in the disease called diabetes. See also GLUCOSE (GLc), GLYCOLYSIS, and AUTOIMMUNE DISEASE.

Isoenzymes See ISOZYMES.

Isoflavins See ISOFLAVONES.

Isoflavones A group of phytochemicals (including genistein and daidzein) that are produced within the seeds of the soybean plant. Evidence suggests that consumption of isoflavones by humans can help to lower blood content of low-density lipoproteins (LDLP), can help to prevent prostate enlargement, and can help to prevent certain types of cancer (e.g., endometrial cancer). See GENISTEIN, SOYBEAN PLANT, PHYTOCHEMICAL, LOW-DENSITY LIPOPROTEINS (LDLP), PROSTATE-SPECIFIC ANTIGEN (PSA), and CANCER.

Isoleucine (ile) A monocarboxylic amino acid occurring in most dietary proteins.

Isomer One of the two or more chemical substances having the same elementary percentage composition (i.e., same atoms) and molecular weight, but differing in structure and therefore in properties. There are many ways in which such structural differences (between the two or more isomeric molecules) occur. One example is *n*-butane [$CH_3(CH_2)_2CH_3$] and isobutane [$CH_3CH(CH_3)_2$]. See also STEREOISOMERS.

Isomerase An enzyme-catalyzing transformation of a compound into its positional isomer. See also ISOMER.

Isoprene The five-carbon hydrocarbon molecule: 2-methyl-1,3 butadiene. It is a recurring structural unit of the terpenoid molecules, which are either linear or cyclic. There exists a very large number of terpenes and many are major components of essential plant oils. See GTPases.

Isotope Refers to one of the several "varieties" of atoms that exist, of the same element, that differ from each other in the number of neutrons in the atom's nucleus. For example, the element chlorine exists primarily in two forms (isotopes) in nature—with 18 neutrons (76% of the time) and with 20 neutrons (24% of the time). The chemical properties of isotopes of a given element are virtually identical. See also ATOMIC WEIGHT.

Isozymes (isoenzymes) Multiple forms of an enzyme that differ from each other in their substrate (substance acted upon) affinity, in their maximum activity, or in their regulatory properties. See also ENZYME, SUBSTRATE (CHEMICAL), and RIBOZYMES.

J

Japan Bio-Industry Association An association of the largest Japanese companies that are engaged in at least some form of genetic engineering research or production. Similar to America's Biotechnology Industry Organization (BIO), it is headquartered in Tokyo. See also BIOTECHNOLOGY INDUSTRY ORGANIZATION (BIO), BIOTECHNOLOGY, GENETIC ENGINEERING, RECOMBINANT DNA (rDNA), SENIOR ADVISORY GROUP ON BIOTECHNOLOGY (SAGB), INTERNATIONAL FOOD BIOTECHNOLOGY COUNCIL.

Jumping Genes Genes that move (change positions) within the genome. Genes associated with transposable elements. A segment fragment of deoxyribonucleic acid (DNA) that can move from one position in the genome to another. See also GENE, GENOME, DEOXYRIBONUCLEIC

ACID (DNA), GENETIC CODE, TRANSPOSITION, *TRANSPOSON,* TRANSLOCATION, INTROGRESSION, and HOT SPOTS.

K

Karyotype A size-order alignment of an organism's chromosome pairs in the format of a chart. It enables the connecting of chromosomes to symptoms (e.g., of genetic diseases in the organism) and traits. See also CHROMOSOMES, GENE, GENOTYPE, TRAIT, LINKAGE, LINKAGE GROUP, MUSCULAR DYSTROPHY (MD), CHROMATIDS, and CHROMATIN.

Karyotyper A scientist (or more frequently an automated analytical machine) that:

- takes a video picture of a given cell under a microscope
- digitizes that picture within a computer
- "cuts out" the individual chromosomes contained within that cell's genome
- arranges the cell's chromosomes in pairs by size order into a chart (called a karyotype)

See also CHROMOSOMES, GENOME, and KARYOTYPE.

Kb An abbreviation for 1,000 (kilo) base pairs of deoxyribonucleic acid (DNA). See also DEOXYRIBONUCLEIC ACID (DNA) and KILOBASE PAIRS (Kbp).

Kd An abbreviation for kilodalton. See KILODALTON (Kd).

Kefauver Rule A 1962 American law that mandates that the Food and Drug Administration (FDA) requires proof of pharmaceutical efficacy for drugs to be sold in the United States.

Keratins Insoluble protective or structural proteins consisting of parallel polypeptide chains arranged in an α-helical or β conformation.

Ketose A simple monosaccharide having its carbonyl groups at other than a terminal position. See also MONOSACCHARIDES.

Killer T Cell See CYTOTOXIC T CELLS.

Kilobase Pairs (Kbp) A unit of DNA equals to 1,000 base pairs. See also BASE PAIR (bp) and DEOXYRIBONUCLEIC ACID (DNA).

Kilodalton (Kd) A unit of mass equal to 1,000 Daltons. See also DALTON.

Koseisho The Japanese government agency that must approve new pharmaceutical products for sale with Japan. It is the equivalent of the U.S. Food and Drug Administration. See also NDA (TO KOSEISHO), FOOD AND DRUG ADMINISTRATION (FDA), COMMITTEE FOR PROPRIETARY

MEDICINAL PRODUCTS (CPMP), COMMITTEE ON SAFETY IN MEDICINES, MEDICINES CONTROL AGENCY (MCA), EUROPEAN MEDICINES EVALUATION AGENCY (EMEA), and BUNDESGESUNDHEITSAMT (BGA).

Krebs Cycle See CITRIC ACID CYCLE.

L

L-Selectin Also known as the homing receptor. See SELECTINS, LECTINS, and ADHESION MOLECULES.

La Jolla Institute for Allergy and Immunology (LIAI) A biotechnology research institute in San Diego, CA, that was created as a nonprofit center by Kirin Brewery Company of Tokyo.

Label (radioactive) A radioactive atom, introduced into a molecule in order to enable observation of that molecule's metabolic transformation (in an organism). See also AUTORADIOGRAPHY.

Lachrymal Fluid (tears) A salty solution produced by the tear glands to bathe and lubricate the eye. Possesses antimicrobial properties.

Lac Operon An operon in *Escherichia coli (E. coli)* that codes for three enzymes involved in the metabolism of lactose. See also OPERON, CODING SEQUENCE, and *ESCHERICHIA COLI (E. COLI)*.

LACI See LIPOPROTEIN ASSOCIATED COAGULATION (CLOT) INHIBITOR.

Lambda Phage A bacteriophage that infects *Escherichia coli (E. coli)*. It is commonly used as a vector in recombinant DNA (deoxyribonucleic acid) research. See also PHAGE (BACTERIOPHAGE) and *ESCHERICHIA COLIFORM (E. COLI)*.

Langerhans Cells See DENDRITIC LANGERHANS CELLS and ISLETS OF LANGERHANS (ALSO CALLED BETA CELLS).

Laurate A medium chain length (i.e., C12) fatty acid that is naturally produced by coconut trees, oil palm trees, and certain species of wild plants. In 1992, canola plants were genetically engineered so that they could also produce (desirable) laurate in their seeds. See also FATTY ACID, CANOLA, GENETIC ENGINEERING, GENETIC CODE, LPAAT PROTEIN, and LAUROYL-ACP THIOESTERASE.

Lauroyl-ACP Thioesterase The enzyme that is required for the synthesis (creation) of laurate in plants. For example, the presence of this enzyme in the California bay tree causes its seed oil to contain as much as 45% laurate. See also LAURATE, ENZYME, and LPAAT PROTEIN.

Lazaroids A class of drugs being developed to "bring back from the dead" tissues that have been (almost) killed due to a lack of oxygen (e.g., caused by a clot blocking a vital artery). See also HUMAN SUPEROXIDE DISMUTASE (hSOD), FIBRIN, and REPERFUSION.

LDLP See LOW-DENSITY LIPOPROTEINS.

LDLP Receptors See LOW-DENSITY LIPOPROTEINS (LDLP).

Leader See LEADER SEQUENCE.

Leader Sequence The nontranslated sequence at the 5′ end of mRNA that precedes the initiation codon. See also MESSENGER RNA (mRNA) and CODON.

Leaky Mutants A mutant in which the mutated gene product, such as an enzyme, still possesses a fraction of its normal biological activity. See also MUTATION, GENE, PROTEIN, BIOLOGICAL ACTIVITY, and ENZYME.

LEAR See CANOLA.

Lecithin A by-product of the refining process for soybean oil (deoiled lecithin from processed soybeans is composed of approximately 20–25% phosphatidyl choline by weight). Historically, lecithin has often been used commercially in food processing as an emulsifier, instantizing agent, and lubricating agent. Lecithin (also known as phosphatidylcholine) is a source of choline when digested; and is a critical component of the lipoproteins which transport fat and cholesterol molecules in the bloodstream (e.g., from the digestive system, to body cells, to the liver, etc.).

Lecithin (choline) promotes synthesis of high-density lipoproteins (i.e., HDLP, also known as "good" cholesterol) by the liver, when it is consumed by humans. Phosphatidyl choline (PC) is involved in cell signal transduction.

Some other common dietary sources of lecithin include eggs, red meats, spinach and nuts. See also LIPOPROTEIN, LIPIDS, CONJUGATED PROTEIN, HIGH-DENSITY LIPOPROTEINS (HDLP), LOW-DENSITY LIPOPROTEINS (LDLP), SOYBEAN PLANT, SOYBEAN OIL, CHOLINE, and SIGNAL TRANSDUCTION.

Lectins A class of proteins that have the capability to rapidly (and reversibly) combine with *specific* sugar molecules (e.g., those sugar molecules or glycoproteins on the surface of adjacent cells, within an organism). Lectins are a common component of the surface (membranes) of plant and animal cells; and are so specific (regarding sugar molecules that they will or won't combine with/(attach to) that they discriminate between different monosaccharides *and* different oligosaccharides (i.e., on the surfaces of adjacent cells within an organism).

This capability to reversibly combine with sugar (i.e., carbohydrate) molecules (on the surface of adjacent cells) is utilized by:

- bacteria and other microorganisms, to adhere to (sugar molecules on surface of) host cells, as the first step in the process of infecting those host cells

- white blood cells (e.g., lymphocytes), to adhere to the walls of blood vessels (endothelium), as the first step to leaving the bloodstream to go fight infection (pathogens, trauma) in tissue adjacent to that blood vessel. The lectin (glycoprotein) that adheres to the (endothelial sugar molecule on) blood vessel wall is called L-selectin, or the homing receptor. The two sugar molecules (glycoproteins) on blood vessel wall (endothelium) are called P-selectin and E-selectin (also known as ELAM-1)
- cancerous tumor cells, to adhere to the walls of blood vessels (endothelium) as part of the tumor-proliferation process known as metastasis (i.e., new tumors are "seeded" throughout the body via this process).

See also PROTEIN, SUGAR MOLECULES, GLYCOPROTEINS, LEUKOCYTES, LECTINS, SELECTINS, LYMPHOCYTES, MONOCYTES, NEUTROPHILS, ENDOTHELIAL CELLS, ENDOTHELIUM, CANCER, and SIGNAL TRANSDUCTION.

Leptin A protein hormone that is produced by fat cells (adipose tissue) in the body. When leptin is produced and travels to cells whose surface bears leptin receptors (e.g., in the brain), those (brain) cells receive signal (transduction) indicating fullness/satiety.

Leptin has been found to be present in the bloodstream of obese humans at a concentration of approximately four times the concentration-found in bloodstreams of lean humans. High levels of leptin present in the bloodstream disrupt some of the activities of insulin (hormone which regulates blood sugar levels), and may lead to diabetes. See also HORMONE, PROTEIN, BIOLOGICAL ACTIVITY, and INSULIN.

Leptin Receptors Cellular receptors which are specific to leptin. In 1996, H. Ralph Snodgrass discovered that leptin receptors are involved in the "sorting" of immature blood cells (from bone marrow) to create subpopulations. See LEPTIN and RECEPTORS.

Lethal Mutation Mutation of a gene to yield no, or a totally defective gene product (protein), thereby making it unable to function, and hence unable to sustain the life of the organism.

Leucine (leu) A monocarboxylic essential amino acid. See also ESSENTIAL AMINO ACIDS.

Leukocytes (white blood cells) A diverse family of nucleated cells that has many immunological functions. See also NEUTROPHILS, EOSINOPHILS, BASOPHILS, LYMPHOCYTE, B LYMPHOCYTES, MONOCYTES, and GRANULOCYTES (POLYMORPHONUCLEAR GRANULOCYTES).

Leukotrienes Chemicals released by certain cells (T cells) which "signal" leukocytes (white blood cells). When thus activated, the leukocytes migrate to the site of infection to combat the pathogens. See also LEUKOCYTES, MAST CELLS, SIGNALLING, SIGNAL TRANSDUCTION, T CELLS, and PATHOGEN.

Levorotary (L) Isomer An isomer of an optically active compound; rotates (when illuminated) the plane of plane-polarized light to the left. See also STEREOISOMERS and DEXTROROTARY (D) ISOMER.

LH See LUTEINIZING HORMONE.

Library A set of cloned DNA fragments together representing the entire genome. See also DEOXYRIBONUCLEIC ACID (DNA) and GENOME.

Ligand (in biochemistry) In general, a molecule or ion that can bind to (interact with) a protein molecule. For example, a pharmaceutical that binds to a receptor protein molecule on the surface of a cell may be called a ligand. See also PROTEIN, RECEPTORS, T CELL RECEPTORS, ENDOCYTOSIS, CD4 PROTEIN, INVASIN, LIGAND (IN CHROMATOGRAPHY), and CHELATION.

Ligand (in chromatography) A term used to describe a substance (the ligand) that has the capacity for specific and noncovalent (reversible) binding to some protein. A ligand may be a coenzyme for a specific enzyme. The ligand can be covalently attached (immobilized) by means of the appropriate chemical reaction to the surface of certain porous column material. When a mixture of proteins containing the enzyme to be isolated is passed through the column, the enzyme, which is capable of tightly binding to the ligand, does so, and is in this manner held to the column. The other proteins present, which have no specific affinity for the ligand, pass on through the column. The protein/ligand complex is then dissociated and the enzyme eluted from the column, which may be accomplished by passing more free (unbound) coenzyme through the column. The ligand may be hormones (i.e., used to isolate receptor molecules) or any other type of molecule that is capable of binding specifically and reversibly to the desired protein or protein complex. See also AFFINITY CHROMATOGRAPHY, SUBSTRATE (IN CHROMATOGRAPHY), CHROMATOGRAPHY, PROTEIN, PEPTIDE, ANTIBODY, and MONOCLONAL ANTIBODIES (MAb).

Ligase An enzyme used to catalyze the joining of single-stranded DNA segments. See also DEOXYRIBONUCLEIC ACID (DNA).

Ligation The formation of a phosphodiester bond to link two adjacent bases separated by a nick in one strand of a double helix of DNA (deoxyribonucleic acid). The term can also be applied to blunt-end ligation and to the joining of RNA (ribonucleic acid) strands. See also DEOXYRIBONUCLEIC ACID (DNA) and LIGASE.

Light-Chain Variable (VL) Domains The regions (domains) of the antibody (molecule's) "light chain" that vary in their amino acid sequence. The "chains" (of atoms) comprising the antibody (immunoglobulin) molecule consist of a region of variable (V) amino acid sequence and a region in which the amino acid sequence remains constant (C). An antibody molecule possesses two antigen binding sites, and it is the variable domains of the light (VL) and heavy (VH) chains which contribute to this (antigen binding ability). See also ANTIBODY, IMMUNOGLOBULIN, PROTEIN, SEQUENCE (OF A PROTEIN MOLECULE), ANTIGEN, AMINO ACID, COMBINING SITE, DOMAIN (OF A PROTEIN), and HEAVY-CHAIN VARIABLE (VH) DOMAINS.

Lignocellulose A complex biopolymer comprising the bulk of woody plants. It consists of polysaccharides and polymer phenols. See also POLYSACCHARIDES.

Linkage A phenomenon discovered by Thomas Hunt Morgan in the early 1900s via his experiments with fruit flies. This term describes the tendency of genes to be inherited together as a result of their locations being physically close to each other on the same chromosome; measured by percent recombination between loci. Because the locus (i.e., location of gene on the chromosome) determines the likelihood that two genes will go together into offspring, "marker genes" that are linked to a gene (e.g., for a given trait or disease) of interest can be utilized to predict the presence of that (trait or disease-causing) gene. See also GENE, LOCUS, CHROMOSOMES, LINKAGE GROUP, MARKER (GENETIC MARKER), MAP DISTANCE, and LINKAGE MAP.

Linkage Group Includes all loci (in DNA molecule) that can be connected (directly or indirectly) by linkage relationships; equivalent to a chromosome. See also LOCUS, CHROMOSOMES, LINKAGE, CHROMATIDS, CHROMATIN, LINKAGE MAP, DEOXYRIBONUCLEIC ACID (DNA).

Linkage Map A depiction of gene loci (on chromosomes) based on the frequency of recombination (of linked genes) in the offspring's genome. See also LINKAGE, LINKAGE GROUP, GENE, LOCUS, and MARKER (GENETIC MARKER).

Linker A short synthetic duplex oligonucleotide containing the target site for some restriction enzyme. It may be added to the ends of a DNA (deoxyribonucleic acid) fragment prepared by cleavage with some other enzyme reconstructions of recombinant DNA.

Linking The process of "attaching" a drug or a toxin to a monoclonal antibody, or another homing molecule of the immune system. Because this attachment must be reversible, so that the homing molecule can release the drug or toxin after delivering that drug or toxin to the desired site in the body (e.g., delivery of a toxin to a tumor, to kill the tumor), linking is a difficult process to reliably achieve. See also IMMUNOTOXIN, CONJUGATE, MONOCLONAL ANTIBODIES (MAb), and TOXIN.

Lipase An enzyme (one of a class of enzymes) that catalyzes the hydrolytic cleavage of lipid molecules (triglycerides) to yield free fatty acids. Lipase was the first enzyme to be produced via genetic engineering and marketed. Lipase also occurs naturally in cow's milk, and in the intestines of many animals. See ENZYME, HYDROLYTIC CLEAVAGE, TRIGLYCERIDES, FATS, and FATTY ACID.

Lipid Bilayer See LIPIDS.

Lipid Vesicles See LIPOSOMES.

Lipids Water-insoluble (fat) biomolecules that are highly soluble in organic solvents such as chloroform. Lipids serve as fuel molecules, highly concentrated energy stores, "signal" molecules, and components of membranes. Membrane lipids are relatively small molecules that have both a hydrophilic (i.e., "water loving") and a hydrophobic (i.e., "water hating") moiety. These (membrane) lipids spontaneously form closed bimolecular sheets in aqueous media (water) which are barriers to the free movement (flow) of polar molecules. See also MOIETY.

Lipophilic A "fat loving" molecule or portion of a molecule. Relating to or having strong affinity for fats or other lipids. See LIPIDS and FATS.

Lipopolysaccharide (LPS) See ENDOTOXIN.

Lipoprotein A conjugated protein containing a lipid or a group of lipids. For example, low-density lipoproteins (also known as "bad" cholesterol) are a "package" of cholesterol (lipid) surrounded by a hydrophilic protein. See also PROTEIN, LOW-DENSITY LIPOPROTEINS (LDLP), CONJUGATED PROTEIN, HYDROPHILIC, and LIPIDS.

Lipoprotein-Associated Coagulation (Clot) Inhibitor (LACI) A protein that prevents formation of blood clots. This occurs because LACI inhibits the controlled series of zymogen activations (enzymatic cascade) which causes the formation of fibrinogen (precursor to fibrin), leading subsequently to clot formation. See also FIBRIN, FIBRONECTIN, and ZYMOGENS.

Liposomes Also called lipid vesicles or vesicle. Aqueous (i.e., watery) compartments enclosed by a lipid bilayer. They can be formed by suspending a suitable lipid, such as phosphatidyl choline, in an aqueous medium. This mixture is then sonicated (i.e., agitated by high-frequency sound waves) to give a dispersion of closed vesicles (i.e., compartments) that are quite uniform in size. Alternatively, liposomes can be prepared by rapidly mixing a solution of lipid in ethanol with water, which yields vesicles that are nearly spherical in shape and have a diameter of 500 Å (Angstroms). Larger vesicles (10,000 Å or 1 μm, or 0.00003937 inch in diameter) can be prepared by slowly evaporating the organic solvent from a suspension of phospholipid in mixed solvent system. Liposomes can be made to contain certain drugs for protective, controlled release delivery to targeted tissues. For example, pharmaceuticals which tend to be rapidly degraded in the bloodstream could be enclosed within liposomes so that more of the nondegraded pharmaceutical would remain by the time it reached the targeted tissue. The controlled release property enables larger doses (of drugs possessing toxic side effects) to be prescribed, knowing that the drug will be released in the body over an extended period of time. See also LIPIDS, MICRON, and ANGSTROM (Å).

Lipoxidase See LIPOXYGENASE (LOX).

Lipoxygenase (LOX) An enzyme that is naturally produced within its seeds (soybeans) by the soybean plant (botanical name *Glycine max (L.) Merrill*). In the presence of heat and moisture, lipoxygenase enzyme catalyzes a chemical reaction in which objectionable "beany" (rancid) flavor is produced from certain components of the soybean. That "beany" flavor decreases the suitability of resultant soybean raw materials for manufacture of human foods.

Prevention of the reactions that create the "beany" flavor can be accomplished via heat denaturation (of lipoxygenase present in the soybeans) or via creation of soybeans that do not contain any lipoxygenase (known as "LOX null" soybeans). Lipoxygenase enzyme also catalyzes a reaction in which certain volatile chemicals are produced that inhibit growth of any *Aspergillus flavus*. See also ENZYME, SOYBEAN PLANT, LOX NULL and AFLATOXIN.

Lipoxygenase Null See LOX NULL.

Living Modified Organism (LMO) See GMO.

LMO Living modified organism. See GMO.

Loci The plural of locus. See also LOCUS.

Locus The position of a gene on a chromosome. See also GENE and CHROMOSOMES.

Loop A single-stranded region at the end of a hairpin in RNA (or single-stranded DNA). It corresponds to the sequence between inverted repeats in duplex DNA. See also RIBONUCLEIC ACID (RNA), DEOXYRIBONUCLEIC ACID (DNA), and SEQUENCE (OF A DNA MOLECULE).

LOSBM Low-oligosaccharide soybean meal. See LOW-STACHYOSE SOYBEANS and SOYBEAN PLANT.

Low-Density Lipoproteins (LDLP) So-called "bad" cholesterol, which carries cholesterol molecules from the digestive system (e.g., intestine) to body cells and can clog arteries over time (a disease called atherosclerosis, or coronary heart disease). Since cholesterol does not dissolve in water (which constitutes most of the volume of blood), the body makes LDL cholesterol (derived from the digestion of fatty foods) into little "packages" surrounded by a hydrophilic protein. That protein "wrapper" is known as apolipoprotein B-100, or apo B-100, and it enables LDL cholesterol to be transported in the bloodstream because the apolipoprotein B-100 is attracted to water molecules in the blood. Part of the apolipoprotein B-100 molecule also will bind to special LDLP receptor molecules in the liver, which then clears those (bound) cholesterol packages out of the body as part of regular liver functions. See also HIGH-DENSITY LIPOPROTEINS (HDLP), HYDROPHILIC, RECEPTORS, PROTEIN, SITOSTANOL, ISOFLAVONES, and WATER SOLUBLE FIBER.

Low-Stachyose Soybeans Those soybean varieties that contain lower levels of the relatively indigestible stachyose carbohydrate (and thus higher levels of easily digestible other nutrients) than traditional varieties of soybeans. Low-stachyose soybeans are particularly useful for feeding of monogastric animals (e.g., swine, poultry). See also STACHYOSE, CARBOHYDRATES (SACCHARIDES), VALUE-ENHANCED GRAINS, SOYBEAN PLANT, and HIGH-SUCROSE SOYBEANS.

Low-Tillage Crop Production A methodology of crop production in which the farmer utilizes a minimum of mechanical cultivation (i.e., only two to four passes over the field instead of the conventional five passes for traditional crop production). The plant residue remaining on field's surface helps to control weeds and reduce soil erosion, but it also provides sites for insects to shelter and reproduce, leading to a need for increased insect control via methods such as inserting a *Bacillus thuringiensis (B.t.)* gene into crop plants. See INTEGRATED PEST MANAGEMENT (IPM), CORN, SOYBEAN PLANT, *BACILLUS THURINGIENSIS (B.t.)*, GENETIC ENGINEERING, EUROPEAN CORN BORER (ECB), *HELICOVERPA ZEA (H. ZEA)*, CORN ROOTWORM, GENE, and COLD HARDENING.

LOX Null Refers to soybeans that do not contain lipoxygenase enzymes (thus, they result in a "null" test reading). See LIPOXYGENASE (LOX), SOYBEAN PLANT, and ENZYME.

LPAAT Protein A protein consisting of lysophosphatidic acid acyl transferase (enzyme), which (when present in plant) causes production of triglycerides (in the seeds) possessing saturated fatty acids in the "middle position" of the triglycerides' molecular (glycerine) "backbone." For example, canola (rapeseed) plants genetically engineered to contain LPAAT protein are able to produce high levels of saturated fatty acids (including laurate) in their oil. See also PROTEIN, LAURATE, ENZYME, TRIGLYCERIDES, SATURATED FATTY ACIDS, MONOUNSATURATED FATS, CANOLA, and GENETIC ENGINEERING.

LPS See ENDOTOXIN.

Lumen The interior (opening through which blood flows) e.g., within a blood vessel. See ENDOTHELIUM.

Luminesce See BIOLUMINESCENCE.

Luminescence See BIOLUMINESCENCE.

Lupus An autoimmune disease of the body, in which anti-DNA antibodies bind to DNA. The resulting complexes (of DNA and antibodies) travel to the kidneys via the bloodstream, and become lodged in kidneys, where they cause inflammatory reactions (that can lead to kidney failure). See also ANTIBODY, DEOXYRIBONUCLEIC ACID (DNA), AUTOIMMUNE DISEASE, and SUPERANTIGENS.

Lupus Erythematosus See LUPUS.

Luteinizing Hormone (LH) A reproductive hormone that acts upon the ovaries to stimulate ovulation. It is secreted by the pituitary gland. See also HORMONE, PITUITARY GLAND, and ENDOCRINE HORMONES.

Lymphocyte A type of cell found in the blood, spleen, lymph nodes, etc. of higher animals. They are formed very early in fetal life, arising in the liver by the sixth week of human gestation. There exist two subclasses of lymphocytes: B lymphocytes and T lymphocytes. B lymphocytes make antibodies (immunoglobins) of which there are five classes: IgM, IgA, IgG, IgD and IgE. The antibodies circulate in the bloodstream. T lymphocytes recognize and reject foreign tissue, modulate B cell activity, kill tumor cells, and kill host cells infected with virus. T-lymphocytes are also called T cells. See also B LYMPHOCYTES, T CELLS, ANTIBODY, HELPER T CELLS (T4 CELLS), BLAST CELL, CYTOTOXIC T CELLS, and ANTIGEN.

Lymphokines Peptides and proteins secreted by (immune system response) stimulated T cells. These hormone-like (peptide and protein) molecules direct the movements and activities of other cells in the immune system. Some examples of lymphokines are interleukin-1, interleukin-2, tumor necrosis factor (TNF), gamma interferon, colony stimulating factors, macrophage chemotactic factor, and lymphocyte growth factor. The suffix "-kine" comes from the Greek word *kinesis*, meaning movement.

Lyochrome See FLAVIN.

Lyophilization The process of removing water from a frozen biomaterial (e.g., a microbial culture or an aqueous protein solution) via application of a vacuum. It is a drying method for long-term preservation of proteins in the solid state, and for long-term storage of live microbial cultures. See also CULTURE and PROTEIN.

Lyse To rupture a membrane (cell). The act of lysis (rupturing a membrane). See also LYSIS.

Lysine (lys) An essential, basic amino acid obtained from many proteins by hydrolysis. See also ESSENTIAL AMINO ACIDS, PROTEIN, and OPAGUE-2.

Lysis The process of cell disintegration; membrane rupturing; breaking up the cell wall. See also CYTOLYSIS, CELL, LYSOZYME, and MEMBRANE TRANSPORT.

Lysosome A membrane-surrounded organelle in the cytoplasm of eucaryotic cells which contains many hydrolytic enzymes. The lysosome internalizes and digests foreign proteins as well as cellular debris. The protein fragments (epitopes) are "presented" to T cells by the major histocompatibility complex (MHC) proteins on the surface of the eucaryotic cell. See also ANTIGEN, MAJOR HISTOCOMPATIBILITY COMPLEX (MHC), and T CELLS.

Lysozyme An enzyme, naturally produced by some animals, which possesses antibacterial (i.e., bacteria killing) properties. Discovered in

1922 by Alexander Fleming, in his nasal mucus, Mr. Fleming named it (from the Greek) *lyso*—due to its ability to lyse (cut) bacteria and *zyme*—due to its being an enzyme.

Lysozyme lyses certain kinds of bacteria, by dissolving the polysaccharide components of the bacteria's cell wall. When that cell wall is weakened, the bacteria cell then bursts because osmotic pressure (inside that bacteria cell) is greater than the weakened cell wall can contain.

Tears and egg whites both contain significant amounts of lysozyme, as agents to prevent bacterial infections (e.g., against bacteria entering body via eye openings, against bacteria entering chicken embryo through the eggshell). See also ENZYME, LYSIS, CELL, CYTOLYSIS, POLYSACCHARIDES, and BACTERIA.

Lytic Infection A viral infection in which the final act of the infection is to lyse (i.e., burst, or destroy) the cell. This then releases the new (progeny) viruses so they can go on to infect other cells. See also LYSE and LYSIS.

M

MAA Marketing Authorization Application It is the European Union (EU) equivalent to a U.S. NDA (New Drug Application). An MAA is an application to the Committee for Proprietary Medicinal Products (CPMP) seeking approval of a new drug that has undergone Phase 2 and Phase 3 clinical trials. See also NDA (TO FDA), CANDA, FOOD AND DRUG ADMINISTRATION (FDA), MAA, NDA (TO KOSEISHO), CPMP, and PHASE I CLINICAL TESTING.

MAb See MONOCLONAL ANTIBODIES.

Macromolecules Large molecules with molecular weights ranging from a few thousand to hundreds of millions. See also MOLECULAR WEIGHT.

Macrophage A phagocytic cell that is the counterpart of a monocyte. A monocyte that has left the bloodstream and has moved into the tissues. Macrophages have basically the same functions as monocytes, but they carry these out in the tissues. In summary, they engulf and kill microorganisms, present antigen to the lymphocytes, and kill certain tumor cells, and their secretions regulate inflammation. In the spleen they engulf and destroy old red blood cells. When they reside in the bone marrow they store iron and then transfer it to red blood cells. In the lungs and GI tract they are scavengers and keep the tissues clean. They also serve as a reservoir for the AIDS virus. They (and other phagocytic cells) are largely responsible for the localization and degradation of foreign materials at inflammatory sites. Macrophages display chemotaxis. See also CELLULAR IMMUNE RESPONSE, CHEMOTAXIS, MONOCYTES, PHAGOCYTE, ADHESION MOLECULE, and LYSOSOME.

Macrophage Colony Stimulating Factor (M-CSF) A colony stimulating factor (CSF) that stimulates production of macrophages in the body. See also COLONY STIMULATING FACTORS (CSFs), and MACROPHAGE.

Magainins Discovered within frog skin tissues by Michael Zasloff in 1987, magainins are antimicrobial, amphopathic peptides that lyse (i.e., burst) certain cells upon contact by "worming" their hydrophobic portion into the cell's membrane, which creates a transmembrane (i.e., through the surface) pore (allowing ions to flow into the cell, causing osmotic bursting). Magainins are selective against bacteria, fungi, and protozoa cells (the word magainin comes from the Hebrew word for "shield"). See also AMPHIPHILIC MOLECULES, CELL, PEPTIDE, BACTERIA, FUNGUS, and ANTIBIOTICS.

Maize See CORN.

Major Histocompatibility Complex (MHC) A chromosomal region (approximately 3,000 Kb) which encodes for three classes of transmembrane (cell) proteins. MHC I proteins (located on the surface of nearly all cells) present foreign epitopes (i.e., fragments of antigens that have been ingested; peptides) to cytotoxic T cells (killer T cells). MHC II proteins (located on the surface of immune system cells and phagocytes) present foreign epitopes to helper T cells, and MHC III proteins are components of the complement cascade. Genes in the MHC must be matched (between an organ donor and organ recipient) to prevent rejection of organ transplants. See also COMPLEMENT CASCADE, GRAFT-VERSUS-HOST DISEASE, Kb, MACROPHAGE, PROTEIN, CELL, T CELL RECEPTORS, ANTIGEN, T CELLS, CYTOTOXIC T CELLS, EPITOPE, GENE, TUMOR-ASSOCIATED ANTIGENS, and HUMAN LEUKOCYTE ANTIGENS.

MALDI-TOF-MS Matrix-Associated Laser Desorption Ionization Time of Flight Mass Spectrometry. A mass spectrometry methodology/technology that can establish, in seconds, the identity, purity, etc. of a sample of an oligonucleotide. Also the identification of gram-positive microorganisms, or characterization of genetic materials (e.g., DNA, RNA, etc.) on hybridization surfaces. See MASS SPECTROMETER, MICROORGANISM, OLIGONUCLEOTIDE, MICROORGANISM, GRAM-POSITIVE, RIBONUCLEIC ACID (RNA), HYBRIDIZATION SURFACES, and DEOXYRIBONUCLEIC ACID (DNA).

MAL (Multiple Aleurone Layer) Gene A gene in corn (maize) that (when present in the DNA of a given plant) causes that plant to produce seed that contains higher-than-normal levels of calcium, magnesium, iron, zinc, and manganese. These higher mineral levels are particularly useful for feeding of swine, since traditional No. 2 yellow (dent) corn does not contain enough for optimal pig growth. See also GENE, DEOXYRIBONUCLEIC ACID (DNA), HIGH-METHIONINE CORN, HIGH-LYSINE CORN, FLOURY-2, and OPAGUE-2.

Mammalian Cell Culture Technology to artificially cultivate cells, of mammal origin, in a laboratory or production-scale device (i.e., *in vitro*). Can be either a batch or continuous process device. The first mammalian cell culture was performed by a neurobiologist named R. G. Harrison in 1907, when he added chopped-up spinal cord tissue to clotted (blood) plasma in a humidified growth chamber. The nerve cells from this spinal cord tissue successfully grew, divided, and extended long fibers into the clot. Many improvements to cell culture process have been made over the years, including special growth media (fluids that bathe the cultured cells with the right amounts of amino acids, salts, and other minerals). See also CONTINUOUS PERFUSION, DISSOCIATING ENZYMES, HARVESTING ENZYMES, *IN VITRO*, PLASMA, CELL, MEDIUM, and AMINO ACID.

Mannanoligosaccharides (MOS) A family of oligosaccharides that can be produced by man in commercial quantities via certain yeast cells.

When consumed (e.g., by livestock), mannose sugars in the MOS stimulate the liver to secrete the mannose-binding protein. Mannose binding protein enters the digestive system and binds to the (mannose-containing) capsule (surface membrane) of pathogenic bacteria. That binding to pathogens triggers the immune system's complement cascade to combat those pathogenic bacteria. See OLIGOSACCHARIDES, FRUCTOSE OLIGOSACCHARIDES, SUGAR MOLECULES, YEAST, COMPLEMENT CASCADE, PATHOGENIC, COMPLEMENT (COMPONENT OF IMMUNE SYSTEM), and CAPSULE.

Map Distance A number proportional to the frequency of recombination between two genes. One map unit corresponds to a recombination frequency of 1 percent. See also GENETICS, GENETIC CODE, GENETIC MAP, LINKAGE, and QUANTITATIVE TRAIT LOCI.

Mapping (of genome) See GENETICS, GENETIC CODE, GENETIC MAP, QUANTITATIVE TRAIT LOCI, and POSITION EFFECT.

Marker (DNA marker) A DNA fragment of known size used to calibrate an electrophoretic gel. See ELECTROPHORESIS and DEOXYRIBONUCLEIC ACID (DNA).

Marker (DNA sequence) A specific sequence of DNA that is virtually always associated with a specified trait, because of "linkage" between that DNA sequence (the "marker") and the gene(s) that cause that particular trait. See DEOXYRIBONUCLEIC ACID (DNA), TRAIT, LINKAGE, LINKAGE GROUP, LINKAGE MAP, GENE, SEQUENCE (OF A DNA MOLECULE), and MARKER ASSISTED SELECTION.

Marker (genetic marker) A trait that can be observed to occur or not to occur in an organism such as, for example, bacteria or plant(s). Genetic markers include such traits as: expression of luciferase in leaf cells (causing leaves to glow), resistance to specific antibiotics, the nature of the cell wall and capsule characteristics, requirements for a particular growth factor, and carbohydrate utilization, to mention a few. For example, if a culture of dividing (growing) bacteria that is not resistant to a particular antibiotic (i.e., lacks the trait of antibiotic resistance) is exposed to only the DNA isolated from bacteria that are resistant to the antibiotic, then a fraction of the cells exposed will directly incorporate this trait (some DNA) into their genome, hence acquiring the trait. The first genetically engineered plants bearing a marker gene were field tested in 1986. See also ALLELE, GENETIC ENGINEERING, POSITIVE AND NEGATIVE SELECTION (PNS), TRANSFORMATION, TRANSFECTION, BIOLUMINESCENCE, and MARKER ASSISTED SELECTION.

Marker Assisted Selection The utilization of DNA sequence "markers" by commercial breeders to select the organisms (e.g., crops, livestock, etc.) which possess gene(s) for a particular performance trait (e.g., rapid growth, high yield, etc.) desired; for subsequent breeding/propagation. See DEOXYRIBONUCLEIC ACID (DNA), SEQUENCE (OF A DNA MOLECULE), MARKER (DNA SEQUENCE), GENE, TRAIT, LINKAGE, LINKAGE GROUP, and LINKAGE MAP.

MAS See MARKER ASSISTED SELECTION.

Mass Spectrometer An analytical device that can be used to determine the molecular weights of proteins and nucleic acids, the sequence of (composition and order of amino acids comprising) protein molecules, the chemical composition of virtually any material, and the rapid identification of intact gram-negative and gram-positive microorganisms (the latter, using matrix-assisted laser desorption ionization time of flight mass spectrometry). See also GRAM-NEGATIVE, GRAM-POSITIVE, MOLECULAR WEIGHT, SEQUENCING (OF DNA MOLECULES), PROTEIN, AMINO ACID, NUCLEIC ACIDS, GENE MACHINE, and MALDI-TOF-MS.

Mast Cells Fixed (noncirculating) cells that are present in many different kinds of body tissues. When two IgE molecules of the same antibody "dock" at adjacent receptor sites on a mast cell, then (the two IgE molecules) capture an allergen (e.g., a particle of pollen) between them, a chemical-energetic signal is sent to the interior (inside mast cell) portion of receptor molecules, which causes that interior portion of molecule to change (i.e., transduction). That signal transduction causes a protein named "syk" to set off a chemical chain reaction inside the mast cell; thereby causing that mast cell to release leukotrienes, histamine, serotonin, bradykinin, and "slow reacting substance." Release of these chemicals into the body causes the blood vessels to become more permeable (leaky) and causes the nose to run, itchy and watery eyes. These chemicals also cause smooth muscle contraction—causing sneezing, breath constriction coughing, wheezing, etc. See also BASOPHILS, ANTIGEN, ANTIBODY, RECEPTORS, SIGNAL TRANSDUCTION, HISTAMINE, and SIGNALLING.

Matrix Metalloproteinases (MMP) A family of enzymes that contain the zinc metal ion (Zn^{2+}) at their active sites. Among this family are the collagenases. See also ENZYME, ION, ACTIVE SITE, CATALYTIC SITE, STROMELYSIN (MMP-3), and COLLAGENASE.

Maximum Residue Level (MRL) Term used for an officially established upper allowable limit, of a given compound (e.g., a synthetic hormone) in a particular product, such as meat. For example, in 1994, the Codex Alimentarius Commission in Rome, Italy decided to establish maximum residue levels for each of five growth promotants that are commonly utilized by the U.S. beef industry. Because the World Trade Organization (WTO) subsequently stated that it would respect MRLs, a WTO member nation cannot legally refuse to allow import of meat products on growth promotant-content basis if the content of the promotant contained in the meat is less than its maximum residue level. See also GROWTH HORMONE, GROWTH FACTOR, CODEX ALIMENTARIUS COMMISSION, and WORLD TRADE ORGANIZATION (WTO).

MCA See MEDICINES CONTROL AGENCY (MCA).

Medicines Control Agency (MCA) The British Government agency that, in concert with the Committee on Safety in Medicines, regulates the approval and sale of pharmaceutical products in the United Kingdom. See also COMMITTEE ON SAFETY IN MEDICINES, FOOD AND DRUG ADMINISTRATION (FDA), COMMITTEE FOR PROPRIETARY MEDICINAL PRODUCTS (CPMP), KOSEISHO, NDA (TO KOSEISHO), IND, and BUNDESGESUNDHEITSAMT (BGA).

Medium A substance used to provide nutrients for cell growth. It may be liquid (e.g., broth) or solid (e.g., agar). See also CULTURE MEDIUM, AGAR, and MAMMALIAN CELL CULTURE.

Megakaryocyte Stimulating Factor (MSF) A colony stimulating factor (protein) involved in the regulation of platelet production, white blood cell production, and red blood cell production from stem cells in bone marrow. See also COLONY STIMULATING FACTORS (CSFs), PLATELETS, and STEM CELLS.

Mega-Yeast Artificial Chromosomes (mega YAC) A large (i.e., greater than 500 base pairs in length) piece of DNA that has been cloned (made) inside a living yeast cell. While most bacterial vectors cannot carry DNA pieces that are larger than 50 base pairs, and "standard" YACs typically cannot carry DNA pieces that are larger than 500 base pairs, mega YACs can carry DNA pieces (chromosomes) as large as one million base pairs in length. See also YEAST, CHROMOSOMES, HUMAN

ARTIFICIAL CHROMOSOMES (HAC), *ARABIDOPSIS THALIANA,* DEOXYRIBONUCLEIC ACID (DNA), CLONE (A MOLECULE), VECTOR, BASE PAIR (bp), and YEAST ARTIFICIAL CHROMOSOMES (YAC).

Meiosis Discovered by Edouard Van Beneden in the 1870s, meiosis is the sequence of complex cell nucleus changes resulting in the production of cells (as gametes) with half the number of chromosomes present in the original cell, and typically involving an actual reduction division in which the chromosomes without undergoing prior splitting join in pairs with homologous chromosomes (of maternal and paternal origin) and then separate so that one member of each pair enters each product cell nucleus and undergoes a second division not involving reduction. Occurs by two successive divisions (meiosis I and II) that reduce the starting number of $4n$ chromosomes to $1n$ in each of four product cells. Product cells may mature to germ cells (sperm or eggs). See also OOCYTES and NUCLEUS.

Melting (of DNA) Melting DNA means to heat-denature it. When this happens the hydrogen bonds holding the DNA molecule together in the normal way are disrupted, allowing a more random polymer structure to exist. See also DENATURED DNA.

Melting (of substance other than DNA) To change from a solid to a nonsolid (e.g., liquid) state by the addition of heat (to the solid substance).

Melting Temperature (of DNA) (T_m) The midpoint of the temperature range over which DNA is denatured. See also MELTING (OF DNA).

Membrane Transport The facilitated transport of a solute across a membrane, usually by a specific membrane protein (e.g., adhesion molecule). See also ENDOCYTOSIS, EXOCYTOSIS, SIGNAL TRANSDUCTION, G-PROTEINS, VAGINOSIS, RECEPTORS, ADHESION MOLECULE, VESICULAR TRANSPORT (OF A PROTEIN), GATED TRANSPORT (OF A PROTEIN), and CALCIUM CHANNEL-BLOCKERS.

Mesophile An organism that grows best in the temperature range of 25°C (77°F) to 40°C (104°F). See THERMOPHILE and PSYCHROPHILE.

Messenger RNA (mRNA) Messenger ribonucleic acid. The intermediary molecule between DNA and ribosomes (in a cell) which synthesize (i.e., make) those proteins coded for by the cell's DNA. Upon receiving the "message" encoded in the DNA, the messenger RNA passes through the ribosomes like a reel of punched paper passes through an old player piano (pianola) giving the ribosomes the specifications for making the coded-for proteins.

This process is aided by transfer RNA (tRNA) molecules, which forage for amino acids that float around in the cell (outside of the cell's nucleus and ribosomes). The transfer RNA (tRNA) molecules attach to, and escort individual amino acids to the ribosome, as and when the messenger RNA (mRNA) directs. Each of the 20 different amino acids has at least one of its own purpose-built tRNA molecules, which possess a three-letter code of nucleotides at the stem of the cloverleaf-shaped rRNA molecule.

The ribosome has room for only two tRNA molecules at a time. The messenger RNA (mRNA) molecule (which itself is passing through the ribosome) calls over the first tRNA molecule, which brings with it the specified amino acid. Short sections of the messenger RNA (mRNA) and transfer RNA (tRNA) molecules lock together inside the ribosome (because where these two molecules meet, their three nucleotides are complementary), the whole (locked together) apparatus shifts along by three notches (i.e., nucleotides), and a second tRNA molecule (bearing another amino acid) slips in next to the first tRNA molecule.

Next, the first amino acid (brought in by the first tRNA molecule) jumps over to the second tRNA molecule; joining to the amino acid that was brought in by the second tRNA molecule, thus making the start of a protein (i.e., a poly-amino acid molecule, also known as polypeptide or protein molecule). The empty (first) tRNA molecule falls out of the ribosome, and the whole (locked together) apparatus (i.e., mRNA plus second tRNA molecule) moves three more notches (i.e., nucleotides) along the mRNA molecule to make room for a third tRNA molecule bearing another amino acid, and so on.

This process of creating ever-longer chains of amino acids continues to repeat itself inside the ribosome until the protein (coded for by the DNA, which code was transferred to mRNA, which transferred it to the ribosome) is completed. See also TRANSCRIPTION, COMPLEMENTARY DNA (c-DNA), CENTRAL DOGMA, DEOXYRIBONUCLEIC ACID (DNA), RIBONUCLEIC ACID (RNA), NUCLEIC ACIDS, CODING SEQUENCE, GENETIC CODE, INFORMATIONAL MOLECULES, CODON, RIBOSOMES, POLYRIBOSOME (POLYSOME), rRNA (RIBOSOMAL RNA), NUCLEOTIDE, POLYMER, TRANSFER RNA (tRNA), PROTEIN, AMINO ACID, POLYPEPTIDE (PROTEIN), and ANTISENSE (DNA SEQUENCE).

Metabolism The entire set of enzyme-catalyzed transformations of organic nutrient molecules (to sustain life) in living cells. Conversion of food and water into nutrients that can be used by the body's cells, *and* the use of those nutrients by those cells (to sustain life, grow, etc.). See also ENZYME, CELL, INTERMEDIARY METABOLISM, METABOLITE, COMBINATORIAL BIOLOGY, CITRIC ACID, AFLATOXIN, *FUSARIUM*, CYTOCHROME P4503A4.

Metabolite A chemical intermediate in the enzyme-catalyzed chemical reactions of metabolism. See also METABOLISM, ENZYME, CELL, INTERMEDIARY METABOLISM, AFLATOXIN, and *FUSARIUM*.

Metalloenzyme An enzyme having a metal ion as its prosthetic group.

Methionine (met) An essential amino acid; furnishes (to organism) both labile methyl groups and sulfur necessary for normal metabolism. See also ESSENTIAL AMINO ACIDS, METABOLISM, CYSTINE, and HIGH-METHIONINE CORN.

Methylated Refers to a DNA molecule that is saturated with methyl groups (i.e., methyl submolecule groups, —CH_3, have attached themselves to the DNA molecule at all possible locations). Generally, when a DNA molecule is methylated, the genes comprising that DNA molecule are "turned off" (i.e., inactivated). See also DEOXYRIBONUCLEIC ACID (DNA), TRANSCRIPTION, MESSENGER RNA (mRNA), GENE, and GENETIC CODE.

MHC See MAJOR HISTOCOMPATIBILITY COMPLEX.

Micelle The spherical structure formed by the association of a number of amphiphilic molecules dissolved in water. Structurally, the outer surface of the micelle (sphere) is covered with the polar domains (head groups) which are directed towards (stick into) the water while the interior of the micelle contains the nonpolar domains (tails) which self-associate to create an "oil droplet" microenvironment. Micelles may be used to solubilize nonwater (oil) soluble or sparingly water soluble molecules in water. They may be formed by ionic or nonionic surfactants. See also AMPHIPHILIC MOLECULES, SUPERCRITICAL CARBON DIOXIDE, CRITICAL MICELLE CONCENTRATION, REVERSE MICELLE (RM), SURFACTANT, FATS, and SELF-ASSEMBLY (OF A LARGE MOLECULAR STRUCTURE).

Microaerophile An organism that grows best in the presence of a small amount of oxygen.

Microbe A microscopic organism; applied particularly to bacteria. The word "microbe" was coined by Monsieur Sedillot, a colleague of Louis Pasteur. See also BACTERIA, and GENETICALLY ENGINEERED MICROBIAL PESTICIDES (GEMP).

Microbial Physiology The cell structure, growth factors, metabolism, and genetics of microorganisms. See also MICROORGANISM, CELL, METABOLISM, GENETICS, and MICROBIOLOGY.

Microbicide Any chemical that will kill microorganisms. Used synonymously with biocide and bactericide. See also MICROORGANISM.

Microbiology The science dealing with the structure, classification, physiology, and distribution of microorganisms, and with their technical and medical significance. The term microorganism is applied to the simple unicellular and structurally similar representatives of the plant and animal kingdoms. With few exceptions, the unicellular organisms are invisible to the naked eye and generally have dimensions of between a fraction of a micron and 200 microns. See also MICRON.

Microfilaments Very thin filaments found in the cytoplasm of cells.

Microgram 10^{-6} gram, or 2.527×10^{-8} ounce (avoirdupoir).

Micron Also called micrometer. A unit of length convenient for describing cellular dimensions; the Greek letter μ is used as its symbol. A micron is equal to 10^{-3} mm (millimeter) or 10^4 Å (Angstroms) or 0.00003937 inch.

Microorganism Any organism of microscopic size (i.e., requires a microscope to be seen). See also MICROBIOLOGY and CAPSULE.

Microphage See POLYMORPHONUCLEAR LEUKOCYTES.

Micropropagation A technique used by man to replicate (mass-produce) a given (e.g., valuable) plant by making genetic clones ("copies") of that original plant. See also CLONE (AN ORGANISM) and GENETICS.

Microsatellite DNA Pieces of the same small segment (i.e., a DNA sequence) which are "repeated" (appear repeatedly in sequence within the DNA molecule) adjacent to a specific gene within the DNA molecule. Thus, these "microsatellites" are linked to that specific gene. See DEOXYRIBONUCLEIC ACID (DNA), LINKAGE, SEQUENCE (OF A DNA MOLECULE), SATELLITE DNA, GENE, and LINKAGE GROUP.

Mimetics See BIOMIMETIC MATERIALS.

Minimized Domains See MINIMIZED PROTEINS.

Minimized Proteins The domain/active site of a (former) native protein after all or most of its extraneous (unneeded) portions (peptides) have been removed. In 1995, Brian Cunningham and James A. Wells reduced the 28-residue (peptide) protein (hormone) Atrial Natriuretic Factor to 15-residues (peptides) size without reducing its potency (biological activity).

Minimized proteins—that retain their potency—hold the potential for medicines possessing a greater serum lifetime (when injected into a patient's body), and as "models" for the creation of organic-chemical-synthesized mimetic drugs possessing the same therapeutic effect as the native protein did. See also PROTEIN, PEPTIDE, ACTIVE SITE, ENZYME, CATALYTIC SITE, DOMAIN (OF A PROTEIN), HORMONE, ATRIAL NATRIURETIC FACTOR, BIOMIMETIC MATERIALS, SERUM LIFETIME, and BIOLOGICAL ACTIVITY.

Minimum Tillage See LOW-TILLAGE CROP PRODUCTION and NO-TILLAGE CROP PRODUCTION.

"Miniprotein Domains" See MINIMIZED PROTEINS.

"Miniproteins" See MINIMIZED PROTEINS.

Mitochondria Granular or rod-shaped bodies in a cell that contain the zyme systems required in the citric acid cycle, electron transport, and oxidative phosphorylation. See also ZYME SYSTEMS.

Mitogen A substance (e.g., growth factor, hormone, etc.) that initiates cell division within the body. For example, most Angiogenic Growth Factors (e.g., fibroblast growth factor) stimulate cell division of the endothelial cells which line blood vessel walls. See also MITOSIS, GROWTH FACTOR, HORMONE, ANGIOGENIC GROWTH FACTORS, and ENDOTHELIAL CELLS.

Mitosis A process of cell duplication, or reproduction, during which one cell gives rise to two identical daughter cells. See also MITOGEN.

Mixed-Function Oxygenases Enzymes catalyzing simultaneous oxidation of two substances by oxygen, one of which is usually NADPH or NADH. See also NADPH, NADH, OXIDATION, and ENZYME.

Moiety Referring to a part or portion of a molecule, generally complex, having a characteristic chemical or pharmacological property. See also ANALOGUE and PHARMACOPHORE.

Mold See FUNGUS.

Mole An Avogadro's number (6.023×10^{23}) of whatever units are being considered. One gram molecular weight of an element or a compound (i.e., same number of grams of an element or a compound as that substance's molecular weight, equal to 6.023×10^{23} molecules). See also MOLECULAR WEIGHT.

Molecular Biology A term coined by Vannevar Bush during the 1940s that eventually came to mean the study and manipulation of molecules that constitute, or interact with, cells. Molecular biology as a distinct scientific discipline originated largely as a result of a decision to provide "support for the application of new physical and chemical techniques to biology" during the 1930s by Warren Weaver, director of the biology (funding) program at America's Rockefeller Foundation (a philanthropic organization). See also MOLECULAR GENETICS, GENETICS, GENETIC ENGINEERING, BIOLOGICAL ACTIVITY, BIOPOLYMER, BIOGENESIS, BIOCHEMISTRY, DEOXYRIBONUCLEIC ACID (DNA), MITOSIS, and MEIOSIS.

Molecular Chaperones See CHAPERONES and PROTEIN FOLDING.

Molecular Diversity Sometimes referred to as "irrational drug design," this refers to the drug design technique of generating large numbers of diverse candidate molecules (e.g., pieces of DNA, RNA, proteins, or other organic moieties) at random (via a variety of methods). These diverse candidate molecules are then tested to see which is best at working against a disease/condition (e.g., fitting a cell receptor, or category of receptors relevant to the disease in question). Molecular candidates that show promise (e.g., via a "pretty good fit" to receptor) are then produced in larger quantities (e.g., via Polymerase Chain Reaction techniques) along with additional molecules that are similar though slightly different in structure (e.g., via site-directed mutagenesis) in an attempt to create a molecule that is a "perfect fit" (e.g., to receptor). See also RATIONAL DRUG DESIGN, DEOXYRIBONUCLEIC ACID (DNA), RIBONUCLEIC ACID (RNA), RECEPTORS, RECEPTOR FITTING (RF), RECEPTOR MAPPING (RM), MOIETY, POLYMERASE CHAIN REACTION (PCR), SITE-DIRECTED MUTAGENESIS, DIVERSITY BIOTECHNOLOGY CONSORTIUM, COMBINATORIAL CHEMISTRY, and COMBINATORIAL BIOLOGY.

Molecular Fingerprinting See COMBINATORIAL CHEMISTRY.

Molecular Genetics The science dealing with the study of the nature and biochemistry of the genetic material. Includes the technologies of genetic engineering. See also GENETICS, GENETIC ENGINEERING, MOLECULAR BIOLOGY, BIOLOGICAL ACTIVITY, BIOPOLYMER, BIOGENESIS, BIOCHEMISTRY, DEOXYRIBONUCLEIC ACID (DNA), MITOSIS, MEIOSIS, MOLECULAR DIVERSITY, and CENTRAL DOGMA.

Molecular Pharming™ A trademark of the Groupe Limagrain company, it refers to the production of pharmaceuticals and other chemicals (e.g., intermediates utilized to manufacture pharmaceuticals) in agronomic plants (which have been genetically engineered). See ANTIBIOTIC, GENETIC ENGINEERING, PHYTOCHEMICALS, EDIBLE VACCINES, and CORN.

Molecular Weight The sum of the atomic weights of the constituent atoms in a molecule. See also ATOMIC WEIGHT.

Monoclonal Antibodies (MAb) Discovered and developed in the 1970s by Cesar Milstein and Georges Kohler, monoclonal antibodies are the name for antibodies derived from a single source or clone of cells that recognize only one kind of antigen. Made by fusing myeloma cancer cells (which multiply very fast) with antibody-producing cells, then spreading the resulting conjugate colony so thin that each cell can be grown into a whole, separate colony (i.e., cloning).

In this way, one gets whole batches of the same (monoclonal) antibody, which are all specific to the same antigen. Monoclonal antibodies have found markets in diagnostic kits and show potential for use in drugs, imaging agents, and purification processes. One example of diagnostic use is the invention in 1997 by Bruno Oesch of a monoclonal antibody–based rapid test to detect the prion (PrP^{Sc}) that causes bovine spongiform encephalopathy (BSE) in cattle. See also ASCITES, MYELOMA, CORN, IMMUNOTOXIN, BLAST CELL, ANTIGEN, ANTIBODY, SINGLE-DOMAIN ANTIBODIES (dAbs), MURINE, CATALYTIC ANTIBODY, SEMISYNTHETIC CATALYTIC ANTIBODY, BSE, and PRION.

Monocytes Also called monocyte macrophages. The round-nucleated cells that circulate in the blood. In summary they engulf and kill microorganisms, present antigen to the lymphocytes, kill certain tumor cells, and are involved in the regulation of inflammation.

These cells are often the first to encounter a foreign substance or pathogen or normal cell debris in the body. When they do, the material is taken up (engulfed) and degraded by means of oxidative and hydrolytic enzymatic attack. Peptides that result from the degradation of foreign protein are then bound to a monocyte protein called class II MHC (major histocompatibility complex) and this self-foreign complex then migrates to the surface of the cell where it is embedded into the cell membrane in such a way as to present the peptide to the outside of the cell. This positioning allows T lymphocytes to recognize (inspect) the peptide.

Whereas self-peptides derived from normal cellular debris are ignored, foreign peptides activate precursors of helper T cells to further mature into active, lymphokine-secreting helper T lymphocytes, also known as T_H cells. When monocytes move out of the bloodstream and into the tissues they are then called macrophages. See also MACROPHAGE, CELLULAR IMMUNE RESPONSE, PATHOGEN, and MHC.

Monomer The basic molecular subunit from which, by repetition of a single reaction, polymers are made. For example, amino acids (monomers) link together via condensation reactions to yield polypeptides or proteins (polymers). A monomer is analogous to a link (monomer) in a metal chain (polymer). See also POLYMER.

Monosaccharides The chemical building blocks of carbohydrates, hence known as "simple sugars." They are classified by the number of carbon atoms in the (monosaccharide) molecule. For example, pentoses have five and hexoses have six carbon atoms. They normally form ring structures. The empirical formula for monosaccharides is $(CH_2O)_n$. See also OLIGOSACCHARIDES, CARBOHYDRATES (SACCHARIDES), and SUGAR MOLECULES.

Monounsaturated Fats Fat molecules possessing less than the maximum possible number of hydrogen atoms. Diets that are high in monounsaturated fat content have been shown to reduce low-density lipoproteins ("bad" cholesterol) blood content, while leaving blood levels of high-density lipoproteins ("good" cholesterol) essentially unchanged. See also FATTY ACID, SATURATED FATTY ACIDS, DEHYDROGENATION, UNSATURATED FATTY ACID, LOW-DENSITY LIPOPROTEINS, HIGH-DENSITY LIPOPROTEINS, OLEIC ACID, and FATS.

Morphogenetic An adjective referring to formation and differentiation of tissues and organs in an organism. See MORPHOLOGY, STEM CELLS, and TOTIPOTENT STEM CELLS.

Morphology First used in print by the poet Johann Wolfgang von Goethe, this word is utilized to refer to the form/structure of an organism or any of its parts. See also TRAIT and PHENOTYPE.

MOS See MANNANOLIGOSACCHARIDES.

MRA See MUTUAL RECOGNITION AGREEMENTS and MUTUAL RECOGNITION ARRANGEMENTS.

MRL See MAXIMUM RESIDUE LEVEL.

mRNA See MESSENGER RNA.

MSF See MEGAKARYOCYTE STIMULATING FACTOR.

Multi-Copy Plasmids Plasmids that are present inside bacteria in quantities greater than one plasmid per (host) cell. See also PLASMID, VECTOR, and COPY NUMBER.

Multienzyme System A sequence of related enzymes participating in a given metabolic (chemical reaction) pathway.

Murine Of, or pertaining to, mice. For example, the first monoclonal antibodies were produced using cells from mice. This frequently caused adverse immune responses to monoclonal antibodies when they were injected into the human body (e.g., thus limiting their use in therapeutic purposes). However, researchers have recently discovered how to make monoclonal antibodies in human cells. See also MONOCLONAL ANTIBODIES (MAb).

Muscular Dystrophy (MD) A genetic disease caused by a defect in the X chromosome; first recognized by G. A. B. Duchenne in 1858. Afflicts males almost exclusively because males have only one X chromosome, whereas females inherit two copies of the X chromosome and have a backup in case one X chromosome is damaged (as is the case for MD victims). In 1981, Kay E. Davies used DNA probes (genetic probes) to discover that the Duchenne muscular dystrophy (DMD) gene must lie somewhere between two unique (to MD victims) segments on the upper, shorter arm of the X chromosome. See also DNA PROBE, CHROMOSOMES, KARYOTYPE, CHROMATIDS, and CHROMATIN.

Mutagen A chemical substance capable of producing a genetic mutation (change), by causing changes in the DNA of living organisms. For example, Dr. Gary Shaw discovered in 1996 that women who smoke cigarettes during their pregnancies are twice as likely to have babies with the genetic deformity known as cleft lip and palate. If those women have a particularly susceptible (to smoke) gene variant (allele) within their DNA, they are as much as eight times as likely to have babies with cleft lip and palate. See also MUTATION, GENETICS, HEREDITY, GENETIC CODE, ALLELE, DEOXYRIBONUCLEIC ACID (DNA), ONCOGENES, and MUTANT.

Mutant An altered cell or organism resulting from mutation (an alteration) of the original wild (normal) type. A change from the normal to the unique or abnormal. See also MUTAGEN, HEREDITY, and WILD TYPE.

Mutase An enzyme catalyzing transposition of a functional group in the substrate (substance acted upon by the enzyme). Intramolecular transfer of a chemical group from one position (i.e., carbon atom) to another within the same molecule. An example of a mutase is phosphoglucomutase. It has a molecular weight of about 60,000 Daltons with about 600 amino acid residues (monomers). The mutase can interchange (move) a phosphate unit between the 1 and 6 position. The 1 refers to a carbon atom designated as "#1" and the 6 refers to a different carbon atom designated as "#6."

Mutation From the Latin term *mutare*, meaning "to change." Any change that alters the sequence of the nucleotide bases in the genetic material (DNA) of an organism or cell; with alteration occurring either by displacement, addition, deletion, cross-linking, or other destruction. The alteration to the DNA sequence would alter its meaning, that is, its ability to produce the normal amount or normal kind of protein, so the (organism or cell) is itself altered. Such an altered organism is called a mutant. See also MUTANT, INFORMATIONAL MOLECULES, HEREDITY, GENETIC CODE, GENETIC MAP, PROTEIN, and DEOXYRIBONUCLEIC ACID (DNA).

Mutual Recognition Agreements (MRAs) Legal agreements (e.g., treaties) between two or more nations, to recognize and respect each other's approval process (e.g., for new crops derived via biotechnology). See GMO, COMMITTEE FOR VETERINARY MEDICINAL PRODUCTS (CVMP), ORGANIZATION FOR ECONOMIC COOPERATION AND DEVELOPMENT (OECD), EVENT, EUROPEAN MEDICINES EVALUATION AGENCY (EMEA), COMMITTEE FOR PROPRIETARY MEDICINAL PRODUCTS (CPMP), and UNION FOR PROTECTION OF NEW VARIETIES OF PLANTS (UPOV).

Mutual Recognition Arrangements See MUTUAL RECOGNITION AGREEMENTS (MRAs).

Mycotoxins Toxins produced by fungi. More than 350 different mycotoxins are known to man. Almost all mycotoxins possess the capacity to harmfully alter the immune systems of animals. Consumption by animals (including humans) of certain mycotoxins (e.g., via eating infected corn, nuts, peanuts, cottonseed products, etc.) can result in liver toxicity, gastrointestinal lesions, cancer, muscle necrosis, etc. See TOXIN, FUNGUS, *FUSARIUM*, AFLATOXIN, VOMITOXIN, *FUSARIUM MONILIFORME*, and FUMONISINS.

Myeloma A tumor cell line derived from a lymphocyte. It usually produces a single type of immunoglobulin. See also HYBRIDOMA, LYMPHOCYTE, and AGING.

Myristoylation Transformation of proteins in cells in such a manner that these cells then cause cancer. See also CANCER.

N

N Glycosylation See GLYCOSYLATION (TO GLYCOSYLATE).

n-3 Fatty Acids Also known as "omega-3" fatty acids. See POLYUNSATURATED FATTY ACIDS (PUFA).

n-6 Fatty Acids Also known as "omega-6" fatty acids. See POLYUNSATURATED FATTY ACIDS (PUFA).

NAD (NADH, NADP, NADPH) Nicotinamide-adenine dinucleotide, also known as diphosphopyridine nucleotide, codehydrogenase 1, coenzyme 1, and coenzymase by its discoverers, Harden and Young. $C_{21}H_{27}O_{14}N_7P_2$. An organic coenzyme (molecule) that functions as a distinct yet integral part of certain enzymes. NAD plays a role in certain enzymes concerned with oxidation/reduction reactions. Meanings: NADH, nicotinamide-adenine dinucleotide, reduced; NADP, nicotinamide-adenine dinucleotide phosphate; and NADPH, nicotinamide-adenine dinucleotide phosphate, reduced. See also ENZYME, COENZYME, OXIDATION-REDUCTION REACTION, and NITRIC OXIDE SYNTHASE.

NADA (New Animal Drug Application) An application to the U.S. Food and Drug Administration (FDA) to begin testing/studies of a new drug for animals (e.g., livestock), that might (eventually) lead to its FDA approval. See also IND.

NADH Nicotine-adenine dinucleotide, reduced. See NAD.

NADP Nicotine-adenine dinucleotide phosphate. See NAD and NITRIC OXIDE SYNTHASE.

NADPH Nicotinamide-adenine dinucleotide phosphate, reduced. See NAD.

"Naked" Gene A bare gene (strand of DNA) that has been extracted-from or derived (e.g., synthesized from sequencing data) from a pathogen. During the 1990s, it was discovered that inserting such "naked" genes into certain tissues in the (usual disease host) organism would sometimes cause those tissues to take up the "naked" genes and express some of the cell-surface proteins indigenous to that pathogen. When that happens, and the (host's) immune system mounts an immune response (to those cell-surface proteins, and thus to the pathogen), these "naked" genes are referred to as "DNA vaccines." See also GENE, DEOXYRIBONUCLEIC ACID (DNA), PATHOGEN, PROTEIN, SYNTHESIZING (OF DNA MOLECULES), SEQUENCING (OF DNA MOLECULES), EXPRESS, DNA VACCINES, IMMUNE RESPONSE, CELLULAR IMMUNE RESPONSE, HUMORAL IMMUNITY, ANTIBODY, and DNA VECTOR.

Nanocrystal Molecules Coined by researchers A. Paul Alivisatos and Peter G. Schultz, it is a term used to describe double-stranded DNA molecules that have attached to them several multi-atom clusters of gold. As of 1996, these researchers were working to try to create nanometer-scale electrical circuits, semiconductors, etc.

A separate methodology, researched by Chad A. Mirkin et al., utilizes strands of DNA to reversibly assemble gold nanoparticles (nanometer-scale multi-atom particles) into supramolecular (many molecule) agglomerations, in which the gold particles are separated from each other by a distance of approximately 60 Angstroms. The aggregation of these DNA-metal nanoparticles causes a visible color change to occur. As of 1996, these researchers were working to try to create simple and rapid tests that would indicate the presence of a virus (e.g., HIV-1 or HIV-2) via a visible color change. Such a test would use two noncomplementary DNA sequences, each of which has attached to it a gold nanoparticle (via a thiol group). The two sequences would be selected for their ability to latch onto a target sequence in the desired virus, but they would be unable to combine with each other, since they are noncomplementary. When double-stranded DNA molecules possessing two "sticky ends" (that are complementary to the sequences attached to virus) are added, the resultant color change indicates virus presence. See also DOUBLE HELIX, DEOXYRIBONUCLEIC ACID (DNA), ANGSTROM (Å), NANOMETERS (nm), HYBRIDIZATION SURFACES, BASE PAIR (bp), SELF-ASSEMBLY (OF A LARGE MOLECULAR STRUCTURE), NANOTECHNOLOGY, STICKY ENDS, HYBRIDIZATION (MOLECULAR GENETICS), SEQUENCE (OF A DNA MOLECULE), VIRUS, and BIOSENSORS (CHEMICAL).

Nanogram (ng) 10^{-9} gram, or 3.527×10^{-11} ounce (avoirdupoir).

Nanometers (nm) 10^{-9} meter. Often used to express wavelengths of light (e.g., in a spectrophotometer). See SPECTROPHOTOMETER.

Nanoparticles See NANOCRYSTAL MOLECULES and NANOTECHNOLOGY.

Nanotechnology Literally means "dwarf technology." A futuristic, conceptual technology in which man manipulates objects whose dimensions are approximately 1 to 100 nanometers. For example, it is possible that in the future a variety of man-made "nano-assemblers" [i.e., tiny (molecular) machines smaller than a grain of sand] would manufacture those things that are produced today in factories. For example, enzyme molecules function essentially as jigs and machine tools to shape large molecules as they are formed in biochemical reactions. The technology of manipulating atoms and molecules in order to form (build) bigger, but still vanishingly small functional structures and machines. See also ENZYME, GENOSENSORS, NANOMETERS (nm), BIOSENSORS (ELECTRONIC), NANOCRYSTAL MOLECULES, and BIOSENSORS (CHEMICAL).

Napole Gene See REDEMENT NAPOLE (RN) GENE.

NAS See NATIONAL ACADEMY OF SCIENCES.

National Academy of Sciences (NAS) A private, self-perpetuating society of distinguished scholars in scientific and engineering research, dedicated to the advancement of science and technology and their use for the general welfare. Under the authority of its congressional charter of 1863, the NAS has a working mandate that calls upon it to advise the U.S. Federal Government on scientific and technical matters.

National Cancer Institute (NCI) One of the National Institutes of Health. See also NATIONAL INSTITUTE OF HEALTH (NIH).

National Heart, Lung, and Blood Institute (NHLBI) One of the National Institutes of Health. See NATIONAL INSTITUTES OF HEALTH (NIH).

National Institute of Allergy and Infectious Diseases (NIAID) The main agency of the National Institutes of Health. See NATIONAL INSTITUTES OF HEALTH (NIH).

National Institute of General Medical Sciences (NIGMS) One of the National Institutes of Health. See NATIONAL INSTITUTES OF HEALTH (NIH).

National Institutes of Health (NIH) The major U.S. Government sponsor of biotechnology research. It is composed of a group of government institutes that each focus on specific medical areas. See also RECOMBINANT DNA ADVISORY COMMITTEE (RAC).

Native Conformation The normal, biologically active conformation (i.e., the three-dimensional arrangement of its atoms) of a protein molecule. See also CONFORMATION.

Natural Killer Cells These cells are involved in tumor surveillance. They also kill virus-laden cells.

NCI See NATIONAL CANCER INSTITUTE (NCI).

NDA (to FDA) New Drug Application (to the U.S. Food and Drug Administration). A (paper) application to the U.S. Food and Drug Administration (FDA) seeking approval of a new drug that has undergone Phase 2 and Phase 3 clinical trials. An NDA is submitted in the form of (thousands of) pages of (clinical and other) data, along with various analyses (e.g., statistical) of that data for efficacy, safety, etc. See also CANDA, FOOD AND DRUG ADMINISTRATION (FDA), MAA, NDA (TO KOSEISHO), and PHASE I CLINICAL TESTING.

NDA (to Koseisho) New drug application. It is the Japanese equivalent to a U.S. IND (investigational new drug) application; to the Koseisho, the Japanese equivalent of the U.S. Food and Drug Administration (FDA). See also IND, KOSEISHO, and FOOD AND DRUG ADMINISTRATION (FDA).

Necrosis Refers to cell death caused by physical injury to the cell (e.g., exposure to toxin, exposure to ultraviolet light, lack of oxygen, etc.). See CELL, TOXIN, RESPIRATION, and TUMOR NECROSIS FACTOR.

Neem Tree A tropical tree that resists insect (e.g., whiteflies, mealybugs, aphids, mites) depradations and certain fungal diseases (e.g., rusts, powdery mildew, etc.) via secretions of liquids that contain Azadirachtin (an insect-repelling chemical). See also AZADIRACHTIN and FUNGUS.

Negative Supercoiling Comprises the twisting of a duplex of DNA (deoxyribonucleic acid) in space in the opposite sense to the turns of the strands in the double helix. See DOUBLE HELIX.

Neoplasia New growth. See NEOPLASTIC GROWTH.

Neoplastic Growth A new growth of animal or plant tissue resembling (more or less) the tissue from which it arises but having distinct biochemical differences from the parent cell. The neoplastic tissue is a mutant version of the original and appears to serve no physiologic function in the same sense as did the original tissue. It may be benign or malignant (i.e., a tumor). See also TUMOR and CANCER.

Nerve Growth Factor (NGF) A protein produced by the salivary glands (and also in tumors) that greatly increases growth/reproduction of nerve cells and guides the formation of neural networks. In the brain,

NGF is thought to increase the production of the messenger chemical, acetylcholine, by protecting and stimulating those neurons that produce acetylcholine. Because those (acetylcholine-producing) neurons are typically the first to be destroyed in an Alzheimer's disease victim, NGF holds potential to be used to counteract (some of) the effects of the disease. NGF is also necessary for normal development of the hypothalamus, a brain structure that regulates a number of hormones. Human T cells appear to have receptors for NGF, which could explain the "mind-body connection" between a person's emotional well-being and physical health (i.e., NGF may be a go-between for the brain and the immune system). NGF was discovered by Rita Levi-Montalcini in 1954. See also GROWTH FACTOR, EPIDERMAL GROWTH FACTOR (EGF), HYPOTHALAMUS, HORMONE, and PROTEIN.

Neuraminidase (NA) A transmembrane (i.e., through the membrane) glycoprotein enzyme that appears in the membrane of the influenza virus.

Neurotransmitter An organic, low molecular weight compound that is secreted from the terminal end of a neuron (in response to the arrival of an electrical impulse) into a liquid-filled gap that exists between neurons. The transmitter molecule then diffuses across the small gap and attaches to the next neuron. This attachment causes structural changes in the membrane of the neuron and initiates the conductance of an electrical impulse. In this way an electrical impulse is transmitted along a neuron network of which the neurons themselves do not physically touch. A neurotransmitter serves to transmit a nerve impulse between different neurons. See also MOLECULAR WEIGHT.

Neutraceuticals See NUTRACEUTICALS.

Neutriceuticals See NUTRACEUTICALS.

Neutrophils Phagocytic (ingesting, scavenging) white blood cells that are produced in the bone marrow. They ingest and destroy invading microorganisms and facilitate post-infection tissue repair. They can secrete collagenase and plasminogen activator. They are the immune system's "first line" of defense against invading pathogens, and large reserves are called forth within hours of the start of a "pathogen invasion." See also PATHOGEN, COLLAGENASE, and MICROORGANISM.

New Drug Application See NDA (TO KOSEISHO), NDA (TO FDA), MAA, IND, and CANDA.

NIAID See NATIONAL INSTITUTE OF ALLERGY AND INFECTIOUS DISEASES.

Nick A break in one strand of a double-stranded DNA molecule. One of the phosphodiester bonds between two adjacent nucleotides is ruptured. No bases are removed from the strand, it is just opened at that point. See also DEOXYRIBONUCLEIC ACID (DNA).

Nicotine-Adenine Dinucleotide (NAD) See NAD.

Nicotine-Adenine Dinucleotide, reduced (NADH) See NAD.

Nicotine-Adenine Dinucleotide phosphate (NADP) See NAD.

Nicotine-Adenine Dinucleotide phosphate, reduced (NADPH) See NAD.

NIH See NATIONAL INSTITUTES OF HEALTH (NIH).

NIHRAC See RECOMBINANT DNA ADVISORY COMMITTEE (RAC).

Ninhydrin Reaction A color reaction given by amino acids and peptides on heating with the chemical ninhydrin. The technique is widely used for the detection and quantitation (measurement) of amino acids and peptides. The concentration of amino acid in a solution (of hydrochloric acid) is proportional to the optical absorbance of the solution after heating it with ninhydrin. α-Amino acids give an intense blue color, and amino acids (such as proline) give a yellow color. One is able to determine concentration of a protein or peptide and also obtain an idea of the type of protein or peptide that is present. See also ABSORBANCE (A), AMINO ACID, and PEPTIDE.

Nitrate Reduction The reduction of nitrate to nitrite or ammonia by an organism. See also REDUCTION (IN A CHEMICAL REACTION).

Nitric Oxide A molecule, produced in the body, which acts as a signalling molecule, oxidant utilized against pathogens by the immune system, and (destructive) free radical. As a "messenger molecule," nitric oxide is utilized by the body for control of blood pressure, immune system regulation, neural signalling, etc. See SIGNALLING, OXIDIZING AGENT (OXIDANT), PATHOGEN, IMMUNE RESPONSE, HUMAN SUPEROXIDE DISMUTASE (hSOD), SIGNAL TRANSDUCTION, and NITRIC OXIDE SYNTHASE.

Nitric Oxide Synthase An enzyme that catalyzes the reaction which the body utilizes to make nitric oxide from L-arginine (via cleavage, off that molecule). The cofactor for that reaction is nicotine-adenine dinucleotide phosphate (NADP). See ENZYME, NITRIC OXIDE, COFACTOR, NAD (NADH, NADP, NADPH), and ARGININE (arg).

Nitrification The oxidation (of ammonia) to nitrate by an organism. See also OXIDATION.

Nitrogenase System A system of enzymes capable of reducing atmospheric nitrogen to ammonia in the presence of ATP. See also REDUCTION (IN A CHEMICAL REACTION) and ENZYME.

Nitrogen Cycle The cycling of various forms of biologically available nitrogen through the plant, animal, and microbial worlds (kingdoms) as well as the atmosphere and geosphere.

Nitrogen Fixation Conversion of atmospheric nitrogen (N_2 gas) into a soluble, biologically available form. The conversion is carried out by nitrogen-fixing organisms (e.g., *Rhizobium* bacteria) which live symbiotically in the roots of certain plants, e.g., alfalfa or soybeans. This is one of nature's ways of fertilizing. When not enough nitrogen fixation occurs, soil is not able to produce maximum crop yields and farmers may need to spread fixed nitrogen onto the field in the form of the fertilizer ammonium nitrate or sodium nitrate. See also SYMBIOTIC, GENISTEIN (gen), BACTERIA, and SOYBEAN PLANT.

"No-Till" Crop Production See NO-TILLAGE CROP PRODUCTION.

No-Tillage Crop Production A methodology of crop production in which the farmer avoids mechanical cultivation (i.e., only one pass over the field). The plant residue remaining on the field's surface helps to control weeds and reduce soil erosion, but it also provides sites for insects to shelter and reproduce, leading to a need for increased insect control via methods such as inserting a *Bacillus thuringiensis (B.t.)* gene into crop plants. See INTEGRATED PEST MANAGEMENT (IPM), CORN, SOYBEAN PLANT, *BACILLUS THURINGIENSIS (B.t.)*, GENETIC ENGINEERING, EUROPEAN CORN BORER (ECB), *HELICOVERPA ZEA (H. ZEA)*, CORN ROOTWORM, GENE, and COLD HARDENING.

Nonessential Amino Acids Amino acids of proteins that can be made (biochemically synthesized within the body) by humans and certain other vertebrate animals from simple chemical precursors (in contrast to the essential amino acids). These amino acids are thus not required in the diet (of humans and those other vertebrates). See also ESSENTIAL AMINO ACIDS.

Nonheme-Iron Proteins Proteins containing iron but no porphyrin groups (within which iron atoms are held) in their structure. See also HEME.

Nonpolar Group A hydrophobic ("water hating") group on a molecule; usually hydrocarbon (composed of hydrogen and carbon atoms) in nature. These groups are more at home in a nonpolar (oil-like) environment. See also POLAR GROUP, AMPHIPATHIC MOLECULES, and AMPHOTERIC COMPOUND.

Nonsense Codon A triplet of nucleotides that does not code for an amino acid. Any one of three triplets (U-A-G, U-A-A, U-G-A) that cause termination of protein synthesis. U-A-G is known as amber and U-A-A is known as ochre. See GENETIC CODE, CODON, and TERMINATION CODON.

Nonsense Mutation A mutation that converts a codon that specifies an amino acid into one that does not specify any amino acid. A change in the nucleotide sequence of a codon that may result in the termination of a polypeptide chain. See NONSENSE CODON, GENETIC CODE, and CODON.

Nontranscribed Spacer A region between transcription units in a tandem gene cluster. See TRANSCRIPTION, MESSENGER RNA (mRNA), GENETIC CODE, GENE SPLICING, and GENE.

North American Plant Protection Organization (NAPPO) One of the international SPS standard-setting organizations that develops plant health standards, guidelines and recommendations (e.g., to prevent transfer of a disease from one country to another). Subsidiary to the International Plant Protection Convention (IPPC), it covers the countries of North America. Its secretariat is located in Nepean, Canada. See also IN-

TERNATIONAL PLANT PROTECTION CONVENTION (IPPC), EUROPEAN PLANT PROTECTION ORGANIZATION (EPPO), and SPS.

Northern Blotting A research test/methodology used to transfer RNA fragments from an agarose gel (e.g., following gel electrophoresis) to a filter paper without changing the relative positions of the RNA fragments (e.g., re electrophoresis separation grid). See also RIBONUCLEIC ACID (RNA), GEL ELECTROPHORESIS, AGAROSE, CHROMATOGRAPHY, and FIELD INVERSION GEL ELECTROPHORESIS.

Nuclear Matrix Proteins Protein molecules that are present in cancerous cells, but not in normal (non-mutated) cells. See also PROTEIN, CELL, MUTATION, MUTANT, MYRISTOYLATION, NEOPLASTIC GROWTH, and "PARP."

Nuclear Receptors Receptors in a cell's outer membrane that serve to convey a "signal" from outside the cell all the way into the cell's nucleus. See also RECEPTORS, SIGNALLING, SIGNAL TRANSDUCTION, NUCLEUS, ENDOCYTOSIS, VAGINOSIS, CD4 PROTEIN, PROTEIN, and CELL.

Nuclease An enzyme capable of hydrolyzing the internucleotide linkages of a nucleic acid (e.g., DNA or RNA). Nucleases tend to degrade (i.e., hydrolyze, cleave) artificially inserted DNA strands, making genetic targeting more difficult. See also GENETIC TARGETING, DEOXYRIBONUCLEIC ACID (DNA), and RIBONUCLEIC ACID (RNA).

Nucleic Acid Probes See DNA PROBE, NUCLEIC ACIDS, POLYMERASE CHAIN REACTION (PCR), and RAPID MICROBIAL DETECTION (RMD).

Nucleic Acids A nucleotide polymer. A large, chain-like molecule containing phosphate groups, sugar groups, and purine and pyrimidine bases; two types are ribonucleic acid (RNA) and deoxyribonucleic acid (DNA). The bases involved are adenine, guanine, cytosine, and thymine (uracil in RNA). See also GENETIC CODE, DNA, and RNA.

Nucleoid The compact body that contains the genome in a bacterium. See also GENOME.

Nucleolus A round, granular structure situated in the nucleus of eucaryotic cells. It is involved in rRNA (ribosomal RNA) synthesis and ribosome formation. See also RIBOSOMES and NUCLEUS.

Nucleophilic Group An electron-rich group with a strong tendency to donate electrons to an electron-deficient nucleus. See also POLAR GROUP and NONPOLAR GROUP.

Nucleoproteins Complexes made up of nucleic acid and protein. These two substances are apparently not linked by strong chemical bonds, but are held together by salt linkages and other weak bonds. Most viruses consist entirely of nucleoproteins, although some viruses also contain fatty substances. Nucleoproteins also occur in animal and plant cells and in bacteria. See also PROTEIN and VIRUS.

Nucleoside A hybrid molecule consisting of a purine (adenine, guanine) or pyrimidine (thymine, uracil, or cytosine) base covalently linked to a five-membered sugar ring (ribose in the case of RNA and deoxyribose in the case of DNA). See also NUCLEOTIDE.

Nucleoside Diphosphate Sugar A coenzyme-like carrier of a sugar molecule functioning in the enzymatic synthesis of polysaccharides and sugar derivatives. See also POLYSACCHARIDES.

Nucleosome Spherical particles composed of a special class of basic proteins in combination with DNA. The particles are approximately 12.5 mm in diameter and are connected to each other by DNA filaments. Under the electron microscope they appear somewhat like a string of pearls. See also CHROMATIN, HISTONES, and PROTEIN.

Nucleotide An ester of a nucleoside and phosphoric acid. Nucleotides are nucleosides that have a phosphate group attached to one or more of the hydroxyl groups of the sugar (ribose or deoxyribose). In short, a nucleotide is a hybrid molecule consisting of a purine or pyrimidine base covalently linked to a five-membered sugar ring which is covalently linked to a phosphate group. While (polymerized) nucleotides are the structural units of a nucleic acid, free nucleotides that are not an integral part of nucleic acids are also found in tissues and play important roles in the cell, e.g., ATP and cyclic AMP. See also ATP, CYCLIC AMP, NUCLEOSIDE, NUCLEIC ACIDS, MESSENGER RNA (mRNA), RIBONUCLEIC ACID (RNA), and DEOXYRIBONUCLEIC ACID (DNA).

Nucleus The usually spherical body with each living cell that contains its hereditary biological material (e.g., DNA, genes, chromosomes, etc.) and controls the cell's life functions (e.g., metabolism, growth, and reproduction). The nucleus is a highly differentiated, relatively large organelle lying in the cytoplasm of the cell. The nucleus is surrounded by a (nuclear) membrane which is quite similar to the plasma (cell) membrane; except the nuclear membrane contains holes or pores. It is characterized by its high content of chromatin, which contains most of the cell's DNA. That chromatin is normally (when cell is not in process of dividing) distributed throughout the nucleus in a diffuse manner. See also GENOME, CELL, GENE, GENETIC CODE, RNA, HEREDITY, DEOXYRIBONUCLEIC ACID (DNA), CHROMOSOMES, MEIOSIS, METABOLISM, CHROMATIDS, CHROMATIN, PLASMA MEMBRANE, ORGANELLES, and NUCLEAR RECEPTORS.

Nutraceuticals Coined in 1989 by Stephen DeFelice, this term is used to refer to either a food or portion of food (e.g., a vitamin, essential amino acid, etc.) that possesses medical or health benefits (to the organism that consumes that nutraceutical). For example, saponins (present in beans, spinach, tomatoes, potatoes, alfalfa, clover, etc.) possess some cancer-prevention properties.

Also sometimes called pharmafoods, functional foods, or designer foods, these are food products that have been designed to contain specific concentrations and/or proportions of certain nutrients (e.g., vitamins, amino acids, etc.) that are critical for good health. See also ESSENTIAL AMINO ACIDS, AMINO ACID, VITAMIN, FOOD GOOD MANUFACTURING PRACTICE (FGMP), SAPONINS, ESSENTIAL NUTRIENTS, and PHYTOCHEMICALS.

Nutriceuticals See NUTRACEUTICALS.

"Nutrient-Enhanced™" A phrase that is now a trademark of ICI Seeds Inc. (Garst Seed Co.); it refers to plants that have been modified to possess novel traits which make those plants more economically valuable for nutritional uses (e.g., higher than normal protein content in feedgrains). See also VALUE-ENHANCED GRAINS, HIGH-OIL CORN, PROTEIN, GENETIC ENGINEERING, HIGH-LYSINE CORN, HIGH-METHIONINE CORN, PLANT'S NOVEL TRAIT (PNT), and HIGH-PHYTASE CORN/SOYBEANS.

O

O Glycosylation See GLYCOSYLATION (TO GLYCOSYLATE).

OAB (Office of Agricultural Biotechnology) A unit of the U.S. Department of Agriculture that is in charge of a part of the federal regulatory process for biotechnology (e.g., field tests of transgenic plants). See also TOXIC SUBSTANCES CONTROL ACT (TSCA), RECOMBINANT DNA ADVISORY COMMITTEE (RAC), FOOD AND DRUG ADMINISTRATION (FDA), and TRANSGENIC (ORGANISM).

OD See OPTICAL DENSITY.

Odorant Binding Protein A protein that enhances people's ability to smell odorants in trace quantities much lower than those needed to activate olfactory (i.e., smelling) nerves. The protein accomplishes this by latching onto (odorant) molecules and enhancing their aroma. Hence, it acts as a kind of "helper" entity in bringing about the ability to smell certain odorants present in low concentration. See PROTEIN.

OECD See ORGANIZATION FOR ECONOMIC COOPERATION AND DEVELOPMENT.

Office International des Epizootics See INTERNATIONAL OFFICE OF EPIZOOTICS (OIE).

OH43 Gene in plants (e.g., corn/maize) that causes production of a seed coat which is more resistant to tearing. Greater tear-resistance results in a lower incidence of fungi infestation in seed, which results in less mycotoxin production in seed. See also GENE, FUNGUS, AFLATOXIN, and MYCOTOXINS.

OIE Office International des Epizootics. See INTERNATIONAL OFFICE OF EPIZOOTICS (OIE).

OIF See OSTEOINDUCTIVE FACTOR.

Oleic Acid A fatty acid that is naturally present in the fat of animals and also in oils extracted from oilseed plants (e.g., soybean, canola, etc.). For example, the soybean oil produced from traditional varieties of soybeans tends to contain 24% oleic acid. See MONOUNSATURATED FATS, FATTY ACID, CANOLA, SOYBEAN PLANT, SOYBEAN OIL, and HIGH-OLEIC OIL SOYBEANS.

Oligionucleotide See OLIGONUCLEOTIDE.

Oligomer A relatively short (the prefix "oligo-" means few, slight) chain molecule (polymer) that is made up of repeating units (e.g., XAXAXAXA or XXAAXXAAXXAA, etc.). Short polymers consisting of only two repeating units are called dimers, those of three repeating units are called trimers. Longer units are called polymers (i.e., many units). As a rule of thumb, oligomers consisting of eleven or more repeating units are called polymers.

Oligonucleotide Synonymous with oligodeoxyribonucleotide, they are short chains of nucleotides (i.e., single-stranded DNA or RNA) that have been synthesized (i.e., made) by chemically linking together a number of specific nucleotides. Oligonucleotides (also called, simply "oligos") are used as synthetic (i.e., man-made) genes, DNA probes, and in site-directed mutagenesis. See also NUCLEOTIDE, GENE, DNA PROBE, OLIGOMER, SITE-DIRECTED MUTAGENESIS, GENE MACHINE, DEOXYRIBONUCLEIC ACID (DNA), RIBONUCLEIC ACID (RNA), and SYNTHESIZING (OF DNA MOLECULES).

Oligonucleotide Probes Short chain fragments of DNA that are used in various gene analysis tests (e.g., the single base change in DNA that causes sickle-cell anemia). See also OLIGONUCLEOTIDE, DEOXYRIBONUCLEIC ACID (DNA), DNA PROBE, and GENE MACHINE.

Oligopeptide A relatively short chain molecule that is made up of amino acids linked by peptide bonds. See also PEPTIDE, POLYPEPTIDE (PROTEIN), OLIGOMER, and AMINO ACID.

"Oligos" See OLIGONUCLEOTIDE.

Oligosaccharides Relatively short molecular chains made up to 10–100 simple sugar (saccharide) units. These sugar (i.e., carbohydrate) chains are frequently attached to protein molecules. When this happens, the resulting molecule is known as a glycoprotein—that is, a hybrid molecule that is part protein and part sugar. The oligosaccharide portion affects a protein's conformation(s) and biological activity. The oligosaccharide (carbohydrate) portion of a glycoprotein functions as a mediator of cellular uptake of that glycoprotein. Glycosylation thus affects the length of time the molecule resides in the bloodstream before it is taken out of circulation (serum lifetime). It is thought that blood group (e.g., A, B, O, etc.) is based upon an oligosaccharide concept. For example, different oligosaccharide "branches" on a given glycoprotein (e.g., tissue plasminogen activator) could cause that glycoprotein to be perceived by the body's immune system to be another (incorrect) blood type, thus provoking an immune response against it. See also POLYSACCHARIDES, CONFORMATION, MONOSACCHARIDES, FURANOSE, PENTOSE, PYRANOSE, GLYCOGEN, GLYCOFORM, FRUCTOSE OLIGOSACCHARIDES, GLYCOPROTEIN, TISSUE PLASMINOGEN ACTIVATOR (tPA), OLIGOMER, SEROLOGY, HUMORAL IMMUNITY, and CELLULAR IMMUNE RESPONSE.

Omega-3 Fatty Acids More properly called "*n*-3 fatty acids." See N-3 FATTY ACIDS.

Omega-6 Fatty Acids More properly called "*n*-6 fatty acids." See N-6 FATTY ACIDS.

Oncogenes Genes within a cell's DNA that "turn on" the process of cell division (replication) at appropriate time(s) during the life of each cell in an organism. When oncogenes are mutated (e.g., via exposure to cigarette smoke or ultraviolet light), those oncogenes can become cancer-causing genes, some of which (e.g., erythroblastosis virus gene) are almost identical to the gene for epidermal growth factor (EGF) receptor (i.e., oncogene is a "deformed copy" of that gene). Such mutated oncogenes code for (i.e., cause to be made) proteins (e.g., protein kinases, protein phosphorylating enzymes, etc.) that trigger uncontrolled cell growth. They sometimes may consist of a human chromosome that has viral nucleic acid material incorporated into it and is a permanent part of that chromosome. See also GENE, CELL, DEOXYRIBONUCLEIC ACID (DNA), ras GENE, MEIOSIS, CARCINOGEN, RIBOSOMES, PROTEIN, TUMOR, CANCER, PROTO-ONCOGENES, GENETIC CODE, RECEPTORS, and MUTAGEN.

Oocytes The cells, produced by ovaries, that eventually become an ovum ("egg cell") via meiosis. See also MEIOSIS and CELL.

Opague-2 A gene in corn (maize) that (when present in the DNA of a given plant) causes that plant to produce seed that contains higher-than-normal levels of lysine, calcium, magnesium, iron, zinc, and manganese. See also LYSINE, HIGH-METHIONINE CORN, ESSENTIAL AMINO ACIDS, CORN, VALUE-ENHANCED GRAINS, HIGH-LYSINE CORN, DEOXYRIBONUCLEIC ACID (DNA), GENE, and MAL (MULTIPLE ALEURONE LAYER) GENE.

Open Reading Frame (ORF) Region of a gene (DNA) that contains a series of triplet (bases) coding for amino acids without any termination codons. The sequence is potentially translatable into protein. See also CODING SEQUENCE, GENETIC CODE, and TRANSLATION.

Operator Also known as the "o locus." The site on the DNA to which a repressor molecule binds to prevent the initiation of transcription. The operator locus is a distinct entity and exists independently of the structural genes and the regulatory gene. It is the structural/biochemical "switch" with which the operon is turned on or off, and it controls the transcription of an entire group of coordinately induced genes. One type of mutation of the operator locus is called operator constitutive mutants. Constitutive mutants continually churn out the protein characteristic for that operon because the operon unit cannot be turned off by the repressor molecule. See also OPERON, PROMOTER, REGULATORY GENES, REPRESSION (OF GENE TRANSCRIPTION/TRANSLATION), REPRESSOR (PROTEIN), and STRUCTURAL GENE.

Operon A gene unit consisting of one or more genes that specify a polypeptide and an operator unit that regulates the structural gene, that is, the production of messenger RNA (mRNA) and hence, ultimately, of a number of proteins. Generally an operon is defined as a group of functionally related structural genes mapping (that is, being) close to each other in the chromosome and being controlled by the same (one) operator. If the operator is "turned on" then the DNA of the genes comprising the operon will be transcribed into mRNA, and down the line specific proteins are produced. If, on the other hand, the operator is "turned off" then transcription of the genes does not occur and the production of the operon-specific proteins does not occur. See also OPERATOR and TRANSCRIPTION.

Optical Activity The capacity of a substance to rotate the plane of polarization of plane-polarized light when examined in an instrument known as a polarimeter. All compounds that are capable of existing in two forms that are nonsuperimposable mirror images of each other exhibit optical activity. Such compounds are called stereoisomers (or enantiomers or chiral molecules) and the two forms arise because compounds having asymmetric carbon atoms to which other atoms are connected may arrange themselves in two different ways. See also STEREOISOMERS, ENANTIOMERS, and CHIRAL COMPOUND.

Optical Density (OD) The absorbance of light of a specific wavelength by molecules normally dissolved in a solution. Light absorption depends upon the concentration of the absorbing compound (chemical entity) in the solution, the thickness of the sample being illuminated, and the chemical nature of the absorbing compound. An analytical instrument known as a spectrophotometer is used to (quantitatively) express the amount of a substance (dissolved) in a solution. Mathematically, this is accomplished using the Beer-Lambert Law. See also SPECTROPHOTOMETER and ABSORBANCE (A).

Optimum pH The pH (level of acidity) at which maximum growth occurs or maximal enzymatic activity occurs or at which any reaction occurs maximally. See also ENZYME.

Optimum Temperature The temperature at which the maximum growth occurs or maximal enzymatic activity occurs or at which any reaction occurs maximally. See also ENZYME and ENSILING.

Optrode A fiberoptic sensor made by coating the tip of a (glass) optic fiber with an antibody that fluoresces when the antibody comes in contact with its corresponding antigen. Alternatively, the fiber tip is sometimes coated with a dye that fluoresces when the dye comes in contact with specific chemicals (e.g., oxygen, glucose, etc.). Functionally, a beam of light is sent down the fiber and strikes ("pumps") the fluorescent complex, which then fluoresces (releases light of a specific wavelength). The light produced by fluorescence travels back up the same optic fiber and is detected by a spectrophotometer upon its return. By application of the Beer-Lambert Law, quantitative detection/measurement of the antigen or chemical *in vivo* in, for example, a patient's bloodstream is possible. See also ANTIGEN, *IN VIVO*, ANTIBODY, GLUCOSE (GLc), and SPECTROPHOTOMETER.

Organelles Membrane-surrounded structures found in eucaryotic cells; they contain enzymes and other components required for specialized cell function (e.g., ribosomes for protein synthesis, or lysosomes for enzymatic hydrolysis). Some organelles such as mitochondria and chloroplasts contain DNA and can replicate autonomously (from the rest of the cell). See also NUCLEUS, EUCARYOTE, ENZYME, RIBOSOMES, and LYSOSOME.

Organization for Economic Cooperation and Development (OECD) An international organization comprised of the world's wealthiest (most developed) nations. In 1991, the OECD's Group of National Experts on Safety in Biotechnology (GNE) completed a document entitled Report on the Concepts and Principles Underpinning Safety Evaluations of Food Derived from Modern Biotechnology. The "aim of that document was to elaborate the scientific principles to be considered (i.e., by OECD member nations' regulatory agencies) in evaluating the safety of new foods and food components" (e.g., genetically modified soybeans, corn/maize, potatoes, etc.). See BIOTECHNOLOGY, SOYBEAN PLANT, GNE, CANOLA, and MUTUAL RECOGNITION AGREEMENTS (MRA).

Organogenesis The production of entire organs, usually from basic cells, such as fibroblasts, and structural material such as collagen. See also COLLAGEN and FIBROBLASTS.

Origin Point or region where DNA (deoxyribonucleic acid) replication is begun. Often abbreviated "Ori." See also REPLICATION (OF VIRUS) and REPLICATION FORK.

Orphan Drug The name of the legal status granted by the Food and Drug Administration's Office of Orphan Products Development (to certain pharmaceuticals). This classification provides the sponsors of those pharmaceuticals with special tax and other financial incentives (e.g., market monopoly for a limited time). If companies feel that they possess a cure (drug) for a certain disease, but the number of potential patients is below a certain number and there is potential competition from rival companies, then the high cost of developing and shepherding the drug through the FDA would be such that the company would not be able to regain its development costs and make a profit. Hence, orphan drug status was designed to encourage drug development efforts for otherwise noneconomic pharmaceuticals with less than 200,000 patients a year.

Orthophosphate Cleavage Enzymatic cleavage of one of the phosphate ester bonds of ATP to yield ADP and a single phosphate molecule known as orthophosphate (designated as P_i). The cleavage of the phosphate bond is energy-yielding and is (except in the case of a futile cycle) coupled enzymatically to reactions that utilize the energy to run the cell. An orthophosphate cleavage reaction releases relatively less energy than does a corresponding pyrophosphate cleavage reaction. See also ADENOSINE DIPHOSPHATE (ADP), ADENOSINE TRIPHOSPHATE (ATP), FUTILE CYCLE, and PYROPHOSPHATE CLEAVAGE.

Osmosis Bulk flow of water through a semipermeable (or more accurately, differentially permeable) membrane into another (aqueous) phase containing more of a solute (dissolved compound). As an example, let us set up an osmotically active system. There are two solutions, A and B. Solution A has less salt dissolved in it than solution B and, furthermore, the two solutions are separated by a differentially permeable membrane (this looks like a plastic film). Water molecules (and only water molecules) will flow from solution A through the membrane and into solution B. The reason for this is that the membrane allows free passage only to water molecules. The bulk flow of water has the effect of diluting solution B, while concentrating solution A. Water will flow from region A to region B until the salt concentrations of both solutions are equal. Osmosis is therefore a process in which water passes from regions of low salt concentration to regions of high salt concentration. The process can be viewed as equalizing the number of water and solute molecules on both sides of the membrane. See also OSMOTIC PRESSURE.

Osmotic Pressure May be defined as the hydrostatic pressure which must be applied to a solution on one side of a semipermeable membrane (solution B in the example for osmosis) in order to offset the flow of solvent (water) from the other side (solution A in the example for osmosis). It is a measure of the tendency or "strength" of water to flow from a region of low salt concentration (and conversely high water concentration) to regions of high salt concentration (and conversely low water concentration). See also OSMOSIS.

Osteoinductive Factor (OIF) A protein that induces the growth of both cartilage-forming cells and bone-forming cells (e.g., after a bone has been broken). When applied in the presence of transforming growth factor-beta, type 2 (another protein), osteoinductive factor first causes connective tissue cells to grow together to form a matrix of cartilage (e.g., across the bone break), then bone cells slowly replace that cartilage. Osteoinductive factor also seems to thwart a type of cell that tears down bone formation, so OIF may someday be used to combat osteoporosis. See also GROWTH FACTOR, TRANSFORMING GROWTH FACTOR-BETA (TGF-BETA), FIBROBLASTS, and FIBROBLAST GROWTH FACTOR (FGF).

Outcrossing The transfer of a given gene or genes (e.g., one synthesized by man and inserted into a plant via genetic engineering) from a domesticated organism (e.g., crop plant) to wild type (relative of plant). See GENE, SYNTHESIZING (OF DNA MOLECULES), GENETIC ENGINEERING, and WILD TYPE.

Overwinding Positive supercoiling. Winding which applies further tension in the direction of winding of the two strands about each other in the duplex. See also DEOXYRIBONUCLEIC ACID (DNA), SUPERCOILING, DOUBLE HELIX, and DUPLEX.

Oxalate A salt or ester of oxalic acid. See also CALCIUM OXALATE.

Oxidant See OXIDIZING AGENT (OXIDANT).

Oxidation Loss of electrons from a compound (or element) in a chemical reaction. When one compound is oxidized, another compound is reduced. That is, the other compound must "pick up" the electrons which the first has lost. See also OXIDATION-REDUCTION REACTION and HYDROGENATION.

Oxidation-Reduction Reaction A chemical reaction in which electrons are transferred from a donor to an acceptor molecule or atom. See also OXIDATION, OXIDIZING AGENT (OXIDANT), and REDUCTION (IN A CHEMICAL REACTION).

Oxidative Phosphorylation The enzymatic phosphorylation of ADP to ATP coupled to electron transport from a substrate to molecular oxygen. The synthesis (production) of ATP from the starting materials of ADP and inorganic phosphate (orthophosphate). See also ADENOSINE DIPHOSPHATE (ADP), ADENOSINE TRIPHOSPHATE (ATP), and ORTHOPHOSPHATE CLEAVAGE.

Oxidizing Agent (oxidant) The acceptor of electrons in an oxidation-reduction reaction. The oxidant is reduced by the end of the chemical reaction. That is, the oxidizing agent is the entity that seeks and accepts electrons. Electron acceptance is, by definition, reduction. See also OXIDATION-REDUCTION REACTION.

Oxygenase An enzyme catalyzing a reaction in which oxygen is introduced into an acceptor molecule.

P

P Element A transposon, whose genes (i.e., within this transposon) resist rearrangement during the process (i.e., transposition) of the P element being incorporated into a new location within an organism's genome (i.e., its deoxyribonucleic acid or DNA). In addition to "carrying" genes to a new location(s) in genome, the P element itself codes for transposase (an enzyme that makes transposition possible). See also TRANSPOSON, GENE, ENZYME, TRANSPOSITION, TRANSPOSASE, DEOXYRIBONUCLEIC ACID (DNA), and GENOME.

P. gossypiella See PECTINOPHORA GOSSYPIELLA.

P-Selectin Formerly known as GMP-140 and PADGEM, it is a selectin molecule that is synthesized by endothelial cells before (adjacent) tissues are infected. Thus "stored in advance," the endothelial cells can present P-selectin molecules on the internal surface of the endothelium within minutes after an infection (of adjacent tissue) begins. This presentation of P-selectin molecules attracts leukocytes to the site of the infection, and draws them out of the bloodstream (the leukocytes "squeeze" between adjacent endothelial cells). See also SELECTINS, LECTINS, ELAM-1, ADHESION MOLECULES, LEUKOCYTES, and ENDOTHELIUM.

p53 Gene A tumor-suppressor gene which controls passage (of a cell) from the "GI" phase to the "s" (i.e., DNA synthesis) phase. The p53 protein that is coded for by p53 gene, is a transcription factor (i.e., it "reads" DNA to determine if damaged, then acts to control cell division, while p53 gene codes for more production of additional p53 protein). Discovered in 1993 by Arnold J. Levine and colleagues, it is believed to be responsible for up to 60% of all human cancer tumors (when the p53 gene is damaged or mutated). Normally, the p53 gene codes for (i.e., causes to be manufactured in cell) the p53 protein, which acts to prevent cells from dividing uncontrollably when the cell's DNA has been damaged (e.g., via exposure to cigarette smoke or ultraviolet light). If, in spite of the presence of p53 protein, a cell begins to divide uncontrollably following damage to its DNA, the p53 gene can cause apoptosis, which is also known as "programmed cell death" (to prevent tumors). See also GENE, TUMOR-SUPPRESSOR GENES, ras GENE, GENETIC CODE, MEIOSIS, DEOXYRIBONUCLEIC ACID (DNA), CARCINOGEN, RIBOSOMES, ONCOGENES, TRANSCRIPTION FACTORS, CANCER, TUMOR, p53 PROTEIN, PROTO-ONCOGENES, PROTEIN, and APOPTOSIS.

p53 Protein A tumor-suppressor protein, sometimes called the "guardian of the genome"; though whose amino acid sequence alterations (resulting from damage or mutation to the p53 gene) are believed to be responsible for up to 60% of all human cancer tumors. The p53 protein has four domains, one of which (i.e., the core domain) binds to a specific sequence(s) of the cell's DNA, in order to prevent cell from dividing uncontrollably when the cell's DNA has been damaged (e.g., via exposure to cigarette smoke, ultraviolet light, or other carcinogen), until the damage to that DNA can be repaired.

As the amount of DNA within a given (damaged) cell increases, the concentration of p53 protein also increases. Because p53 protein is a transcription factor (i.e., "reads" DNA to determine if damaged, then acts to control cell division, while p53 gene codes for production of more p53 protein), p53 is very efficient at preventing/inhibiting tumors. See also GENE, p53 GENE, TUMOR-SUPPRESSOR GENES, ras GENE, ras PROTEIN, GENETIC CODE, MEIOSIS, CARCINOGEN, DEOXYRIBONUCLEIC ACID (DNA), AFLATOXIN, RIBOSOMES, ONCOGENES, CANCER, TUMOR, PROTO-ONCOGENES, PROTEIN, TRANSCRIPTION FACTORS, and DOMAIN (OF A PROTEIN).

Paclitaxel An anticancer compound that was originally isolated from the Pacific yew tree. See also CANCER and TAXOL.

PAGE See POLYACRYLAMIDE GEL ELECTROPHORESIS.

Palindrome A DNA molecule sequence that is the same when one strand of the molecule is read left to right and the other strand is read right to left. See also DEOXYRIBONUCLEIC ACID (DNA) and READING FRAME.

Papovavirus A class of animal viruses, e.g., SV40 and polyoma. See also VIRUS.

Parkinson's Disease A disease of the human brain, in which those nerve cells that are involved in controlling movement (motor control) die. Discovered in 1919 by doctors treating an epidemic of encephalitis lethargica (onset of Parkinson's disease commonly follows that encephalitis, can also be induced by drugs, etc.). The (natural) cause of Parkinson's disease is unknown although it can be induced by drug misuse. When a human brain is functioning normally, cells within a region of the brain called the substantia nigra initiate motor (i.e., muscle) activity by releasing the chemical "messenger" known as dopamine. In the brain of a person suffering from Parkinson's disease, those dopamine-producing cells die off, causing a progressive loss of motor control for that person. See also NEUROTRANSMITTER, CILIARY NEUROTROPHIC FACTOR (CNTF), SIGNALLING, and GLIAL DERIVED NEUROTROPHIC FACTOR (GDNF).

"Parp" An enzyme, produced by genetically engineered hamster cells, which is utilized by man in order to determine/test if a given substance is carcinogenic. See also ENZYME, CARCINOGEN, NUCLEAR MATRIX PROTEINS, and GENETIC ENGINEERING.

Particle Gun See BIOLISTIC® GENE GUN and "SHOTGUN" METHOD [TO INTRODUCE FOREIGN (NEW) GENES INTO PLANT CELLS].

Partition Coefficient A constant (number) that expresses the ratio in which a given solute will be partitioned (i.e., distributed) between two given immiscible liquids (e.g., oil and water) at equilibrium.

Passive Immunity An immune response (to a pathogen) that results from injecting *another* organism's antibodies into the organism that is being challenged by the pathogen. See POLYCLONAL ANTIBODIES (USED IN HUMANS). See also HUMORAL IMMUNITY, ANTIBODY, COMPLEMENT (COMPONENT OF IMMUNE SYSTEM), COMPLEMENT CASCADE, IMMUNOGLOBULIN, PATHOGEN, ANTIGEN, and MONOCLONAL ANTIBODIES (MAb).

PAT Gene A dominant gene which, when inserted into a plant's genome, imparts resistance to glufosinate-ammonium containing herbicides. Because the glufosinate-ammonium herbicides act via inhibition of glutamine synthetase (an enzyme that catalyzes the synthesis of glutamine), this inhibition of enzyme kills plants (e.g., weeds). That is because glutamine is crucial for plants to synthesize critically needed amino acids.

The PAT gene is often used by genetic engineers as a marker gene. See also GENE, GENOME, GENETIC ENGINEERING, MARKER (GENETIC MARKER), DOMINANT ALLELE, ESSENTIAL AMINO ACIDS, HERBICIDE-TOLERANT CROP, GTS, SOYBEAN PLANT, CANOLA, CORN, GLUTAMINE, and GLUTAMINE SYNTHETASE.

Pathogen A virus, bacterium, parasitic protozoan, or other microorganism that causes infectious disease by invading the body of an organism known as the host. It should be noted that infection is not synonymous with disease because infection does not always lead to injury of the host.

Pathogenic Disease-causing. See PATHOGEN.

PBR The intellectual property rights that are legally accorded to plant breeders by laws, treaties, etc. Similar to patent law for inventors. See PLANT BREEDER'S RIGHTS (PBR), PLANT'S NOVEL TRAIT (PNT), PLANT VARIETY PROTECTION ACT (PVP), EUROPEAN PATENT CONVENTION, EUROPEAN PATENT OFFICE (EPO), and U.S. PATENT AND TRADEMARK OFFICE (USPTO).

pBR322 An *Escherichia coli (E. coli)* plasmid cloning vector that contains the ampicillin resistance and tetracycline resistance genes. It consists of a circle of double-stranded DNA. See also *ESCHERICHIA COLI (E. COLI)*, PLASMID, VECTOR, and DEOXYRIBONUCLEIC ACID (DNA).

PC Phosphatidyl choline. See LECITHIN.

PCR See POLYMERASE CHAIN REACTION.

PDGF See PLATELET-DERIVED GROWTH FACTOR.

PDWGF See PLATELET-DERIVED WOUND GROWTH FACTOR.

Pectinophora gossypiella Also known as the pink bollworm, this is one of three insect species that is called "bollworms" (when they are on cotton plants). The holes that they chew in cotton plant's bolls have been shown to enable the *Aspergillas flavus* fungus to infect those (chewed) cotton plants. See *B.t. KURSTAKI*, *HELICOVERPA ZEA*, *HELIOTHIS VIRESCENS*, and BRIGHT GREENISH-YELLOW FLUORESCENCE (BGYF).

PEG-SOD (polyethylene glycol superoxide dismutase) A modified version of the enzyme human superoxide dismutase (hSOD) in which polyethylene glycol (a polymer made up of ethylene glycol monomers) is combined with the hSOD molecule. The PEG seems to wrap around or about the enzyme in such a way that the whole complex is able to exist in the blood for longer periods of time than the unmodified hSOD enzyme. This is because the PEG effectively camouflages the hSOD molecule and hence protects it from being inactivated by the body's own defense mechanisms in the bloodstream. This technology is important in that hSOD is used to fight certain diseases by injecting it into the body. However, the SOD must be present in the body for extended periods of time in order to effectively work, and since the injected SOD is a foreign molecule, the body tries to destroy it (and hence its function) as quickly as possible. See also HUMAN SUPEROXIDE DISMUTASE (hSOD), CATALASE, and ENZYME.

Penicillin G (benzylpenicillin) The original penicillin (antibiotic) molecule, discovered by Alexander Fleming in the 1920s, on spoiled bread (mold). In the 1940s, scientists at the U.S. Department of Agriculture in Peoria, Illinois (in America) discovered how to produce commercial quantities of Penicillin G by utilizing the fungus *Penicillium chrysogenum*, which they found on a cantelope in Peoria, Illinois. Penicillin kills bacteria by blocking an enzyme which is crucial to growth and repair of the bacteria's cell wall (peptidoglycan layer), but penicillin does not harm other species, so it is species-specific to bacteria. See ANTIBIOTIC, FUNGUS, BACTERIA, ENZYME, SPECIES SPECIFIC, and *PENICILLIUM*.

Penicillinases (E.C. 3.5.2.6) Also known as β-lactamases, these are enzymes that hydrolyze (break down) the β-lactam ring (portion) of the penicillin molecule's structure. Some microorganisms (e.g., pathogenic bacteria) have become able to produce these enzymes as a defense to penicillin and cephalosporin antibiotics (drugs). See ENZYME, HYDROLYZE, PENICILLIN G, PATHOGENIC, BACTERIA, and ANTIBIOTIC.

Penicillium Refers to the genus of fungi (mold) that belongs to the category *Deuteromycotina*, and often causes (food) spoilage. Some of this genus have been utilized commercially to produce antibiotics. See GENUS, FUNGUS, ANTIBIOTIC, and PENICILLIN G.

Pentose A simple sugar (monosaccharide molecule) whose backbone structure contains five carbon atoms. There exists many different pentoses. Some examples of pentoses are: ribose, arabinose, and xylose, to mention just a few. See also MONOSACCHARIDES.

Pepsin A crystallizable proteinase (enzyme) that in an acidic medium digests (breaks down) most proteins to polypeptides. It is secreted by glands in the mucous membrane of the stomach of higher animals. In combination with dilute hydrochloric acid it is the chief active principle (component) of gastric juice. Also used in manufacturing peptones and in digesting gelatin for the recovery (i.e., recycling) of silver from photographic film. See also DIGESTION (WITHIN ORGANISMS), PROTEIN, PEPTIDE, and PEPTONE.

Peptidase An enzyme that hydrolyzes (cleaves) a peptide bond. See also PEPTIDE BOND, PEPSIN, PEPTONE, and PEPTIDE MAPPING ("FINGERPRINTING").

Peptide Two or more amino acids covalently joined by peptide bonds.

An oligomer component of a polypeptide. A dipeptide, for example, consists of two (di) amino acids joined together by a peptide bond or linkage. By analogy, this structure would correspond to two joined links of a chain. See also POLYPEPTIDE (PROTEIN), OLIGOMER, and AMINO ACID.

Peptide Bond A covalent bond (linkage) between the α-amino group of one amino acid and the α-carboxyl group of another amino acid. This is the linkage or bond which holds the amino acids (chain links) together in a polypeptide chain. It is the all-important bond which holds the amino acid monomers together to form the polymer known as a polypeptide. See also PEPTIDE, POLYPEPTIDE (PROTEIN), and OLIGOMER.

Peptide Mapping ("fingerprinting") The characteristic pattern of peptides resulting from partial hydrolysis (cleavage, digestion) of a protein. The pattern is obtained by separating the peptides by two-dimensional chromatography in which the peptides are first subjected to chromatography using one solution which separates many but not all peptides. The chromatogram is then turned 90 degrees and is again chromatographed using a second solution which then separates all of the peptides producing the final fingerprint of the protein. See also CHROMATOGRAPHY.

Peptido-Mimetic See BIOMIMETIC MATERIALS.

Peptone A protein that has been partially hydrolyzed (i.e., cleaved) by the peptidase pepsin. See also PROTEIN, HYDROLYTIC CLEAVAGE, PEPTIDASE, PEPSIN, and PEPTIDE MAPPING (FINGERPRINTING).

Perforin A 70 Kd (kilodalton) protein that is instrumental in the lysis of infected cells. A series of reactions occurs on the surface of a cell which results in the polymerization of certain monomers to form transmembrane (i.e., through the membrane) pores 100 Å (Angstroms) wide, which allows ions to rush into the cell (due to osmotic pressure) and thus burst (lyse) that cell, so the (formerly) internal pathogens can be attacked by the body's immune system. Perforin is a protein that is akin to the C9 component of the complement. See also OSMOTIC PRESSURE, COMPLEMENT (COMPONENT OF IMMUNE SYSTEM), COMPLEMENT CASCADE, Kd, CYTOTOXIC T CELLS, CECROPHINS (LYTIC PROTEINS), and MAGAININS.

Periodicity The number of base pairs per turn of the DNA double helix. See also DEOXYRIBONUCLEIC ACID (DNA).

Periodontium Tissue that anchors teeth in the jaw. Regrowth of periodontal tissue can be stimulated by a combination of platelet-derived growth factor and insulin-like growth factor-1. See PLATELET-DERIVED GROWTH FACTOR (PDGF) and INSULIN-LIKE GROWTH FACTOR-1 (IGF-1).

Peritoneal Cavity/Membrane The smooth, transparent, serous membrane that lines the cavity of the abdomen of a mammal.

Peroxidase An enzyme that is naturally produced in soybeans by approximately half of all commercial soybean varieties. Peroxidase very effectively inhibits (stops) growth of any *Aspergillus flavus* fungi that might be present. Peroxidase can be used to replace more toxic and environmentally problematic chemicals in certain industrial processes. Among other applications, peroxidase can replace formaldehyde use in paints, varnishes, glues, and computer chip manufacturing. See also ENZYME.

Persistence The tendency of a compound (e.g., an insecticide) to resist degradation by *biological* means (e.g., metabolism by microorganisms) after that compound has been introduced into the environment (e.g., sprayed onto a field) or by *physical* means (e.g., degradation caused by exposure to sunlight, moisture, etc.). See also METABOLISM, MICROORGANISM, and BIODEGRADABLE.

Peyer's Patches A special set of lymphoid organs found in the intestinal wall. They filter out antigens that enter the intestine in food or come from the bacteria growing naturally in the intestine, and "present" those intact antigens to adjacent lymphoid tissues, via special "M" cells of the Peyer's patches. This activates the lymphocytes in the patches, which then migrate out of the node and into the blood where they float in the tissue spaces just inside the intestinal lining. There they secrete antibodies (primarily IgA), which are then transported into the lumen (contents) of the gut and attack (bind) the antigen. See also HUMORAL IMMUNITY, CELLULAR IMMUNE RESPONSE, GUT-ASSOCIATED LYMPHOID TISSUES (GALT), ALLERGIES (FOODBORNE), and ANTIGEN.

Pfiesteria piscicida A single-celled microscopic algae which has a predator/prey relationship with fish in its ecosystem. During a large portion of its life cycle, *Pfiesteria piscicida* exists in a nontoxic cyst form at the bottom of a river. When those (cysts) detect certain substances (e.g., excreta) emitted by live fish, the *Pfiesteria piscicida* transform into an amoeboid or dinoflagellate form, which secretes a water-soluble neurotoxin into the water (which incapacitates nearby fish). The *Pfiesteria piscicida* next attach themselves to those fish, excrete a lipid-soluble toxin which destroys the epidermal layer of the fish's skin, allowing the *Pfiesteria piscicida* to begin "eating" the fish's tissue. Human exposure to the neurotoxin apparently causes short-term memory loss. See ECOLOGY, CELL, TOXIN, and LIPIDS.

PHA See POLYHYDROXYALKANOIC ACID (PHA).

Phage (bacteriophage) Another name for virus. A virus that attacks bacteria is known as a bacteriophage. Bacteriophages are frequently used as vectors for carrying (foreign) DNA into cells by genetic engineers. See also BACTERIOPHAGE, VECTOR, GENETIC ENGINEERING, and DEOXYRIBONUCLEIC ACID (DNA).

Phagocyte A cell such as a leukocyte that engulfs and digests cells, cell debris, microorganisms, and other foreign bodies in the bloodstream and tissues (phagocytosis). The ingested material is then degraded via enzymes. A whole class of cells is known to be phagocytic. See also

MACROPHAGE, MICROPHAGE, MONOCYTES, T CELLS, POLYMORPHONUCLEAR LEUKOCYTES (PMN), CELLULAR IMMUNE RESPONSE, POLYMORPHONUCLEAR GRANULOCYTES, and LYSOSOME.

Pharmacokinetics (pharmacodynamics) A branch of pharmacology dealing with the reactions between drugs or synthetic food ingredients and living structures (e.g., tissues, organs). The study of the:

- absorption—transport of the pharmaceutical or food ingredient into the bloodstream (e.g., from the intestinal tract, in the case of food ingredients)
- distribution—initial physical disposition/behavior of the substance in the body after the substance enters the body. For example, does the substance preferentially concentrate in the fat cells of the body? etc.
- metabolism—breakdown of the substance (if breakdown does occur) into other compounds, and ultimate disposition of those compounds (or the original substance, if breakdown does not occur). For example, some pharmaceuticals break down into smaller compound(s); one of which then acts upon the relevant body cells (e.g., to relieve pain, lower blood pressure, etc.).
- elimination—the speed and thoroughness with which the substance is excreted or otherwise removed from the body.

In short, pharmacokinetics deals with what happens to a substance that is introduced into a living system. For example, how quickly it is broken down, to what intermediates and metabolites it is broken down, and what the pathway of this breakdown is. See also PHARMACOLOGY, "ADME" TESTS, ABSORPTION, METABOLISM, INTERMEDIARY METABOLISM, DIGESTION (WITHIN ORGANISMS), and PHASE I CLINICAL TESTING.

Pharmacology The study of chemicals (e.g., pharmaceuticals) and their effects on living organisms. See also PHARMACOKINETICS (PHARMACODYNAMICS).

Pharmacophore The portion of a molecule (e.g., a pharmaceutical) that is responsible for its biological activity (i.e., therapeutic action on recipient's tissue, etc.). See also BIOLOGICAL ACTIVITY, ACTIVE SITE, CATALYTIC SITE, and MINIPROTEINS.

Phase I Clinical Testing The first in a series of human tests of new pharmaceuticals mandated by the Food and Drug Administration (FDA). The primary purpose of the Phase I clinical test is to detect if the new pharmaceutical is toxic or otherwise harmful to normal, healthy humans. The conclusion of Phase I testing leads to Phase II and Phase III testing. See also FOOD AND DRUG ADMINISTRATION (FDA), KEFAUVER RULE, KOSEISHO, BUNDESGESUNDHEITSAMT (BGA), COMMITTEE FOR PROPRIETARY MEDICINAL PRODUCTS (CPMP), IND, and IND EXEMPTION.

PHB See POLYHYDROXYLBUTYLATE.

Phenotype The outward appearance (structure) or other visible characteristics of an organism. See also GENOTYPE, MORPHOLOGY and GENE.

Phenylalanine (phe) An essential amino acid. L-Phenylalanine is one of the raw materials used to manufacture NutraSweet® (NutraSweet Co.) synthetic sweetener. See also LEVOROTARY (L) ISOMER, ESSENTIAL AMINO ACIDS, and STEREOISOMERS.

Pheromones From the Greek words "pherein" (to carry) and "hormon" (to excite), they are sex hormones emitted by insects and animals; and spread through the air by the wind and diffusion for the purposes of attracting the opposite sex. Some pheromones have been produced artificially and used in lure traps to attract and catch male insects so as to prevent their mating with females (i.e., a biological pesticide). Pheromone traps for Japanese beetles are commonplace in infested areas (e.g., when utilizing Integrated Pest Management). It is envisioned that commercial exploitation of this area of science will increase. See also HORMONE and INTEGRATED PEST MANAGEMENT (IPM).

Phosphate-Group Energy The decrease in free energy as one mole of a phosphorylated compound at 1.0 M concentration undergoes hydrolysis to equilibrium at pH 7.0 and 25°C (77°F). The energy that is available to do biochemical work. The energy arises from the breakage (cleavage) of a phosphate to phosphate bond. See also FREE ENERGY, HYDROLYSIS, and MOLE.

Phosphatidyl Choline See LECITHIN.

Phosphinothricin Another name for the herbicide active ingredient glufosinate. See GLUFOSINATE.

Phosphorylation The introduction of a phosphate group into a molecule. Formation of a phosphate derivative of a biomolecule, usually by enzymatic transfer of a phosphate group from ATP. See also ADENOSINE TRIPHOSPHATE (ATP).

Phosphorylation Potential (ΔG_p) The actual free-energy change of ATP hydrolysis under a given set of conditions. See also PHOSPHORYLATION, FREE ENERGY, HYDROLYSIS, and ADENOSINE TRIPHOSPHATE (ATP).

Photon A single unit of light energy. See also PHOTOSYNTHESIS and PHOTOSYNTHETIC PHOSPHORYLATION (PHOTOPHOSPHORYLATION).

Photoperiod The optimum length or period of illumination required for the growth and maturation of a plant. The photoperiod is distinct from photosynthesis. See also PHYTOCHROME.

Photophore See BIOLUMINESCENCE.

Photosynthesis The synthesis (production) of bioorganic compounds (molecules) using light energy as the power source. The synthesis of carbohydrates (hexose) occurs via a complicated, multistep process involving reactions that occur both in the light (light reactions) and in the dark

(dark reactions). In eucaryotic cells the photosynthetic machinery necessary to capture light energy and subsequently utilize it is contained in structures called chloroplasts, which contain the molecule that initially captures light energy, called chlorophyll. Chlorophyll looks green. Green plants synthesize carbohydrates from carbon dioxide and water, which are used as a hydrogen source. The synthesis reaction, which is light-driven, liberates oxygen in the process. Other organisms use this oxygen to sustain life. Plants are not the only users of photosynthesis technology. Other organisms such as green sulfur bacteria and purple bacteria also carry out photosynthesis, but they use other compounds besides water as a hydrogen source. See also CARBOHYDRATES (SACCHARIDES), CHLOROPLASTS, EUCARYOTE, and HEXOSE.

Photosynthetic Phosphorylation (photophosphorylation) The formation of ATP from the starting compounds ADP and inorganic phosphate (P_i). The formation is coupled to light-dependent electron flow in photosynthetic organisms. See also PHOTON, PHOTOSYNTHESIS, ADENOSINE TRIPHOSPHATE (ATP), and ADENOSINE DIPHOSPHATE (ADP).

Phylogenetic Constraint The limitations inherent in an organism as a result of what its ancestors were. For example, a horse will never fly and an ape will never speak, because the ancestors of neither possessed those capabilities. See also GENOTYPE, PHENOTYPE, GENOME, and MORPHOLOGY.

Physical Map (of genome) A diagram showing the linear order of genes or genetic markers on the genome, with units indicating the actual distance between the genes or markers. See also GENETIC MAP, GENE, GENOME, and POSITION EFFECT.

Physiology The branch of biology dealing with the study of the functioning of living things. The materials of physiology include all life: animals, plants, microorganisms, and viruses.

Phytase A digestive enzyme which is present in the digestive systems of many plant-eating animals to enable breakdown of phytate (also known as "phytic acid"). Phytase is sometimes present within the plant material consumed by animals. For example, phytase is naturally produced in the seed coat of wheat. See also ENZYME, DIGESTION (WITHIN ORGANISMS), PHYTATE, and HIGH-PHYTASE CORN/SOYBEANS.

Phytate A chemical complex (large molecule) substance that is the dominant (i.e., 60% to 80%) chemical form of phosphorus within cereal grains, oilseeds, and their byproducts. Monogastric animals (e.g., swine) cannot digest and utilize the phosphorus within phytate, because they lack the enzyme known as phytase in their digestive system, so that phosphorus (phytate) is excreted into the environment. When phytase enzyme is present in the ration of a monogastric animal, at a high enough level, the monogastric animal is then able to digest the phytate (thereby releasing that phosphorus for absorption by the animal). See also PHYTASE, ENZYME, DIGESTION (WITHIN ORGANISMS), and HIGH-PHYTASE CORN/SOYBEANS.

Phytic Acid See PHYTATE.

Phytochemicals A term used to refer to certain biologically active chemical compounds that occur in fruits, vegetables, grains, herbs, flowers, bark, etc. Phytochemicals act to repel or control insects, prevent plant diseases, control fungi and adjacent weeds. Phytochemicals also sometimes confer beneficial health effects to the animals (e.g., humans) that consume the plant (portions) containing those applicable phytochemicals.

For example, vitamin C in citrus fruits, beta carotene in carrots and other orange vegetables, *d*-limonene in orange peels, tannins in green tea, capsaicin in chili peppers, omega-3 fatty acids in fish oil, genistein in soybeans, etc.

Beta carotene has been found to aid eyesight and may help prevent lung cancer. *d*-Limonene has been found to protect rats against breast cancer. Tannins appear to prevent stomach cancer. Capsaicin can reduce arthritis pain. Omega-3 fatty acids help to lower triglyceride levels in the blood. Genistein appears to block growth of breast cancer tumors, and to prevent the loss of bone density that leads to the disease osteoporosis. Tocotrienols act as antioxidants, and also inhibit synthesis of cholesterol (in humans). See also CANCER, DEXTROROTARY (D) ISOMER, FATTY ACID, GENISTEIN (Gen), BIOLOGICAL ACTIVITY, MOLECULAR PHARMING, FLAVONOIDS, RESVERATROL, NUTRACEUTICALS, CHOLESTEROL, OMEGA-3 FATTY ACIDS, and GENISTEIN.

Phytochrome A protein plant pigment that serves to direct the course of plant growth and development and differentiation in a plant. The response is independent of photosynthesis, e.g., in the photoperiod (i.e., length of light period) response. See also PHOTOPERIOD, PROTEIN, and PHOTOSYNTHESIS.

Phytohormone See PLANT HORMONE.

Phytophthora Root Rot A plant disease that is caused by phytophthora fungus (*Phytophthora sojae*). Some soybean varieties are genetically resistant to as many as twenty-one races/strains of phytophthora fungus. See also FUNGUS, RPS1c GENE, RPS1k GENE, GENOTYPE, STRAIN, SOYBEAN PLANT, and RPS6 GENE.

Phytophthora sojae See PHYTOPHTHORA ROOT ROT.

Phytoplankton Algae that are floating or freely suspended in the water.

Picogram (pg) 10^{-12} gram or 3.527×10^{-14} ounce (avoirdupoir). See also MICROGRAM.

Picorna A family of the smallest known viruses. The viruses of this family are a cause of the common cold and Hepatitis A in humans, one form of foot and mouth disease in animals, and at least one disease in corn (maize). In 1994, Dr. Asim Dasgupta discovered a cellular molecule within ordinary backer's yeast that prevents picorna virus reproduction. This advance could lead to the creation of a treatment, in the future, to cure one or more of the above-mentioned diseases after infection has begun. See also VIRUS, CLADISTICS, and CLADES.

Pink Bollworm See *PECTINOPHORA GOSSYPIELLA*.

Pink Pigmented Facultative Methylotroph (PPFM) A type of bacteria that is naturally present in virtually all plants. PPFM produces cytokinin, which aids the cell division (growth) process in plants. PPFM also produces a chemical substance similar to vitamin B-12.

In 1996, Joe Polacco discovered that impregnation of aged seeds with PPFM improved the germination (sprouting) rate of those aged seeds. See also BACTERIA, MITOSIS, CELL DIFFERENTIATION, and VITAMIN.

Pituitary Gland One of the endocrine glands, it lies beneath the hypothalamus (at the base of the brain). Along with the other endocrine glands, the pituitary helps to control long-term bodily processes. This control is accomplished via interdependent secretion of hormones along with the other glands comprising the total endocrine system. For example, the pituitary helps to control the body's growth from birth until the end of puberty, by secreting growth hormone (GH). Secretion of GH by the pituitary is itself governed by the hormone known as growth hormone–releasing factor (GHRF), received by the pituitary gland from the hypothalamus.

The pituitary gland also helps to control reproduction (e.g., development and growth of ovaries, timing of ovulation, maturation of oocytes, etc.) by secreting two gonadotropic (reproductive) hormones named luteinizing hormone (LH) and follicle-stimulating hormone (FSH). Secretion of LH and FSH by the pituitary is itself governed by the hormones gonadotropin-releasing hormone (GnRH, received by pituitary from the hypothalamus) and estrogen/progesterone (received by pituitary from the ovaries). See also ENDOCRINE GLANDS, ENDOCRINE HORMONES, HORMONE, ENDOCRINOLOGY, HYPOTHALAMUS, FOLLICLE-STIMULATING HORMONE (FSH), ESTROGEN, GROWTH HORMONE–RELEASING FACTOR (GHRF), and GROWTH HORMONE (GH).

Plant Breeder's Rights (PBR) The intellectual property rights that are legally accorded to plant breeders by laws, treaties, etc. Similar to patent law for inventors. See PLANT'S NOVEL TRAIT (PNT), PLANT VARIETY PROTECTION ACT (PVP), EUROPEAN PATENT CONVENTION, EUROPEAN PATENT OFFICE (EPO), U.S. PATENT AND TRADEMARK OFFICE (USPTO), and UNION FOR PROTECTION OF NEW VARIETIES OF PLANTS (UPOV).

Plant Hormone An organic compound that is synthesized in minute quantities by the plant. It influences and regulates plant physiological processes. Also called a phytochrome. The four general types of hormones that together influence cell division, enlargement, and differentiation are the auxins, gibberellins, kinins, and abscisic acid, See also HORMONE, GIBBERELLINS, and PHYTOCHROME.

Plant Variety Protection Act (PVP) A law passed by the United States' Congress in 1970 that enables intellectual property protection (analogous to patent protection) for new seed plants and seeds in America. See also U.S. PATENT AND TRADEMARK OFFICE (USPTO), EUROPEAN PATENT CONVENTION, EUROPEAN PATENT OFFICE (EPO), PLANT'S NOVEL TRAIT (PNT), PLANT BREEDER'S RIGHTS (PBR), and UNION FOR PROTECTION OF NEW VARIETIES OF PLANTS (UPOV).

Plantibody See ANTIBIOTIC.

Plant's Novel Trait (PNT) The new (novel) trait added to a plant (e.g., crop plant such as cotton, corn/maize, soybean, etc.). Example novel traits are herbicide-tolerance (via CP4 EPSPS, PAT gene, etc.), insect resistance (via *B.t.* gene, cholesterol oxidase gene, etc.), resistance to aluminum toxicity (via CSb gene, etc.), among others. See TRAIT, CORN, SOYBEAN PLANT, CP4 EPSPS, GENE, PAT GENE, *B.t.*, *BACILLUS THURINGIENSIS (B.t.)*, EVENT, and CITRATE SYNTHASE (CSb) GENE.

Plasma A pale, amber-colored fluid constituting the fluid portion of the blood in which are suspended the cellular elements. Plasma contains 8–9 percent solids. Of these, 85 percent are proteins consisting of three major groups, which are: fibrinogen, albumin, and globulin. The other components are the lipids, which include the neutral fats, fatty acids, lecithin, and cholesterol. Also present are sodium, chloride and bicarbonate, potassium, calcium, and magnesium. A most essential function of plasma is the maintenance of blood pressure and the exchange (with tissues) of nutrients for waste. See also ABSORPTION and HOMEOSTASIS.

Plasma Membrane A thin structure that completely surrounds the cell as a "skin." They may be seen with the aid of an electron microscope. The entire membrane appears to be about 100 Angstroms (Å; 0.1 μm) thick and is composed of two dark lines each about 30 Å thick which are, however, separated by a lighter area. This trilaminar "sandwich" structure is referred to as the unit membrane. The plasma membrane is composed of lipoidal (fat-like) material in which proteins and protein complexes and whole functional systems are embedded. In the plasma membrane are incorporated such energy-dependent transport systems as Na^+ and K^+ transporting ATPase and amino acid transport systems. Besides the cell, membranes surround such systems as the endoplasmic reticulum, vacuoles, lysosomes, golgi bodies, mitochondria, chloroplasts, and the nucleus, to mention just a few. The plasma membrane and membranes in general function in part as a permeability barrier to the free movement of substances between the inside and exterior of the cell or organelles that they surround.

Plasmid An independent, stable, self-replicating piece of DNA in bacterial cells that is not a part of the normal cell genome and that never becomes integrated into the host chromosome. This is in contrast to a similar genetic element known as an episome plasmid which may exist independently of the chromosome or may become integrated into the host chromosome. Plasmids are known to confer resistance to antibiotics and may be transferred by cell-to-cell contact (by conjugation via the sex pilus) or by viral-mediated transduction. Plasmids are commonly used in recombinant DNA experiments as acceptors of foreign DNA. Known forms of plasmids include both linear and circular molecules. See also EPISOME (OF A BACTERIUM), VECTOR, COPY NUMBER, MULTI-COPY PLASMIDS, DEOXYRIBONUCLEIC ACID (DNA), CELL, GENOME, CHROMOSOME, and ANTIBIOTIC.

Plasmocyte Another name for a blast cell. See BLAST CELL.

Plastid An independent, stable, self-replicating piece of DNA inside a plant cell that is not a part of the normal cell genome (i.e., in nucleus). Because there can exist up to 10,000 plastids in a given plant cell, the insertion of a gene (e.g., via genetic engineering) into plastids can result in a higher yield (of the protein coded for by that gene) than is achieved via insertion of the gene into the cell's nucleus DNA. See also DEOXYRIBONUCLEIC ACID (DNA), CELL, COPY NUMBER, GENOME, PROMOTER, GENE, and GENETIC ENGINEERING.

Platelet Activating Factor (PAF) See CHOLINE.

Platelet-Derived Growth Factor (PDGF) An angiogenic growth factor produced by the blood's platelet cells which attracts the growth of capillaries into the vicinity of a fresh wound. This action releases still other growth factors, and starts the process of building a fibrin network, to support the subsequent (blood) clot. PDGF is a competence factor (i.e., a growth factor that is required to make a cell able or competent to react to other growth factors). PDGF is normally contained within the platelet cells, so does not circulate in the blood in a form enabling it to be freely available to its "target cells." This "containment" of PDGF in platelets ensures site-specific delivery of the PDGF directly to a wound site, so stimulus (i.e., of capillary growth) is localized to the actual wound site. After PDGF has caused the formation of the initial clot at a wound site, PDGF attracts connective tissue cells into the vicinity of the wound (to start the tissue-repair process). PDGF also acts as a mitogen (substance causing cell to divide and thus multiply) for connective tissue cells, granulocytes, and monocytes (each of which is involved in the wound's healing process). See also ANGIOGENIC GROWTH FACTORS, FIBRIN, FIBRONECTIN, PLATELETS, MITOGEN, GRANULOCYTES, and MONOCYTES.

Platelet-Derived Wound Growth Factor (PDWGF) See PLATELET-DERIVED GROWTH FACTOR (PDGF).

Platelet-Derived Wound Healing Factor (PDWHF) See PLATELET-DERIVED GROWTH FACTOR (PDGF).

Platelets Disk-shaped blood cells that stick to the (microscopically jagged) edges of wounds. The aggregation of platelets at the wound site leads to blood clotting, forming a temporary wound covering. During this blood clotting process, the platelets release platelet-derived growth factor (PDGF) which attracts fibroblasts to the wound area (for subsequent healing process). See also FIBRIN, FIBRONECTIN, PLATELET-DERIVED GROWTH FACTOR (PDGF), and FIBROBLASTS.

Pleiotropic Adjective used to describe a gene that affects more than one trait (apparently unrelated) characteristic of the phenotype (appearance of an organism). For example, biologist David Ho in 1993 discovered a single gene in the barley (*Hordeum vulgare*) plant that controls the traits of the plant's height, drought resistance, strength, and time to maturity. See also GENE, GENETIC CODE, DEOXYRIBONUCLEIC ACID (DNA), INFORMATIONAL MOLECULES and PHENOTYPE.

PNT See PLANT'S NOVEL TRAIT (PNT).

Point Mutation A mutation consisting of a change of only one nucleotide in a DNA molecule. See also MUTATION, HEREDITY, MUTANT, MUTAGEN, DEOXYRIBONUCLEIC ACID (DNA), NUCLEOTIDE, BASE EXCISION SEQUENCE SCANNING (BESS), and SITE-DIRECTED MUTAGENESIS (SDM).

"Points to Consider" Document See POINTS TO CONSIDER IN THE MANUFACTURE AND TESTING OF MONOCLONAL ANTIBODY PRODUCTS FOR HUMAN USE.

Points to Consider in the Manufacture and Testing of Monoclonal Antibody Products for Human Use The Food and Drug Administration's (FDA's) governing rules for IND (investigational new drug) submission for monoclonal antibody (MAb) based pharmaceuticals. See also IND.

Polar Group A hydrophilic ("water loving") portion of a molecule; it may carry an electrical charge. A group that "likes" to be in the presence of water molecules or other polar compounds. See also NONPOLAR GROUP, POLARITY (CHEMICAL), POLAR MOLECULE (OR DIPOLE), AMPHIPATHIC MOLECULES, and AMPHOTERIC COMPOUND.

Polar Molecule (or dipole) A molecule in which the centers of positive and negative (electrical) charge do not coincide, so that one end of the molecule carries a positive (or partial positive) charge and the other end a negative (or partial negative) charge. See also POLARITY (CHEMICAL), POLAR GROUP, ION-EXCHANGE CHROMATOGRAPHY, and NONPOLAR GROUP.

Polar Mutation A mutation in one gene which, because transcription occurs only in one direction, reduces the expression of subsequent genes in the same transcription unit further down the line. See also TRANSCRIPTION, TRANSLATION, EXPRESS, and NUCLEIC ACIDS.

Polarimeter An instrument used for measuring the degree of rotation of plane-polarized light by an optically active compound/solution. See also STEREOISOMERS, OPTICAL ACTIVITY, LEVOROTARY (L) ISOMER, and DEXTROROTARY (D) ISOMER.

Polarity (chemical) The degree to which an atom or molecule bears an electrical charge or a partial electrical charge. In general, the more polar (i.e., separation or partial separation of charge) a molecule is, the more hydrophilic ("water loving") it is. Polarity results from an uneven distribution of electrons between the atoms comprising a molecule. See also POLAR GROUP, HYDROPHILIC, and POLAR MOLECULE (OR DIPOLE).

Polarity (genetic) Having to do with the one way or unidirectionality of gene transcription in an operon unit. That is, the region near the operator is always transcribed before the more distant regions. By analogy, transcription begins at the left end of an operon unit and proceeds (reads, transcribes) toward the right end of the operon unit. The distinction between the 5′ and the 3′ ends of nucleic acids. See also POLAR MUTATION and TRANSCRIPTION.

Polyacrylamide Gel A "sieving" gel, that is used in electrophoresis. See POLYACRYLAMIDE GEL ELECTROPHORESIS (PAGE).

Polyacrylamide Gel Electrophoreis (PAGE) A form of chromatography in which molecules are separated on the basis of size and charge. The stationary phase (the polyacrylamide gel) is a polymerized version of acrylamide monomers. The gel looks and feels like Jello™. On a molecular basis it consists of an intertwined and cross-linked mesh of polyacrylamide strings. As can be imaged, there are holes in the gel (like in a plastic mesh bag) and with enough cross-linking the size of the holes begins to approach the size of the molecules which are to be separated. Since some molecules will be larger and some smaller, some of them will be able to pass through the gel matrix more easily than others. This is part of the basis for separation. It should be noted at this point that if the gel is cross-linked enough and because of this the holes in that gel are smaller than the molecules to be separated, then the molecules will not be able to penetrate into the gel and no separation can occur. The charge on the molecule also plays a role in the separation. Functionally, the gel serves to hold and separate the molecules. Although details are not presented here, after the gel has been prepared (poured and cross-linked) a small amount of the solution containing the molecules to be separated is placed into wells (grooves to hold the liquid) on the gel and the system is subjected to an electric current. Over the course of minutes to hours molecules bearing different charge/mass separate. See also BIOLUMINESCENCE, CHROMATOGRAPHY, FIELD INVERSION GEL ELECTROPHORESIS, and ELECTROPHORESIS.

Polyadenylation The addition of a sequence of polyadenylic acid to the

3′ end of a eucaryotic mRNA after its transcription (post-transcriptional). See also MESSENGER RNA (mRNA) and TRANSCRIPTION.

Polycistronic Coding regions representing more than one gene in mRNA (i.e., they code for two or more polypeptide chains). Many mRNA molecules in procaryotes are polycistronic. See also RIBOSOMES and PROCARYOTES.

Polyclonal Antibodies (used in humans) A mixture of antibody molecules (that are specific for a given antigen) that has been purified from an immunized (to that given antigen) animal's blood. Such antibodies are polyclonal in that they are the products of many different populations of antibody-producing cells (within the animal's body). Hence they differ somewhat in their precise specificity and affinity for the antigen. Years ago, antibodies (then called antitoxin) that were purified from an immunized animal's blood (e.g., a horse) were injected into humans suffering from certain diseases (e.g., diphtheria). In these cases the pathogen had caused disease by secreting large amounts of toxin into the victim's bloodstream. The antitoxin combined quantitatively (e.g., 1:1, 2:1, 1:2, 1:3, 3:1, etc.) with, and neutralized the toxin (for those few diseases for which it was applicable). Vaccines are now used instead, because of the adverse immune response caused by the horse's blood (antigens), See also ANTIBODY, PASSIVE IMMUNITY, MONOCLONAL ANTIBODIES (MAb), ANTIGEN, PATHOGEN, and TOXIN.

Polyclonal Response (of immune system to a given pathogen) Because a given pathogen generally has several antigenic sites on its surface, the B lymphocytes (activated by helper T cells in response to a pathogen invading the body) synthesize several (subtly different) antibodies against that pathogen. And since the antibodies are made by different cells the response is known as poly(many)clonal. See also PATHOGEN, ANTIGEN, ANTIBODY, HAPTEN, EPITOPE, HELPER T CELLS (T4 CELLS), LYMPHOCYTE, B LYMPHOCYTES, and LYMPHOKINES.

Polyethylene-Glycol Superoxide Dismutase (PEG-SOD) See PEG-SOD (POLYETHYLENE GLYCOL SUPEROXIDE DISMUTASE) and HUMAN SUPEROXIDE DISMUTASE (hSOD).

Polygalacturonase (PG) An enzyme (e.g., present in tomatoes) that starts the breakdown (softening) of the fruit tissue. Recent advances make it possible to significantly delay the softening (i.e., spoilage) process by reducing the production of polygalacturonase through genetic engineering of the plant. In 1986, William Hiatt of the American company Calgene discovered the gene for polygalacturonase, which led to that company commercializing a tomato variety that had been genetically engineered to reduce production of polygalacturonase in that variety's tomatoes (in 1994). See also EPSP SYNTHASE, GENETIC ENGINEERING, ANTISENSE (DNA SEQUENCE), ENZYME, GENE, and ACC SYNTHASE.

Polygenic (trait, product, etc.) A trait or end product (e.g., in a grain-produced crop) that requires simultaneous expression of more than one gene. For example, the level of protein produced in soybeans is controlled by five genes. See also POLYHYDROXYLBUTYLATE (PHB), PROTEIN, SOYBEAN PLANT, GENE, TRAIT, SOYBEAN OIL, BCE4, *ARABIDOPSIS THALIANA*, and PLASTID.

Polyhydroxyalkanoates See POLYHYDROXYALKANOIC ACID (PHA).

Polyhydroxyalkanoic Acid (PHA) An "energy storage" substance that is naturally produced by certain bacteria (90 strains known). When PHA is removed from the bacteria and purified, this substance has physical properties quite similar to thermoplastics like polystyrene. PHA can quickly be broken down by soil microorganisms, so PHA is a biodegradable plastic. See also STARCH, BACTERIA, BIOPOLYMER, POLYHYDROXYLBUTYLATE (PHB), and MICROORGANISM.

Polyhydroxylbutylate (PHB) An "energy storage" substance that is naturally produced by certain bacteria, yeasts, and plants. When removed from the bacteria and purified, this substance has physical properties quite similar to thermoplastics like polystyrene. PHB can quickly be broken down by soil microorganisms, so PHB is a biodegradable plastic. Three separate enzymes are utilized by the organism in order to make the PHB molecule. In 1994, researchers succeeded in transferring genes for PHB production into the weed plant *Arabidopsis thaliana* and the crop plant rapeseed (canola). See also STARCH, BACTERIA, BIOPOLYMER, ENZYME, POLYGENIC, MICROORGANISM, POLYHYDROXYALKANOIC ACID (PHA), CANOLA, and *ARABIDOPSIS THALIANA*.

Polymer A molecule possessing a regular, repeating, covalently bonded arrangement of smaller units called monomers. By analogy, a chain (polymer) that is composed of links (monomer) hooked together. See OLIGOMER and PROTEIN.

Polymerase An enzyme that catalyzes the assembly of nucleotides into RNA (RNA polymerase) and of deoxynucleotides into DNA (DNA polymerase). See DNA POLYMERASE, RNA POLYMERASE, REVERSE TRANSCRIPTASES, DNA, RNA, and *TAQ*.

Polymerase Chain Reaction (PCR) A reaction that uses the enzyme DNA polymerase to catalyze the formation of more DNA strands from an original one by the execution of repeated cycles of DNA synthesis. Functionally, this is accomplished by heating and melting double-stranded (hydrogen bonded) DNA into single-stranded (nonhydrogen bonded) DNA and producing an oligonucleotide primer complementary to each DNA strand. The primers bind to the DNA and mark it in such a way that the addition of DNA polymerase and deoxynucleoside triphosphates cause a new strand of DNA to form which is complementary to the target section of DNA. The process described previously is repeated again and again to produce millions of copies of the desired strand of DNA. PCR and its registered trademarks are the property of F. Hoffmann-La Roche & Co. AG, Basel, Switzerland. See also POLYMERASE CHAIN REACTION (PCR) TECHNIQUE, DEOXYRIBONUCLEIC ACID (DNA), DNA PROBE, PROBE, Q-BETA REPLICASE TECHNIQUE, COCLONING (OF MOLECULES), and POSITIVE AND NEGATIVE SELECTION (PNS).

Polymerase Chain Reaction (PCR) Technique Developed in 1984 and 1985 by Kary B. Mullis, Randall K. Saiki, Stephen J. Scharf, Fred A. Faloona, Glenn Horn, Henry A. Erlich, and Norman Arnheim, the PCR technique is an *in vitro* method that greatly amplifies (makes millions of copies of) DNA sequences that otherwise could not be detected or studied. It can be utilized to amplify a given DNA sequence that constitutes less than one part per million of initial sample (e.g., a 100-base-pair target DNA sequence within the genome of one of the higher organisms, which can contain up to 500 million base pairs). The procedure alleviates the necessity of *in vivo* replication of a target DNA sequence, or of replication of one-of-a-kind tiny DNA samples (e.g., from a crime scene). See also *IN VITRO, IN VIVO*, POLYMERASE CHAIN REACTION (PCR), DEOXYRIBONUCLEIC ACID (DNA), BASE PAIR (bp), GENOME, SEQUENCE (OF A DNA MOLECULE), *TAQ*, and DNA POLYMERASE.

Polymorphism (chemical) The property of a chemical substance crystallizing (or simply existing) in two or more forms having different structures. For example, diamond and graphite are two different structures (manifestations) of the element carbon. Deoxyribonucleic acid (DNA) is a polymorphic compound because the polymer can take on different forms (see A-DNA, B-DNA, and Z-DNA). See also DEOXYRIBONUCLEIC ACID (DNA), DNA PROFILING, and POLYMORPHISM (GENETIC).

Polymorphism (genetic) A name applied to a condition in which a species of plant or animal is represented by several distinct, nonintegrating forms or types unrelated to age or sex. The differences are often in coloration, though any characteristic of the organism may be involved (e.g., nuclei shape for polymorphonuclear leukocytes). See also POLYMORPHONUCLEAR LEUKOCYTES (PMN), POLYMORPHONUCLEAR GRANULOCYTES, and POLYMORPHISM (CHEMICAL).

Polymorphonuclear Granulocytes Neutrophils, eosinophils, and basophils are collectively known as polymorphonuclear granulocytes. This is due to the fact that collectively their nuclei are segmented into lobes and they have granule-like inclusions within their cytoplasm. See also GRANULOCYTES, BASOPHILS, EOSINOPHILS, NEUTROPHILS, and CYTOPLASM.

Polymorphonuclear Leukocytes (PMN) Formerly named microphages. Phagocytic (i.e., foreign particle–ingesting) white blood cells that have a lobed nucleus. For example, during an attack of the common cold (when virus first invades mucous membranes of the human nose), the body responds by making Interleukin-8 (IL-8); a glycoprotein that attracts large quantities of polymorphonuclear leukocytes to the mucous membranes of the nose (to try to combat the infection). See also CELLULAR IMMUNE RESPONSE, LEUKOCYTES (WHITE BLOOD CELLS), POLYMORPHISM (GENETIC), VIRUS, GLYCOPROTEIN, and INTERLEUKIN-8 (IL-8).

Polypeptide (protein) A molecular chain of amino acids linked by peptide bonds. Synonymous with protein. Via the synthesis (of this "chain") performed by ribosomes, each polypeptide (protein) in nature is the ultimate expression product of a gene. All of the amino acids commonly found in proteins have an asymmetric carbon atom, except the amino acid glycine. Thus, the polypeptide is potentially chiral in nature. See also PROTEIN, AMINO ACID, GENE, PEPTIDE, STEREOISOMERS, CHIRAL COMPOUND, EXPRESS, RIBOSOMES, POLYRIBOSOME (POLYSOME), and MESSENGER RNA (mRNA).

Polyribosome (polysome) A complex of a messenger RNA (mRNA) molecule on which ribosomes (ribosomal RNA; rRNA) are anchored. A number of ribosomes bound to only a single mRNA molecule. One mRNA molecule hence functions as a template for a number of polypeptide chains at one time. See RIBOSOMES, rRNA, and MESSENGER RNA (mRNA).

Polysaccharides Linear and/or branched (structure) macromolecules (i.e., large molecules) composed of many monosaccharide units (monomers such as glucose) linked by glycosidic bonds. See also GLYCOSIDE, MONOSACCHARIDES, AMYLOSE, and AMYLOPECTIN.

Polysome See POLYRIBOSOME.

Polyunsaturated Fatty Acids (PUFA) Unsaturated fatty acids, possessing more than one double bond, that impart a variety of health benefits to humans that consume them. For example, the "omega-3" (*n*-3) PUFAs possess antithrombotic effects and also reduce blood concentrations of triglycerides. High dietary levels (in human diet) of the "omega-6" (*n*-6) PUFAs have been related to decreased coronary heart disease (CHD). See also UNSATURATED FATTY ACIDS, ESSENTIAL FATTY ACIDS, THROMBOSIS, and TRIGLYCERIDES.

Porcine Somatotropin (PST) A hormone, produced in the pituitary gland of pigs, that increases a swine's muscle tissue production efficiency. Injecting this hormone causes a faster growing, leaner pig.

Porphyrins Complex nitrogenous compounds containing four substituted pyrroles covalently joined into a ring structure. When complexed with a central metal atom it is called a metalloporphyrin.

Position Effect A change in the expression of a gene that is brought about by its translocation to a new site in the genome. For example, a previously active gene may become inactive if placed on a new site in the genome. See also GENOME, TRANSLATION, GENETIC MAP, MAP DISTANCE, and PROMOTER.

Positional Cloning A technique used by researchers to zero in on the gene(s) responsible for a given trait or disease. A genetic map of the organism's genome is used to make an educated guess as to the precise location of the gene of interest (e.g., near marker __ or __, etc.). Then those guessed genes are cloned, inserted into living organisms or cells, and tested to see if the guessed gene causes expression of the protein of interest (e.g., a protein that causes the disease that the researcher is attempting to cure). See also CLONE (A MOLECULE), GENE, GENE AMPLIFICATION, GENE DELIVERY (GENE THERAPY), DNA PROBE, GENE MACHINE, GENETIC ENGINEERING, GENETIC MAP, GENETIC MARKER, GENOME, MAP DISTANCE, FUNCTIONAL GENOMICS, POSITION EFFECT, and EXPRESS.

Positive and Negative Selection (PNS) A separation technique; a technique to speed up the task of selecting, from thousands of laboratory specimens, the few cells with precisely the desired genetic changes induced (via genetic engineering). The thousands of genetically altered cells are brought about (produced) by genetic engineering experiments. Many genetic alterations are accomplished by injecting or flooding (specimen) cells with fragments of new genetic material (genes). A few cells are produced that have precisely the desired genetic changes among a large number of cells that do not have the desired changes. Sort of like a "needle in a haystack." By analogy, the few cells possessing the desired trait represent the needles while the multitude of cells not possessing the trait represent the hay. In order to isolate the few desired cells the needles must be separated from the hay. PNS gets rid of the nondesired cells and leaves only the cells possessing the desired genetic change. This is accomplished in the following way. The pieces of newly injected genetic material are composed not only of the desired sequence of DNA, but also another piece of DNA (known as a marker) which renders only those cells possessing the desired (genetic) change resistant to certain antibiotic drugs (such as neomycin) and certain antiviral drugs (e.g., gancyclovir). When all of the engineered cells are exposed to the drug (which normally kills all of the cells) only those cells possessing the desired genetic change (and the concomitant piece of DNA providing drug resistance) survive and hence are "selected." The other cells not having the drug resistance are selected against, and die. See also GENETIC ENGINEERING, GENE, MARKER (GENETIC MARKER), Q-BETA REPLICASE TECHNIQUE, and POLYMERASE CHAIN REACTION (PCR) TECHNIQUE.

Positive Supercoiling Occurs in double-stranded cyclic DNA molecules having no breaks at all in either strand. If the double helix (of DNA) is wound further in the same direction as the winding of the two strands of the double helix molecule, then the circular duplex itself takes on superhelical turns. By analogy, supercoiling or superhelicity may be described as follows. A piece of rope composed of two or three smaller strands of rope are wound around each other to yield the finished rope. This is equivalent to the normal double-stranded DNA. If the ends of the rope are then joined or tied together and the resultant circle of rope is again wound in the same direction as the winding that produced the rope in the first place, supercoils will be formed and the rope will become a much thicker (supercoiled) but shorter piece of rope. See also DOUBLE HELIX.

Post-Transcriptional Processing (Modification) of RNAs The enzyme-catalyzed processing or structural modifications that RNAs such as mRNAs, rRNAs, and tRNAs must undergo before they are functionally finished products. For example, in eucaryotes a block of poly A containing at least 200 AMP residues is enzymatically attached to the 3′ end of mRNA in the nucleus of the cell. The mRNAs with the "tail" are then transferred to the cytoplasm and the tail enzymatically removed to form the functional mRNAs. It is believed that the poly A tail aids in the transfer of the complex and/or targets the complex to the cytoplasm. See also POST-TRANSLATIONAL MODIFICATION OF PROTEIN MRNA, rRNA, and tRNA.

Post-Translational Modification of Protein Enzymatic processing of a polypeptide chain after its translation from its mRNA, i.e., addition of carbohydrate moieties to the protein or the removal of a portion of the polypeptide chain in order to produce a functional protein in the correct environment. See also POLYPEPTIDE (PROTEIN), MOIETY, MESSENGER RNA (mRNA), ENZYME, RIBOSOMES, CARBOHYDRATES, and GLYCOPROTEIN.

PPFM See PINK PIGMENTED FACULTATIVE METHYLOTROPH.

Pribnow Box The consensus sequence T-A-T-A-A-T-G centered about 10 base pairs before the starting point of bacterial genes. It is a part of the promoter and is especially important in binding RNA polymerase. See also RNA POLYMERASE, TATA HOMOLOGY, HOMEOBOX, PROMOTER, and BASE PAIR (bp).

Primary Structure The sequence of amino acids in a protein chain. See also POLYPEPTIDE (PROTEIN).

Primer (DNA) A short sequence deoxyribonucleic acid (DNA) that is paired with one strand of the template DNA. It is the growing end of the DNA chain and it simply provides a free 3′—OH end at which the enzyme DNA polymerase adds on deoxyribonucleotide units (monomers). Which deoxyribonucleotide is added is dictated by base pairing to the template DNA chain. Without a DNA primer sequence a new DNA chain cannot form since DNA polymerase is not able to initiate DNA chains. See also POLYMERASE, POLYMERASE CHAIN REACTION (PCR), and POLYMERASE CHAIN REACTION (PCR) TECHNIQUE.

Prion Proteinaceous structures (molecules) found in the membrane (surface) of cells, in the brains of all vertebrate animals. In 1982, Dr. Stanley Prusiner discovered that misshapen (mutated) versions could cause the neurodegenerative disease Bovine Spongiform Encephalopathy (BSE) in cattle, and the neurodegenerative diseases Creutzfeld-Jakob Disease (CJD), kuru, Gerstmann-Straussler-Scheinker Syndrome, and Fatal Familial Insomnia (FFI) in humans. Dr. Prusiner named these molecules prions for "proteinaceous infected particle," because unlike infectious pathogenic bacteria or viruses, prions do not contain DNA. The dye named Congo Red, and IDX (a derivative of the chemotherapeutic doxorubicin) have shown some ability to slow prion-caused neurodegeneration. See also PROTEIN, MUTANT, BACTERIA, DEOXYRIBONUCLEIC ACID (DNA), PROTEIN STRUCTURE, BSE, PROTO-ONCOGENES, STRESS PROTEINS, and MONOCLONAL ANTIBODIES.

Probe A relatively small molecule that can be used to sense the presence and condition of a specific protein or nucleic acid by a unique interaction with that macromolecule. See also DNA PROBE and HYBRIDIZATION (MOLECULAR BIOLOGY).

Process Validation (for production of a pharmaceutical) Defined by America's Food and Drug Administration (FDA) as "Establishing documented evidence which provides a high degree of assurance that a specific process will consistently produce a (pharmaceutical) product meeting pre-determined specifications and quality characteristics." See FOOD AND DRUG ADMINISTRATION (FDA), GOOD MANUFACTURING PRACTICES (GMP), GOOD LABORATORY PRACTICES (GLP), and cGMP.

Procaryotes Simple organisms that lack a distinct nuclear membrane and other organelles. Many structural systems are different between procaryotes and eucaryotes including the DNA arrangement, composition of membranes, the respiratory chain, the photosynthetic apparatus, ribosome size, the presence or lack of cytoplasmic streaming, the cell wall, flagella, the mode of sexual reproduction, and the presence or lack of vacuoles. Some representative procaryotes are the bacteria and blue-green algae. See EUCARYOTE.

Progesterone A female sex hormone, secreted by the ovaries, that supports pregnancy and lactation (i.e., milk production). See also HORMONE, PITUITARY GLAND, and ESTROGEN.

Programmed Cell Death See p53 GENE and APOPTOSIS.

Promoter The region on DNA to which RNA polymerase binds and initiates transcription. The promoter "promotes" the transcription (expression) of that gene. A region of DNA (deoxyribonucleic acid) which lies "upstream" of the transcriptional initiation site of a gene. The promoter controls *where* (e.g., which portion of a plant, which organ within an animal, etc.) and *when* (e.g., which stage in the lifetime of an organism) that the gene is expressed. For example, the promoter named "Bce4" is "seed-specific" [i.e., it only "promotes" the expression of a given gene's product (e.g., protein, fatty acid, amino acids, etc.) within a plant's seed]. See also POLYMERASE, GENE, EXPRESS, RNA POLYMERASE, BCE4, PLASTID, DEOXYRIBONUCLEIC ACID (DNA), POLYGENIC (TRAIT, PRODUCT, ETC.), and TRANSCRIPTION.

Proof-Reading Any mechanism for correcting errors in nucleic acid synthesis that involves scrutiny of individual (chemical) units after they have been added to the (molecular) chain. This function is carried out by a 3′ to 5′ exonuclease, among others. Proof-reading dramatically increases the fidelity of the base pairing mechanism. See also SEQUENCING (OF DNA MOLECULES).

Prostate-Specific Antigen (PSA) An antigen whose concentration increases significantly five to ten years prior to the diagnosis of prostate cancer. This means that PSA level measurements could lead to a diagnosis of prostate cancer before symptoms appear. However, a series of tests is required in order to accurately gauge the probability of cancer because PSA levels can also be elevated when a man develops a noncancerous enlarged prostate. See also ANTIGEN, TUMOR, and TUMOR-ASSOCIATED ANTIGENS.

Prosthetic Group A heat-stable metal ion or an organic group (other than an amino acid) that is covalently bonded to the apoenzyme protein. It is required for enzyme function. The term is now largely obsolete. See also ION, AMINO ACID, PROTEIN, ENZYME, APOENZYME and COENZYME.

Protease An enzyme that catalyzes the hydrolytic cleavage (breakdown) of proteins. By analogy, the enzyme breaks the link (peptide bond) holding a chain together. Proteases represent a whole class of protein-degrading enzymes. See also HYDROLYTIC CLEAVAGE and PEPTIDE BOND.

Protease Nexin I (PN-I) A protein that acts as an inhibitor of protease. See also PROTEASE, PROTEIN, and PROTEASE NEXIN II (PN-II).

Protease Nexin II (PN-II) A protein that is thought to regulate important activities in the body and brain by inhibiting specific enzymes and interacting with certain body cells. PN-II is formed from a precursor molecule known as beta-amyloid, via metabolic processing of the beta-amyloid. Recent research indicates that incorrect metabolic processing of beta-amyloid by the body results in amyloid plaques in the brain. The amyloid plaques are generally found in victims of Alzheimer's disease, and directly correlate (in number) with the degree of dementia. See also PROTEASE NEXIN I (PN-I), REGULATORY ENZYME, PROTEIN, ENZYME, INHIBITION, and METABOLISM.

Protein From the Greek word *proteios*, which means "the first" or "the most important." Any of a class of high molecular weight polymer compounds composed of a variety of α-amino acids joined by peptide linkages. Via the synthesis (of this "chain") performed by ribosomes, each protein is the ultimate expression product of a gene. Proteins are the "workhorses" of living systems and include enzymes, antibodies, receptors, peptide hormones, etc. All of the amino acids commonly found in proteins have an asymmetric carbon atom, except the amino acid glycine. Thus the protein is potentially chiral in nature. See also AMINO ACID, GENE, PEPTIDE, ABSOLUTE CONFIGURATION, STEREOISOMERS, CHIRAL COMPOUND, EXPRESS, OLIGOMER, PROTEIN FOLDING, MESSENGER RNA (mRNA), RIBOSOMES, and POLYRIBOSOME (POLYSOME).

Protein C An anticlotting (glyco) protein that prevents post-operative arterial clot formation when administered intravenously. May be synergistic (in its anticlotting effect) with tissue plasminogen activator (tPA). See also THROMBOMODULIN, TISSUE PLASMINOGEN ACTIVATOR (tPA), PROTEIN, and GLYCOPROTEIN.

Protein Engineering The selective, deliberate (re)designing and synthesis of proteins. This is done in order to cause the resultant proteins to carry out desired (new) functions. Protein engineering is accomplished by changing or interchanging individual amino acids in a normal protein. This may be done via chemical synthesis or recombinant DNA technology (i.e., genetic engineering).

"Protein engineers" (actually genetic engineers) use recombinant DNA technology to alter a particular nucleoside or triplet (codon) in the DNA (genes) of a cell. In this way it is hoped that the resulting DNA codes for the different (new) amino acid in the desired location in the protein produced by that cell. See also PROTEIN, POLYPEPTIDE (PROTEIN), GENE, CODON, GENETIC ENGINEERING, AMINO ACID, ESSENTIAL AMINO ACIDS, and SYNTHESIZING (OF PROTEINS).

Protein Folding The complex interactions of a polypeptide molecular chain with its environment and itself and other protein entities, which cause the polypeptide molecule to fold up into a highly organized, tightly packed, three-dimensional structure. Proven to occur spontaneously, by Christian B. Anfinsen during the 1960s; for protein molecules outside of living cells. This ability of polypeptide chains to fold into a great variety of topologies, combined with the large number of sequences (in the molecular chain) that can be derived from the 20 common amino acids in proteins, confers on protein molecules their great powers of recognition and selectivity. How a protein folds up determines its chemical function. During the 1990s, it was discovered that inside living cells, "chaperone" molecules are needed for proper protein folding to occur. These chaperones are protein molecules (e.g., certain heat-shock proteins) that form a loosely bound complex to suppress incorrect protein folding as the protein molecule is emerging from the cell's ribosome, so protein folding is both complete and correct as soon as the newly formed protein molecule is released from the cell's ribosome. See also AMINO ACID, PROTEIN, POLYPEPTIDE (PROTEIN), RIBOSOMES, CHAPERONES, PRION, ABSOLUTE CONFIGURATION, CONFORMATION, ENZYME, and PROTEIN STRUCTURE.

Protein Inclusion Bodies See REFRACTILE BODIES (RB).

Protein Kinases Enzymes capable of phosphorylating (covalently bonding a phosphate group to) certain amino acid residues in specific proteins. See also PHOSPHORYLATION.

Protein Quality See AMINO ACID PROFILE.

Protein Sequencer See SEQUENCING (OF PROTEIN MOLECULES), GENE MACHINE, and SEQUENCING (OF DNA MOLECULES).

Protein Structure A polypeptide chain may take on a certain structure in and of itself because of the amino acid monomers it contains and their location within the chain. The chain may furthermore interact with other polypeptide chains to form larger proteins known as oligomeric proteins. In what follows, the levels of protein structure normally encountered will be highlighted:

- Primary structure—refers to the backbone of the polypeptide chain and to the sequence of the amino acids of which it is comprised.
- Secondary structure—refers to the shape (recurring arrangement in space in one dimension) of the individual polypeptide chain. In some cases, because of its primary structure, the chain may take on an extended or longitudinally coiled conformation.
- Tertiary structure—refers to how the polypeptide chain (the primary structure) is bent and folded in three-dimensional space in order to form the normal tightly folded and compact structure.
- Quaternary structures—refers to how, in larger proteins made up of two or more individual polypeptide chains, the individual polypeptide chains are arranged relative to each other. These large multipolypeptide proteins are called oligomeric proteins and the individual chains are called subunits. An example of such a protein is hemoglobin.

See also CONFORMATION, PROTEIN FOLDING, POLYPEPTIDE (PROTEIN), and CHAPERONES.

Proteolytic Enzymes Enzymes which catalyze the hydrolysis (breakdown) of proteins or peptides. Proteins (enzymes) that destroy the structure (by peptide bond cleavage) and hence the function of other proteins. These other proteins may or may not themselves be enzymes. See also PROTEASE and UBIQUITIN.

Proto-Oncogenes Cellular genes that can become cancer-producing. Proto-oncogenes are activated to oncogenes via different mechanisms, including point mutation, chromosome translocation, insertional mutation, and amplification. See also ONCOGENES, AMPLIFICATION, and MUTATION.

Protoplasm A general term referring to the entire contents of a living cell; living substance.

Protoplast A structure consisting of the cell membrane and all of the intracellular components, but devoid of a cell wall. This (removal of cell's outer wall) can be done to plant cells via treatment with cell-wall-degrading enzymes or electroporation. Under specific conditions (e.g., electroporation), certain DNA sequences (genes) prepared by man, can enter protoplasts. The cell then incorporates some or all of that DNA into its genetic complement (genome), and produces whatever product the newly introduced gene codes for.

In the case of plant protoplasts, whole plants can be regenerated from the (genetically engineered) protoplasts, resulting in plants that produce whatever product(s) the introduced gene(s) codes for. See also CELL, ENZYME, ELECTROPORATION, GENE, GENETIC ENGINEERING, DEOXYRIBONUCLEIC ACID (DNA), CODING SEQUENCE, PROTEIN, SOYBEAN PLANT, CORN, and CANOLA.

Protoxin A chemical compound that only becomes a toxin after it is altered in some way.

For example, the *B.t.* protoxins (e.g., Cry9C, Cry1A (b), Cry1A (c), etc.) only become toxic after they are chemically altered by the alkaline environment inside the gut of certain insects. See *BACILLUS THURINGIENSIS (B.t.)*, *B.t. KURSTAKI*, CRY PROTEINS, CRY1A (b) PROTEIN, CRY1A (c) PROTEIN, CRY9C PROTEIN, *B.T. ISRAELENSIS*, and *B.T. TENEBRIONIS*.

Protozoa A microscopic, single-celled animal form. A unicellular organism without a true cell wall, that obtains its food phagotropically. See also PHAGOCYTE.

PRR See PHYTOPHTHORA ROOT ROT.

PSA See PROSTATE-SPECIFIC ANTIGEN (PSA).

Pseudomonas aeruginosa See CITRATE SYNTHASE (CSb) GENE.

Pseudomonas fluorescens A normally harmless soil microorganism (bacteria) that colonizes the roots of certain plants. At least one company has incorporated the gene for a protein that is toxic to insects (taken from *Bacillus thuringiensis*) into a *Pseudomonas fluorescens*. This was done in order to confer insect resistance to the plants the roots of which the genetically engineered *Pseudomonas fluorescens* has colonized. See also *BACILLUS THURINGIENSIS (B.t.)*, BACTERIA, WHEAT TAKE-ALL DISEASE, GENETIC ENGINEERING, and ENDOPHYTE.

PST See PORCINE SOMATOTROPIN.

Psychrophile An organism that requires 0°C (32°F) for growth. See MESOPHILE and THERMOPHILE.

Pure Culture A culture containing only one species of microorganism. See also CULTURE and CULTURE MEDIUM.

Purine A basic nitrogenous heterocyclic compound found in nucleotides and nucleic acids; it contains fused pyrimidine and imidazole rings. Adenine and guanine are examples.

PVP See PLANT VARIETY PROTECTION ACT.

PVPA See PLANT VARIETY PROTECTION ACT.

PVR Plant Variety Rights. See PLANT VARIETY PROTECTION ACT.

PWGF See PLATELET-DERIVED WOUND GROWTH FACTOR and GROWTH FACTOR.

Pyralis An insect that is also known as the European corn borer (*ostrinia nubialis*). See EUROPEAN CORN BORER (ECB).

Pyranose The six-membered ring forms of sugars are called pyranoses. This is because they are derivatives of the heterocyclic compound pyran. See SUGAR MOLECULES.

Pyrexia Fever; elevation of the body temperature above normal. See PYROGEN.

Pyrimidine A heterocyclic organic compound containing nitrogen atoms at (ring) positions 1 and 3. Naturally occurring derivatives are components of nucleic acids and coenzymes, uracil, thymine, and cytosine.

Pyrogen A substance capable of producing pyrexia (i.e., fever). See PYREXIA.

Pyrophosphate Cleavage The enzymatic removal of two phosphate groups (designated as PP_i) from ATP in one piece leaving AMP as another product. This cleavage releases more energy, which can be used in certain reactions that require more of a "push" to get them going. See also ATP and ORTHOPHOSPHATE CLEAVAGE.

Q-beta Replicase A viral RNA polymerase secreted by a bacteriophage that infects *Escherichia coli* bacteria. Q-beta replicase can copy a naturally occurring RNA (molecule) sequence (e.g., from bacteria, viruses, fungi, or tumor cells) at a geometric (i.e., very fast) rate. See also POLYMERASE, BACTERIOPHAGE, RIBONUCLEIC ACID (RNA), and Q-BETA REPLICASE TECHNIQUE.

Q-beta Replicase Technique An RNA assay (test) that "amplifies RNA probes" that a researcher is seeking. For instance, by using the Q-beta replicase technique to assay for the presence of RNA that is specific to the AIDS virus, it is possible to detect an AIDS infection in a patient's blood sample long before that infection has progressed to the point where antibodies would appear in the blood. See also Q-BETA REPLICASE, RNA PROBES, RIBONUCLEIC ACID (RNA), POSITIVE AND NEGATIVE SELECTION (PNS), ASSAY, IMMUNOASSAY, ANTIBODY, POLYMERASE CHAIN REACTION (PCR) TECHNIQUE, COCLONING (OF MOLECULES), and WESTERN BLOT TEST.

QSAR See QUANTITATIVE STRUCTURE-ACTIVITY RELATIONSHIP (QSAR).

QTL See QUANTITATIVE TRAIT LOCI (QTL).

Quantitative Structure-Activity Relationship (QSAR) A computer modeling technique that enables researchers (e.g., drug development chemists) to predict the likely activity (e.g., effect on tissue) of a new compound before that compound is actually created. QSAR is based on decades of research investigating the impact on "activity" of the chemical structures of thousands of thoroughly studied molecules. See also BIOLOGICAL ACTIVITY, PHARMACOPHORE, PHARMACOKINETICS (PHARMACODYNAMICS), PHARMACOLOGY, ANALOGUE, and RATIONAL DRUG DESIGN.

Quantitative Trait Loci (QTL) Individual specific DNA sequences that are related to known traits (e.g., litter size in animals, egg production in birds, yield in crop plants.) See MARKER (DNA SEQUENCE), TRAIT, LINKAGE, DEOXYRIBONUCLEIC ACID (DNA), LINKAGE GROUP, LINKAGE MAP, GENE, SEQUENCE (OF A DNA MOLECULE), MARKER ASSISTED SELECTION, CORN, and HIGH-OIL CORN.

Quaternary Structure The three-dimensional structure of an oligomeric protein; particularly the manner in which the subunit chains fit together. See also PROTEIN, OLIGOMER, CONFIGURATION, and NATIVE CONFORMATION.

Quick-Stop The term used to describe how DNA mutants of *Escherichia coli* cease replication immediately when the temperature is increased to 42°C (108°F). See also *ESCHERICHIA COLI (E. COLI)*.

R

RAC See RECOMBINANT DNA ADVISORY COMMITTEE.

Racemate An equimolar (i.e., equal number of molecules) mixture of the D and L stereoisomers of an optically active compound. A solution of dextrorotary (D) isomer (enantiomer) will rotate the plane in which the light was polarized a specific number of degrees to the right (dextro) while a solution containing the same number of levorotary (L) isomer molecules will rotate the plane in which the light was polarized the same number of degrees (as in the D isomer case) to the left (levo). The difference between D and L enantiomers is that the rotations of the plane of plane-polarized light are equal in magnitude, but opposite in sign. Hence, a 50:50 mixture of both enantiomers (known as a racemic mixture) shows no optical activity. That is, a solution containing a 50:50 mixture of enantiomers will not rotate the plane of plane polarized light when it is passed through the solution. See also ENANTIOMERS, STEREOISOMERS, LEVOROTARY (L) ISOMER, and DEXTROROTARY (D) ISOMER.

Racemic (mixture) See RACEMATE.

Radioactive Isotope An isotope with an unstable nucleus that spontaneously emits radiation. The radiation emitted includes alpha particles, nucleons, electrons, and gamma rays. See also ISOTOPE.

Radioimmunoassay A very sensitive method of quantitating a specific antigen using a specific radiolabeled antibody. Functionally, the antibody is made radioactive by the covalent incorporation of radioactive iodine. The radioimmuno probe thus prepared is exposed to its antigen (which may be a protein, or a receptor, etc.) in excess (the exact amount will have to be determined). The radiolabeled probe then binds to the antigen and the unbound, free probe is washed away. The radioactivity is then determined (counted) and by comparison to a standard plot which has been constructed previously, the amount of antigen (binding) is determined. See also ANTIBODY, ASSAY, HORMONE, and RADIOIMMUNOTECHNIQUE.

Radioimmunotechnique A method of using a radiolabeled antibody to quantitate a known antigen. See also RADIOIMMUNOASSAY, ANTIGEN, and ANTIBODY.

Random Amplified Polymorphic DNA (RAPD) Technique A genetic mapping methodology that utilizes as its basis the fact that specific DNA sequences (polymorphic DNA) are "repeated" (i.e., appear in sequence) with gene of interest. Thus, the polymorphic DNA sequences are linked to that specific gene. Their linked presence serves to facilitate genetic mapping (i.e., "location" of specific gene(s) on an organism's genome). See GENETIC MAP, SEQUENCE (OF A DNA MOLECULE), RESTRICTION FRAGMENT LENGTH POLYMORPHISM (RFLP) TECHNIQUE, LINKAGE, DEOXYRIBONUCLEIC ACID (DNA), PHYSICAL MAP (OF GENOME), LINKAGE GROUP, MARKER (GENETIC MARKER), LINKAGE MAP, TRAIT, GENOME, and GENE.

RAPD See RANDOM AMPLIFIED POLYMORPHIC DNA TECHNIQUE.

Rapid Microbial Detection (RMD) A broad term used to describe the various testing products/technologies that can be utilized to quickly detect the presence of microorganisms (e.g., pathogenic bacteria in a food processing plant). These testing products are based on immunoassay, DNA probe, electrical conductance and/or impedance, bioluminescence, and enzyme-induced reactions (e.g., which produce fluorescence or a color change to indicate the presence of specific microorganism). See also BIOLUMINESCENCE, MICROBE, BACTERIA, PATHOGEN, IMMUNOASSAY, ENZYME, PROBE, DNA PROBE, ELECTROPHORESIS, and HAZARD ANALYSIS AND CRITICAL POINTS (HACCP).

ras Gene An oncogene that is believed to be responsible for up to 90% of all human pancreatic cancer, 50% of human colon cancers, 40% of lung cancers, and 30% of leukemias. The ras gene is present in the DNA of all human tissues, and codes for ras proteins, which help to signal each cell to divide and grow at appropriate time(s). When the ras gene has been damaged or mutated (e.g., via exposure to cigarette smoke or ultraviolet light), it codes for (i.e., causes to be manufactured in the cell's ribosome) a mutated version of the ras gene that can cause the cell to become cancerous (i.e., divide and grow uncontrollably). See also GENE, ONCOGENES, p53 GENE, GENETIC CODE, MEIOSIS, DEOXYRIBONUCLEIC ACID (DNA), CARCINOGEN, RIBOSOMES, CANCER, TUMOR, ras PROTEIN, PROTO-ONCOGENES, and PROTEIN.

ras Protein A transmembrane (i.e., through the cell membrane) protein that is coded for by the ras gene. The ras protein end that is outside the cell membrane acts as a receptor for applicable growth factors (e.g., fibroblast growth factor), and conveys that signal (i.e., to divide/grow) into the cell when that chemical signal (i.e., the growth factor) touches "receptor end" of the ras protein. When the ras gene has been damaged or mutated (e.g., via exposure to cigarette smoke or ultraviolet light), that gene causes excess ras proteins to be manufactured, which causes over-signalling of the cell to divide and grow (i.e., cell becomes cancerous). See also GENE, ras GENE, ONCOGENES, GENETIC CODE, PROTEIN, p53 PROTEIN, MEIOSIS, CARCINOGEN, RIBOSOMES, DEOXYRIBONUCLEIC ACID (DNA), CANCER, TUMOR, PROTO-ONCOGENES, RECEPTORS, CD4 PROTEIN, SIGNALLING, and SIGNAL TRANSDUCTION.

Rational Drug Design The engineering (building) of chemically synthesized drugs based on knowledge of receptor modeling and drug/target interaction with the aid of supercomputers/interactive graphics/etc.); the educated, creative design of the three-dimensional structure of a drug atom by atom, that is, "from the ground up." This approach represents a major advance over the prior practice of first synthesizing large numbers of compounds (or finding them in nature), followed by thousands of tedious screenings to test for efficacy against a given disease. The approach of rational drug design has, however, not yet been perfected and optimized due, in part, to gaps in our knowledge of drug/receptor interaction and to gaps in our knowledge in general. See also RECEPTORS, RECEPTOR MAPPING (RM), ANALOGUE, and MOLECULAR DIVERSITY.

RB See REFRACTILE BODIES.

rDNA See RECOMBINANT DNA.

Reading Frame The particular nucleotide sequence that starts at a specific point and is then partitioned into codons. The reading frame may be shifted by removing or adding a nucleotide(s). This would cause a new sequence of codons to be read. For example, the sequence CATGGT is normally read as the two codons: CAT and GGT. If another adenosine nucleotide (A) were inserted between the initial C and A, producing the sequence CAATGGT, then the reading frame would have been shifted in such a way that the two new (different) codons would be CAA and TGG, which would code for something completely different. See also CODON, GENETIC CODE, FRAMESHIFT, DEOXYRIBONUCLEIC ACID (DNA), and MUTATION.

Reassociation (of DNA) The pairing of complementary single strands (of the molecule) to form a double helix (structure). See also DOUBLE HELIX.

RecA The product of the RecA locus (in a gene of) *Escherichia coli.* It is a protein with dual activities, acting as a protease and also able to exchange single strands of DNA (deoxyribonucleic acid) molecules. The protease activity controls the SOS response. The nucleic acid handling facility (i.e., ability to exchange single strands of DNA) is involved in recombination/repair pathways. See also SOS RESPONSE, LOCUS, PROTEIN, RIBOSOMES, and *ESCHERICHIA COLI (E. COLI)*.

Receptor Fitting (RF) A research method used to determine the macromolecular structure that a chemical compound (e.g., an inhibitor) must have in order to fit (in a lock-and-key fashion) into a receptor. For example, a pain inhibitor compound blocking a pain receptor on the surface of a cell. See also CD4 PROTEIN, T CELL RECEPTORS, RECEPTORS, RECEPTOR MAPPING (RM), INTERLEUKIN-1, RECEPTOR ANTAGONIST (IL-1ra), and RATIONAL DRUG DESIGN.

Receptor Mapping (RM) A method used to guess at (determine) the three-dimensional structure of a receptor binding site extrapolating from the known structure of the molecule binding to it. This approach can be carried out because of the complementary shape of the receptor and the binding molecule. Functionally, the researcher projects the (guessed-at) properties of the receptor ligands into a mathematical model in which the profile of the receptor is predicted by complementariness (to known chemical molecular structures). The receptor mapping process requires repetitive refinement of the mathematical model to fit properties continually being discovered via the use/interaction of chemical reagents bearing the known molecular structures. See also CD4 PROTEIN, T CELL RECEPTORS, RECEPTORS, and RECEPTOR FITTING (RF).

Receptor-Mediated Endocytosis See ENDOCYTOSIS.

Receptors Functional proteinaceous structures typically found in the membrane (surface) of cells that tightly bind specific molecules (organic, proteins or viruses). Some (relatively rare) receptors are located inside the cell's membrane (e.g., free-floating receptor for Retin-A). Both (membrane, internal) types of receptors are a functional part of information transmission to the cell. A general overview is that once bound, both the receptor and its "bound entity" as a complex is internalized by the cell via a process called endocytosis, in which the cell membrane in the vicinity of the bound complex invaginates. This process forms a membrane "bubble" on the inside of the cell, which then pinches off to form an endocytic vesicle. The receptor then is released from its bound entity by cleavage in the cell's lysosomes. It is recycled (returned) to the surface of the cell (e.g., low-density lipoprotein receptors). In some cases the receptor, along with its bound molecule may be degraded by the powerful hydrolytic enzymes found in the cell's lysosomes (e.g., insulin receptors, epidermal growth factor receptors, and nerve growth factor receptors). Endocytosis (internalization of receptors and bound ligand such as a hormone) removes hormones from the circulation and makes the cell temporarily less responsive to them because of the decrease in the number of receptors on the surface of the cell. Hence the cell is able to respond (to new signal). A receptor may be thought of as a butler who allows guests (in this case molecules that bind specifically to the receptor) to enter the house (cell) and who accompanies them as they enter.

Another mode of "reception" occurs when, following binding, a transmembrane protein (e.g., one of the G proteins) activates the portion of the transmembrane (i.e., through the cell membrane) protein lying inside the cell. That "activation" causes an effector inside cell to produce a "signal" chemical inside the cell which causes the cell to react to the original external chemical signal (that bound itself to the receptor portion of the transmembrane protein). See also CD4 PROTEIN, T CELL RECEPTORS, RECEPTOR FITTING (RF), RECEPTOR MAPPING (RM), LYSOSOMES, INTERLEUKIN-1 RECEPTOR ANTAGONIST (IL-1ra), CD95 PROTEIN, TRANSFERRIN, VAGINOSIS, SIGNAL TRANSDUCTION, ENDOCYTOSIS, G PROTEINS, CELL, SIGNALLING, PROTEIN, NUCLEAR RECEPTORS, and HUMAN IMMUNODEFICIENCY VIRUS TYPES 1 & 2.

Recessive Allele Discovered by Gregor Mendel in the 1860s, this refers to an allelic gene whose existence is obscured in the phenotype of a heterozygote by the dominant allele. In a heterozygote the recessive allele does not produce a polypeptide; it is switched off. In this case the dominant allele is the one producing the polypeptide chain. See also ALLELE, DOMINANT (GENE), HOMOZYGOUS, and HETEROZYGOTE.

Recombinant DNA (rDNA) DNA formed by the joining of genes (genetic material) into a new combination. See also RECOMBINATION and GENETIC ENGINEERING.

Recombinant DNA Advisory Committee (RAC) The former standing U.S. national committee set up in 1974 by the National Institutes of Health (NIH) to advise the NIH director on matters regarding policy and safety issues of recombinant DNA research and development. Over time, it had evolved to become part of the American government's regulatory process for recombinant DNA research and product approval. The RAC was terminated by the director of the NIH in 1996 because the "human health and environmental safety concerns expressed at the inception (of genetic engineering/biotechnology) had not materialized." See also GENE TECHNOLOGY OFFICE, GENETIC ENGINEERING, ZKBS (CENTRAL COMMISSION ON BIOLOGICAL

SAFETY), NATIONAL INSTITUTES OF HEALTH (NIH), RECOMBINANT DNA (rDNA), BIOTECHNOLOGY, RECOMBINATION, INDIAN DEPARTMENT OF BIOTECHNOLOGY, and COMMISSION OF BIOMOLECULAR ENGINEERING.

Recombination The joining of genes, sets of genes, or parts of genes, into new combinations, either biologically or through laboratory manipulation (e.g., genetic engineering). See also GENETIC ENGINEERING, GENE, and RECOMBINANT DNA (rDNA).

Red Blood Cells See ERYTHROCYTES.

Redement Napole (RN) Gene A swine gene that causes animals (possessing at least one negative allele of this gene) to produce meat which is more acidic than average, and thus that meat has a lower "water-holding" capacity. The RN gene was first identified in the Hampshire breed of swine in France. The Hampshire breed has been known to produce meat that is more acidic than average since the 1960s. See also GENE, ALLELE, and ACID.

Reduction (biological) The decomposition of complex compounds and cellular structures by heterotrophic organisms. In a given ecological system, this heterotrophic decomposition serves the valuable function of recycling organic materials. This occurs because the heterotrophs absorb some of the decomposition products (for nourishment) and leave the balance of the (decomposed) substances for consumption (recycling) by other organisms. For example, bacteria break down fallen leaves on the floor of a forest, thus releasing some nutrients to be utilized by plants. See also HETEROTROPH.

Reduction (in a chemical reaction) The gain of (negatively charged) electrons by a chemical substance. When one substance is reduced by another, the other compound is oxidized (loses electrons) and is called the reducing agent. See also OXIDATION-REDUCTION REACTION and OXIDIZING AGENT (OXIDANT).

Redundancy A term used to describe the fact that some amino acids have more than one codon (that codes for production of that amino acid). There are approximately 64 possible codons available to code for 20 amino acids. Therefore, some amino acids will be specified by more than one codon. These (extra) codons are redundant. See also CODON, GENETIC CODE, and RIBOSOMES.

Refractile Bodies (RB) Dense, insoluble (i.e., not easily dissolved) protein bodies (i.e., clumps) that are produced within the cells of certain microorganisms. The refractile bodies function as a sort of natural storage device for the microorganism). They are called refractile bodies because their greater density (than the rest of the microorganism's body mass) causes light to be refracted (bent) when it is passed through them. This bending of light causes the appearance of very bright and dark areas around the refractile body and makes them visible under a microscope.

Relatively rare in natural occurrence, refractile bodies can be induced (i.e., caused to occur) in procaryotes (e.g., bacteria) when the procaryotes are genetically engineered to produce eucaryotic (e.g., mammal) proteins. The proteins are stored in refractile bodies. For example, the *Escherichia coli* bacterium can be genetically engineered to produce bovine somatotropin (BST, a cow hormone) which is stored within refractile bodies in the bacterium. After some time of growth when a significant amount of BST has been synthesized the *Escherichia coli* cells are disrupted (i.e., broken open), and the refractile bodies are removed by centrifugation and washed. They are then dissolved in appropriate solutions to release the protein molecules. This step denatures (unfolds, inactivates) the BST molecules and they are refolded to their native conformation (i.e., restored to the natural conformation found within the cow) in order to regain their natural activity. The protein is then formulated in such a way as to be commercially viable as a biopharmaceutical.

Refractile bodies are also known as inclusion bodies, protein inclusion bodies, and refractile inclusions.

One point of interest is that the prerequisite for the generation of a mammalian protein by (in) a living foreign system such as *E. coli* is that the system used to generate the protein (1) must not have an immune system capable of destroying the foreign protein it is making or (2) the foreign protein made must be camouflaged or protected from any defense mechanisms possessed by the synthesizing organism.

See also PROTEIN, GENETIC ENGINEERING, GENETIC CODE, PROCARYOTES, EUCARYOTE, *ESCHERICHIA COLI (E. COLI)*, BOVINE SOMATOTROPIN (BST), ULTRACENTRIFUGE, CONFORMATION, NATIVE CONFORMATION, and PROTEIN FOLDING.

Regulatory Enzyme A highly specialized enzyme having a regulatory (controlling) function through its capacity to undergo a change in its catalytic activity. There exist two major types of regulatory enzymes: (1) covalently modulated enzymes, and (2) allosteric enzymes.

Covalently modulated enzymes are enzymes that can be interconverted between active and inactive (or less active) forms by the covalent attachment (or removal) of a modulating metabolite by other enzymes. Hence the activity of one enzyme can, under certain conditions, be regulated by other enzymes. Glycogen phosphorylase, an oligomeric protein with four major subunits (tetramer), is a classic example of a covalently modulated enzyme. The enzyme occurs in two forms: (1) phosphorylase a, the more active form, and (2) phosphorylase b, the less active form. In order for the enzyme to possess maximal catalytic activity (i.e., be phosphorylase a) certain serine residue on all four subunits must have a phosphate covalently attached. If, due to other regulatory signals it has received, the enzyme phosphorylase phosphatase hydrolytically cleaves and removes the phosphate group from the four subunits, the tetramer dissociates into the inactive (or much less active) dimer, phosphorylase b. Another enzyme, phosphorylase kinase, is able to rephosphorylate the four specific serine residues of the four subunits at the expense of ATP and regenerate the active phosphorylase a tetramer.

Allosteric enzymes are enzymes that possess a special site on their surfaces that is distinct from the enzyme's catalytic site and to which specific metabolites (called effectors or modulators) are reversibly and noncovalently bound. The allosteric binding site is as specific for a particular metabolite as is the catalytic site, but it cannot catalyze a reaction, only bind the effector. The binding of the effector causes a conformation change in the enzyme such that its catalytic activity is impaired or stopped. Allosteric enzymes are normally the first enzymes in, or are near the beginning of, a multienzyme system. The very last product produced by the multienzyme system (the end product) may act as a specific inhibitor of the allosteric enzyme by binding to that enzyme's allosteric site. The binding consequently causes a conformation change to occur in the enzyme, which inactivates it. A classic example of an allosteric enzyme in a multienzyme sequence is the enzyme L-threonine dehydratase, which is the initial enzyme in the enzyme sequence that catalyzes the conversion of L-threonine to L-isoleucine. This reaction occurs in five enzyme-catalyzed steps. The end product, L-isoleucine, strongly inhibits L-threonine dehydratase, the first enzyme in the five-enzyme sequence. No other intermediate in the sequence is able to inhibit the enzyme. This kind of repression is called feedback or end-product inhibition.

It should be noted that allosteric control may be negative (as in the example above) or positive. In positive control the effector binds to an allosteric site and stimulates the activity of the enzyme. Furthermore, some allosteric enzymes respond to two or more specific modulators with each modulator having its own specific binding site on the enzyme. An allosteric enzyme that has only one specific modulator is called monovalent whereas an enzyme responding to two or more specific modulators is called polyvalent. Combinations of the above possibilities could lead to very fine tuning of the enzymes involved in the synthesis and/or degradation of metabolites.

Note that in the two examples above, the common denominator is the structural change that occurs upon execution of the mechanism. See also METABOLITE and REPRESSIBLE ENZYME.

Regulatory Genes Genes whose primary function is to control the state of synthesis of the products of other genes.

Regulatory Sequence A DNA sequence involved in regulating the expression of a gene, e.g., a promoter or operator region (in the DNA molecule). See also OPERATOR and PROMOTER.

Remediation The cleanup or containment (if chemicals are moving) of a hazardous waste disposal site to the satisfaction of the applicable regulatory agency [e.g., the Environmental Protection Agency (EPA)]. Such cleanup can sometimes be accomplished via use of microorganisms that have been adapted (naturally or via genetic engineering) to consume those chemical wastes that are present in the disposal site. See also ACCLIMATIZATION.

Renaturation The return to the natural structure of a protein or nucleic acid from a denatured (more random coil) state. For example, a protein may be denatured [lose its native (natural) structure] by exposure to surfactants such as SDS or to changes in the pH of the medium, etc. If the surfactant is slowly removed or the pH is slowly readjusted to the optimum for the protein, it will refold (snap) back into its original (native) form. See also NATIVE CONFIGURATION, DENATURATION, and SDS.

Renin A proteolytic enzyme that is secreted by the juxtaglomerular cells of the kidney. Its release is stimulated by decreased arterial pressure and renal blood flow resulting from decreased extracellular fluid volume. It catalyzes the formation of angiotensin I from hypertensinogen. Angiotensin I is then converted to angiotensin II by another enzyme located in the endothelial cells of the lungs. Angiotensin II then causes the increase in the force of the heartbeat and constricts the arterioles. This scenario causes a rise in the blood pressure and is thus a cause of hypertension (high blood pressure). See also HOMEOSTASIS, RENIN INHIBITORS, and ATRIAL PEPTIDES.

Renin Inhibitors Those chemicals that act to block the hypertensive (i.e., high blood pressure–inducing) effect of the enzyme, renin. See also HOMEOSTASIS, RENIN INHIBITORS, and ATRIAL PEPTIDES.

Rennin See CHYMOSIN.

Reovirus A virus containing double-stranded RNA. It is isolated from the respiratory and intestinal tracts of humans and other mammals. The prefix "reo-" is an acronym for respiratory enteric orphan. See also RETROVIRUSES.

Reperfusion The restoration of blood flow to an occluded (i.e., blocked) blood vessel. May be done biochemically (e.g., via tissue plasminogen activator) or via surgery. See HUMAN SUPEROXIDE DISMUTASE (hSOD) and LAZAROIDS.

Replication (of DNA) Reproduction of a DNA molecule (inside a cell). This process can be viewed as occurring in stages, in which the first stage consists of an enzyme "unwinding" the double helix of the DNA molecule at a replication origin, forming a replication fork. At the replication fork, the two separated (DNA) strands serve as templates for new DNA synthesis. That new DNA synthesis is accomplished on each strand via enzymes known as DNA polymerase, which travel along each (single) strand making a second complementary strand by catalyzing the addition of DNA bases (to the new, growing strands).

The end result is two new double helices (DNA molecules), each of which has one chain from the original DNA molecule and one chain that was newly synthesized by the DNA polymerase enzymes. See also DEOXYRIBONUCLEIC ACID (DNA), DNA POLYMERASE, ENZYME, REPLICATION FORK, DUPLEX, DOUBLE HELIX, and BASE PAIR (bp).

Replication (of virus) Reproduction of the original virus. This process can be viewed as occurring in stages, in which the first stage consists of the adsorption of the virus to the host cell, followed by penetration of the virus (or its nucleic acid) into the cell, the taking over of the cell's biomachinery and harnessing of it to replicate viral nucleic acid along with the synthesis of other virus constituents, the correct assembly of the nucleic acids and other constituents into a functional virus, followed finally by release of the virus from the confines of the cell. See also VIRUS, CELL, and NUCLEIC ACIDS.

Replication Fork The point at which strands of parental duplex DNA are separated in a Y shape. This region represents a growing point in DNA replication. See REPLICATION (OF DNA), DEOXYRIBONUCLEIC ACID (DNA), and DUPLEX.

Reporter Gene A specific gene that is inserted into the DNA of a cell so that cell will "report" (to researchers) when signal transduction has occurred in that cell. For example, when researchers are testing numerous candidate drugs for their ability to stop cells from (over) producing a hormone or growth factor, the researchers need to *quickly* know when one of the candidate drugs has had the desired effect on the cell of interest. By prior insertion into that cell of a gene (e.g., that causes a certain chemical to be produced by the cell when signal transduction has taken place), that cell "reports" (when a candidate drug has had the desired effect on the cell) by producing the chemical (coded for by the reporter gene) which can be rapidly detected by the researcher (e.g., via biosensors placed adjacent to the cell). See also GENE, GENETIC ENGINEERING, GENETIC CODE, CELL, CELL CULTURE, SIGNAL TRANSDUCTION, HORMONE, GROWTH FACTOR, and BIOSENSORS (ELECTRONIC).

Repressible Enzyme An enzyme whose synthesis (rate of production) is inhibited (repressed) when the product that it (or it in a multienzyme sequence) synthesizes is present in high concentrations. It is a way of shutting down the synthesis of an enzyme whose product is not required because so much of it is readily available to the cell. When that enzyme product is no longer available (e.g., because the cell has consumed that product) more of the enzyme is synthesized (to catalyze production of more product). See also REPRESSION (OF AN ENZYME), REGULATORY ENZYME, and ENZYME.

Repression (of an enzyme) The prevention of synthesis of certain enzymes when their reaction products are present. See also REPRESSIBLE ENZYME.

Repression (of gene transcription/translation) The inhibition of transcription (or translation) by the binding of a repressor protein to a specific site on the DNA (or RNA) molecule. The repressor molecule is the product of a repressor gene. See also REPRESSOR (PROTEIN), TRANSCRIPTION, TRANSLATION, and DEOXYRIBONUCLEIC ACID (DNA).

Repressor (protein) The product of a regulatory gene, it is a protein that combines both with an inducer (or corepressor) and with an operator region (e.g., of DNA). See also INDUCERS, COREPRESSOR, OPERATOR, and REPRESSION (OF GENE TRANSCRIPTION/TRANSLATION).

Research Foundation for Microbiological Diseases Also known as Riken. A Japanese institution that performs research on infectious diseases. See also NATIONAL INSTITUTE OF ALLERGY AND INFECTIOUS DISEASES (NIAID) and KOSEISHO.

Residue (of chemical within a foodstuff) See MAXIMUM RESIDUE LEVEL (MRL).

Residue (portion of a protein molecule) See MINIMIZED PROTEINS.

Respiration Oxidative process in living cells in which oxygen or an inorganic compound serves as the terminal (final, ultimate) electron acceptor. Aerobic organisms obtain most of their energy from the oxidation of organic fuels. This process is known as respiration. See also OXIDATION-REDUCTION REACTION, REDUCTION (IN A CHEMICAL REACTION), OXIDATION, and OXIDIZING AGENT (OXIDANT).

Restriction Engdoglycosidases A class of enzymes, each of which cleaves (i.e., cuts) oligosaccharides (e.g., the side chains on glycoprotein molecules) at a specific location within the chain. They are an important tool in carbohydrate engineering, enabling the carbohydrate engineer to sequence (i.e., determine the structure of) existing oligosaccharides, to create different oligosaccharides, and to create different glycoproteins via removal/addition/change of the oligosaccharide chains on glycoprotein molecules. See also OLIGOSACCHARIDES, GLYCOPROTEIN, CARBOHYDRATE ENGINEERING, GLYCOSIDASES, ENDOGLYCOSIDASE, EXOGLYCOSIDASE, GLYCOFORM, GLYCOBIOLOGY, and GLYCOSYLATION (TO GLYCOSYLATE).

Restriction Endonucleases A class of enzymes that cleave (i.e. cut) DNA at a specific and unique internal location along its length. These enzymes are naturally produced by bacteria that use them as a defense mechanism against viral infection. The enzymes chop up the viral nucleic acids and hence their function is destroyed. Discovered in the late 1970s by Werner Arber, Hamilton Smith, and Daniel Nathans, restriction endonucleases are an important tool in genetic engineering, enabling the biotechnologist to splice new genes into the location(s) of a molecule of DNA where a restriction endonuclease has created a gap (via cleavage of the DNA). See also VECTOR, ENZYME, POLYMERASE, GENE, GENETIC ENGINEERING, GENE SPLICING, and ELECTROPHORESIS.

Restriction Enzymes See RESTRICTION ENDONUCLEASES.

Restriction Fragment Length Polymorphism (RFLP) Technique A "genetic mapping" technique that analyzes the *specific* sequence of bases (i.e., nucleotides) in a piece of DNA. Since the specific sequence of bases in their DNA molecules is different for each species, strain, variety, and individual (due to DNA polymorphism), RFLP can be utilized to "map" those DNA molecules (e.g., for plant breeding purposes, for criminal investigation purposes, etc.). See GENETIC MAP, SEQUENCE (OF A DNA MOLECULE), RANDOM AMPLIFIED POLYMORPHIC DNA (RAPD) TECHNIQUE, DEOXYRIBONUCLEIC ACID (DNA), PHYSICAL MAP (OF GENOME), LINKAGE, LINKAGE GROUP, MARKER (GENETIC MARKER), LINKAGE MAP, TRAIT, BASE PAIR (bp), DNA PROFILING, POLYMORPHISM (CHEMICAL), NUCLEIC ACIDS, GENETIC CODE, and INFORMATIONAL MOLECULES.

Restriction Map A pictorial representation of the specific restriction sites (i.e., nucleotide sequences that are cleaved by given restriction endonucleases) in a DNA molecule (e.g., plasmid or chromosome). See also RESTRICTION SITE, RESTRICTION ENDONUCLEASES, and DNA.

Restriction Site A nucleotide sequence (of base pairs) in a DNA molecule that is "recognized," and cleaved by a given restriction endonuclease. See also NUCLEOTIDE, SEQUENCE (OF A DNA MOLECULE), BASE PAIR (bp), DNA, RESTRICTION ENDONUCLEASES, and RESTRICTION MAP.

Resveratrol Also known as 3,5,4 trihydroxy stilbene, it is a phytochemical that is produced by certain plants in response to "wounding" (e.g., by fungal growth on plant) or other stress. Plants that produce resveratrol include red grapes, mulberries, soybeans, and peanuts.

Resveratrol inhibits cell mutations, stimulates at least one enzyme that can inactivate certain carcinogens, and (when consumed by humans) contributes to a low incidence of cardiovascular disease. See PHYTOCHEMICALS, SOYBEAN PLANT, FUNGUS, CARCINOGEN, and MUTATION.

Retinoids A group of biologically active compounds that are chemical derivatives of vitamin A. Among other effects on living cells, some of the retinoid compounds act to deprive cancerous cells of their ability to proliferate endlessly, so these (formerly cancerous) cells then progress to a natural death (after exposure to applicable retinoid). See also CELL, VITAMIN, BIOLOGICAL ACTIVITY, CANCER, and NEOPLASTIC GROWTH.

Retroviral Vectors Certain retroviruses that are used by genetic engineers to carry new genes into cells. These molecules become part of that cell's protoplasm. See also RETROVIRUSES, GENETIC ENGINEERING, VECTOR, GENE, and PROTOPLASM.

Retroviruses (from the Latin word *retrovir*, which means "backward man") Oncogenic (i.e., cancer-producing), single-stranded, diploid RNA (ribonucleic acid) viruses that contain (+) RNA in their virions and propagate through a double-helical DNA intermediate. They are known as retroviruses because their genetic information flows from RNA to DNA (reverse of normal). That is, the viruses contain an enzyme that allows the production of DNA using RNA as a template. Retroviruses can only infect cells in which DNA is replicating, such as tumor cells (since they are constantly replicating) or cells comprising the lining of the stomach (since that lining must replace itself every few days). See also ONCOGENES, DIPLOID, RIBONUCLEIC ACID (RNA), REVERSE TRANSCRIPTASES, and CENTRAL DOGMA.

Reverse Micelle (RM) Also known as reversed micelle or inverted micelle. A spheroidal structure formed by the association of a number of amphipathic (i.e., bearing both polar and nonpolar domains) surfactant molecules dissolved in organic, nonpolar solvents such as benzene, hexane, isooctane and oils such as corn and sesame. The structure of an RM is the reverse of that of a micelle. Reverse micelles may be characterized by a structure in which the polar groups of the surfactant and any water present are centrally located with the surfactant hydrocarbon chains pointing outwards into the surrounding hydrocarbon medium. Reverse micelles may be used to solubilize polar molecules (i.e., water, enzymes) in organic nonpolar solvents and oils. See also AMPHIPATHIC MOLECULES, MICELLE, and SURFACTANT.

Reverse Phase Chromatography (RPC) A method of separating a mixture of proteins or nucleic acids or other molecules by specific interactions of the molecules with a hydrophobic (i.e., "water hating") immobilized phase (i.e., stationary substrate) which interacts with hydrophobic regions of the protein (or nucleic acid) molecules to achieve (preferential) separation of the mixture. See also CHROMATOGRAPHY.

Reverse Transcriptases Also known as RNA-directed DNA polymerases. A class of enzymes first discovered to be present in RNA tumor-virus which allows the synthesis of DNA using the RNA present in the virus as a template. This is the reverse of what normally happens and hence the name. Reverse transcriptases closely resemble the DNA-directed DNA polymerases (DNA polymerases) in that they require the same materials and conditions as the DNA polymerases. See also CENTRAL DOGMA and POLYMERASE.

Reversed Micelle See REVERSE MICELLE (RM). Also see MICELLE for comparison.

RFLP (restriction fragment length polymorphism) See POLYMORPHISM (CHEMICAL), RESTRICTION ENDONUCLEASES, and RESTRICTION FRAGMENT LENGTH POLYMORPHISM (RFLP) TECHNIQUE.

rh Used to denote compounds (human molecules) made through the use of recombinant DNA technology. Recombinant (r) human (h). See also rhTNF, RECOMBINANT DNA (rDNA), RECOMBINATION, and GENETIC ENGINEERING.

Rhizobium (bacteria) See NITROGEN FIXATON.

Rho Factor A protein involved in (chemically) assisting *Escherichia coli* RNA polymerase in the termination of transcription at certain (rho dependent) sites on the DNA molecule. See also TRANSCRIPTION, POLYMERASE, and *ESCHERICHIA COLI (E. COLI)*.

rhTNF Recombinant human TNF. See TUMOR NECROSIS FACTOR (TNF).

RIA See RADIOIMMUNOASSAY.

Ribonucleic Acid (RNA) A long-chain, usually single-stranded nucleic acid consisting of repeating nucleotide units containing four kinds of heterocyclic, organic bases: adenine, cytosine, guanine, and uracil. These bases are conjugated to the pentose sugar ribose and held in sequence by phosphodiester (chemical) bonds. The primary function of RNA is protein synthesis within a cell. However, RNA is involved in various ways in the processes of expression and repression of hereditary information. The three main functionally distinct varieties of RNA molecules are: (1) messenger RNA (mRNA) which is involved in the transmission of DNA information, (2) ribosomal RNa (rRNA) which makes up the physical machinery of the synthetic process, and (3) transfer RNA (tRNA) which also constitutes another functional part of the machinery of protein synthesis. See also HEREDITY, GENETIC CODE, RIBOSOMES, INFORMATIONAL MOLECULES, and NANOTECHNOLOGY.

Ribose D-Ribose, a five-carbon-atom monosaccharide (i.e., a sugar). It is important to life because it and the closely allied compound deoxyribose form a part of the molecules that constitute the backbone of nucleic acids. See also NUCLEIC ACIDS and MONOSACCHARIDES.

Ribosomal RNA See rRNA.

Ribosomes The molecular "machines" within cells that coordinate the interplay of tRNAs, mRNA, and proteins in the complex process of protein synthesis (manufacture). RNA constitutes nearly two-thirds of the mass of these large (mega-Dalton) molecular assemblies. The formation of a ribosome (in a cell) is largely a self-assembly process, because all of the information needed for the correct assembly of this structure is contained in the primary structure of its (molecular) components. The assembly process is ordered and proceeds in stages. Many ribosomes (in a given cell) can simultaneously translate an mRNA molecule. The structure, consisting of a group of ribosomes bound to an mRNA molecule that is actively synthesizing protein, is called a polyribosome or a polysome. The ribosomes in this (polysome) unit operate independently of each other, each synthesizing a complete polypeptide (protein) chain. See also PROTEIN, POLYPEPTIDE (PROTEIN), PROTEIN FOLDING, POLYCISTRONIC, PRIMARY STRUCTURE (under PROTEIN STRUCTURE), TRANSCRIPTION, TRANSCRIPTION UNIT, MESSENGER RNA (mRNA), TRANSFER RNA (tRNA), rRNA, DALTON, and SELF-ASSEMBLY (OF LARGE MOLECULAR STRUCTURES).

Ribozymes Discovered by Thomas Cech and Sidney Altman, they are RNA molecules that act as enzymes, that is, possess catalytic activity and can specifically cleave (cut) other RNA molecules. The ribozyme (RNA) molecule and the other RNA molecule come together; whereupon the ribozyme molecule cuts the other RNA molecule at a specific defined (three-base) site. Because the ribozyme molecule acts as an enzyme in this reaction, the ribozyme molecule is not consumed or destroyed, but goes on to cut other RNA molecules. See also RIBONUCLEIC ACID (RNA), ENZYME, and RIBOSOMES.

Ricin A lethal protein naturally produced in castor beans. In 1994, Robert J. Ferl and Paul C. Sehnke genetically engineered the tobacco plant to produce ricin. Attached to a pharmaceutical "guided missile" such as a monoclonal antibody or the CD4 protein, ricin is potentially useful for treatment against some tumors and has been investigated as a possible treatment against acquired immune deficiency syndrome (AIDS). See also IMMUNOTOXIN, MONOCLONAL ANTIBODIES (MAb), CD4 PROTEIN, GENETIC ENGINEERING, FUSION PROTEIN, FUSION TOXIN, SOLUBLE CD4, and GENISTEIN (Gen).

Rifkin, Jeremy Head of the Foundation on Economic Trends. See FOUNDATION ON ECONOMIC TRENDS.

Riken See RESEARCH FOUNDATION FOR MICROBIOLOGICAL DISEASES.

RMD See RAPID MICROBIAL DETECTION.

RN Gene See REDEMENT NAPOLE (RN) GENE.

RNA See RIBONUCLEIC ACID.

RNA Polymerase An enzyme that catalyzes the synthesis of a complementary mRNA (messenger RNA) molecule from a DNA (deoxyribonucleic acid) template in the presence of a mixture of the four ribonucleotides (ATP, UTP, GTP, and CTP). Also called transcriptase. See also CENTRAL DOGMA, POLYMERASE, DNA POLYMERASE, and PROMOTER.

RNA Probes See DNA PROBE.

RNA Transcriptase See RNA POLYMERASE.

RNA Vectors An RNA (ribonucleic acid) vehicle for transferring genetic information from one cell to another. See VECTOR and RETROVIRAL VECTORS.

Roving Gene See JUMPING GENES, TRANSPOSITION, TRANSPOSASE, GENE, GENOME, and DEOXYRIBONUCLEIC ACID (DNA).

Rps6 Gene A gene that confers to any soybean plant (possessing that gene in its DNA) resistance to some strains/races of phytophthora root rot (PRR) disease. See also GENE, DEOXYRIBONUCLEIC ACID (DNA), PHYTOPHTHORA ROOT ROT, and SOYBEAN PLANT.

Rps1c Gene A gene that confers to any soybean plant (possessing that gene in its DNA) resistance to several strains/races of phytophthora root rot (PRR) disease. See also GENE, DEOXYRIBONUCLEIC ACID (DNA), SOYBEAN PLANT, and PHYTOPHTHORA ROOT ROT.

Rps1k Gene A gene that confers to any soybean plant (possessing that gene in its DNA) resistance to as many as twenty-one strains/races of phytophthora root rot (PRR) disease. See also GENE, DEOXYRIBONUCLEIC ACID (DNA), PHYTOPHTHORA ROOT ROT, and SOYBEAN PLANT.

rRNA (ribosomal RNA) The nucleic acid component of ribosomes, making up two-thirds of the mass of the bacteria *Escherichia coli* ribosome, and about one-half of the mass of mammalian ribosomes. Ribosomal RNA accounts for nearly 80 percent of the RNA content of the bacterial cell. See also RIBOSOMES and *ESCHERICHIA COLI (E. COLI)*.

S

S1 Nuclease An enzyme that specifically degrades (destroys) single-stranded sequences of DNA. See also RESTRICTION ENDONUCLEASES, ENZYME, and DEOXYRIBONUCLEIC ACID (DNA).

SAGB Senior Advisory Group on Biotechnology. See SENIOR ADVISORY GROUP ON BIOTECHNOLOGY.

Salting Out A technique used for forcing (dissolved) proteins out of a solution by increasing the concentration of salt in the solution. The Na^+ and Cl^- ions derived from the salt compete for and "tie up" water molecules that are solubilizing the protein molecules thereby rendering them insoluble or more insoluble.

Sanitary and Phytosanitary (SPS) Agreement See WORLD TRADE ORGANIZATION (WTO) and SPS.

Sanitary and Phytosanitary (SPS) Measures See WORLD TRADE ORGANIZATION (WTO) and SPS.

Saponification Alkaline hydrolysis of triacyl glycerols to yield fatty acid salts. The molecules thus produced are known as surfactants (surface active agents) commonly called soap. The process of soapmaking. See also HYDROLYSIS.

Saponins A group of phytochemicals that is produced by certain plants (e.g., the soybean plant, ginseng, etc.). Evidence suggests that human consumption of saponins produced by soybeans can help to lower blood content of low-density lipoproteins (LDLP) and can help to prevent certain types of cancer. See also PHYTOCHEMICALS, SOYBEAN PLANT, LOW-DENSITY LIPOPROTEINS (LDLP), and CANCER.

Satellite DNA Many tandem repeats (identical or related) of a short basic repeating unit (in the DNA molecule). See also DEOXYRIBONUCLEIC ACID (DNA).

Saturated Fatty Acids (SAFA) Fatty acids containing fully saturated alkyl chains (on a molecule). This means that the carbon atoms comprising the chains are held together by one carbon-to-carbon bond and not two or three. High levels of dietary SAFA have been related to increased coronary heart disease (CHD) in humans. Beef fat typically contains approximately 54% saturated fatty acids. Sheep fat contains approximately 58% saturated fatty acids. Pork fat contains approximately 45% saturated fatty acids. Chicken fat contains approximately 32% saturated fatty acids. In general, fats possessing the highest levels of saturated fatty acids tend to be solid at room temperature; and those fats possessing the highest levels of unsaturated fatty acids tend to be liquid at room temperature. That rule of thumb was the original "dividing line" between "fats" and "oils," respectively. See also FATTY ACID, DEHYDROGENATION, MONOUNSATURATED FATS, SAPONIFICATION, LPAAT PROTEIN, UNSATURATED FATTY ACID, and POLYUNSATURATED FATTY ACIDS (PUFA).

Saxitoxins Paralytic poisons that are produced by certain shellfish. See also RICIN and RABIN.

SBO Soybean oil.

Scab See *FUSARIUM*.

Scale-Up The transition step in moving a (chemical) process from experimental (e.g., "test tube," small, bench) scale to a larger scale producing more or much more product that the bench scale (e.g., production of tons/year in a chemical plant). A process may require a number of scale-ups, which each scale-up producing more product than the last one.

Scanning Tunneling Electron Microscopy See ELECTRON MICROSCOPY (EM).

SCP See SINGLE-CELL PROTEIN.

SDM Site-directed mutagenesis. See SITE-DIRECTED MUTAGENESIS.

SDS Sodium dodecyl sulfate. Also known as sodium lauryl sulfate (SLS). A surfactant commonly used in biochemical and biotechnological applications for the solubilization of membrane components and hard-to-solubilize (dissolve) molecules. For example, it is often utilized at high concentration in water solution (e.g., along with potassium acetate) to dissolve plant DNA samples (e.g., when a scientist wants to sequence that sample of plant DNA). The SDS/PA in water solution helps the scientist to separate out contaminants that are commonly present in samples from plant tissues (i.e., polysaccharides, proteins, etc.) because DNA molecules are much more soluble in SDS/PA solution than are those contaminant molecules. Above a critical concentration (CMC), SDS forms micelles in water which are thought to be responsible for its solubilizing action. SDS is also used in such items as shampoo. See also CRITICAL

MICELLE CONCENTRATION, MICELLE, REVERSE MICELLE (RM), PROTEIN, SURFACTANT, DEOXYRIBONUCLEIC ACID (DNA), POLYSACCHARIDES, SEQUENCING (OF DNA MOLECULES), and HEXADECYLTRIMETHYLAMMONIUM BROMIDE (CTAB).

Seed-Specific Promoter See PROMOTER.

"Seedless" Fruits See TRIPLOID.

Selectins Also called LEC-CAMs (leukocyte-cell adhesion molecules). A class of molecular structurally related lectins that mediate (i.e., control, cause, etc.) the contacts between a variety of cells (e.g., leukocytes and endothelial cells); and function as cellular adhesion receptors. See also RECEPTORS, LECTINS, ADHESION MOLECULES, LEUKOCYTES, ENDOTHELIAL CELLS, ENDOTHELIUM, and SIGNAL TRANSDUCTION.

Self-Assembly (of a large molecular structure) The essentially automatic ordering and assembly of certain molecules into a large structure. Examples of such large molecular structures include micelles, reverse micelles, ribosomes, and Tobacco Mosaic Virus (TMV).

The first discovery of a self-assembling active biological structure occurred in 1955, when Heinz Frankel-Conrat and Robley Williams showed that TMV will reassemble into functioning, infectious virus particles (after TMV has been dissociated into its components via immersion in concentrated acetic acid). See also MICELLE, REVERSE MICELLE (RM), RIBOSOMES, TOBACCO MOSAIC VIRUS (TMV), and NANOCRYSTAL MOLECULES.

Semisynthetic Catalytic Antibody An antibody that is produced (e.g., via monoclonal antibody techniques) in response to a carefully selected antigen (i.e., one of the molecules involved in the chemical reaction that you are trying to catalyze). Such an antibody is then made to be catalytic by "attaching" a (molecular) group that is known to catalyze the desired chemical reaction. This attaching is done either via chemical modification of the antibody, or via genetic engineering of the cell (DNA) that produces that antibody. See also CATALYST, ANTIBODY, CATALYTIC ANTIBODY, SITE-DIRECTED MUTAGENESIS, MONOCLONAL ANTIBODIES (MAb), ANTIGEN, GENETIC ENGINEERING, and ABZYMES.

Senior Advisory Group on Biotechnology (SAGB) An association of approximately 35 of the largest European companies that are engaged in at least some form of genetic engineering research or production. Similar to America's Biotechnology Industry Organization (BIO), the SAGB works with governments and the public to promote safe and rational advancement of genetic engineering and biotechnology. It was formed in 1989 and is based in Brussels, Belgium. See also BIOTECHNOLOGY, GENETIC ENGINEERING, RECOMBINANT DNA (rDNA), JAPAN BIO-INDUSTRY ASSOCIATION, INTERNATIONAL FOOD BIOTECHNOLOGY COUNCIL, and BIOTECHNOLOGY INDUSTRY ASSOCIATION (BIO).

Sense Normal (forward) orientation of DNA sequence (gene) in genome. See GENE SILENCING and ANTISENSE (DNA SEQUENCE).

Sepsis Also known as systemic inflammatory response syndrome, this life-threatening condition ("septic shock") occurs when the body's immune system over-responds to infection (e.g., by gram-negative bacteria) in which release of bacterial endotoxin (lipopolysaccharide, or LPS) occurs. Those immune system cells (e.g., macrophages, etc.) overproduce numerous inflammatory agents (e.g., cytokines), which induce fever, shock, and sometimes organ failure. See GRAM-NEGATIVE (G−), BACTERIA, CYTOKINES, ENDOTOXIN, and MACROPHAGE.

Septic Shock See SEPSIS.

Sequence (of a DNA molecule) The specific nucleic acids that comprise a given segment of a DNA molecule. See also DEOXYRIBONUCLEIC ACID (DNA), GENETIC CODE, GENE, CHROMOSOMES, NUCLEIC ACIDS, SEQUENCING (OF DNA MOLECULES), and STRUCTURAL GENOMICS.

Sequence (of a protein molecule) The specific amino acids (and the order in which they are coupled together) that comprise a given segment of a protein molecule. See also PROTEIN, AMINO ACID, STRUCTURAL GENE, STRUCTURAL GENOMICS, and SEQUENCING (OF PROTEIN MOLECULES).

Sequence Map A pictorial representation of the sequence of amino acids in a protein molecule, the sequence of nucleic acids in a DNA molecule, or the sequence of oligosaccharide components in a glycoprotein/carbohydrate molecule. See also SEQUENCING (OF DNA MOLECULES), SEQUENCING (OF PROTEIN MOLECULES), SEQUENCING (OF OLIGOSACCHARIDES), SEQUENCE (OF A DNA MOLECULE), SEQUENCE (OF A PROTEIN MOLECULE), and RESTRICTION MAP.

Sequencing (of DNA molecules) The process used to obtain the sequential arrangement of nucleotides in the DNA backbone. The cleavage into fragments (followed by separation of those fragments, which can then be sequenced individually) of DNA molecules by one of several methods: (1) a chemical cleavage method followed by polyacrylamide gel electrophoresis (PAGE), (2) a method consisting of controlled interruption of enzymatic replication methods followed by PAGE, (3) a didexyl method utilizing fluorescent "tag" atoms attached to the DNA fragments, followed by use of spectrophotometry to identify the respective DNA fragments by their differing "tags" (which fluoresce at different wavelengths). This (fluorescent tag) variant of the dideoxy method can be automated to "decipher" large DNA molecules (i.e., genomes). Such automated machines are sometimes called "gene machines." See also POLYACRYLAMIDE GEL ELECTROPHORESIS (PAGE), GENE MACHINE, SEQUENCE (OF A DNA MOLECULE), and BASE EXCISION SEQUENCE SCANNING (BESS).

Sequencing (of oligosaccharides) See RESTRICTION ENDOGLYCOSIDASES and SEQUENCE MAP.

Sequencing (of protein molecules) The process used to obtain the sequential arrangement of amino acids in a protein molecule. See also PROTEIN, AMINO ACID, and SEQUENCE (OF A PROTEIN MOLECULE).

Sequon A (potential) site on a protein molecule's "backbone" where a sugar molecule (or a chain of sugar molecules, i.e., an oligosaccharide) may be attached. See also PROTEIN, SUGAR MOLECULES, GLYCOPROTEIN, GLYCOGEN, GLYCOSYLATION (TO GLYCOSYLATE), PROTEIN ENGINEERING, and OLIGOSACCHARIDES.

Serine (ser) A nonessential amino acid; a biosynthetic precursor of several metabolites, including cysteine, glycine, and choline. See also ESSENTIAL AMINO ACIDS.

Seroconversion The development of antibodies (specific to that disease-causing microorganism) in response to vaccination or natural exposure to a disease-causing microorganism. See also SEROLOGY, ANTIBODY, IMMUNOGLOBULIN (IgA, IgE, IgG, and IgM), HUMORAL IMMUNITY, PATHOGEN, POLYCLONAL ANTIBODIES (USED IN HUMANS), and PASSIVE IMMUNITY.

Serologist See SEROLOGY.

Serology A subdiscipline of immunology, concerned with the properties and reactions of blood sera. It includes the diverse techniques used for the "test tube" measurement of antibody-antigen reactions, including blood typing (e.g., for transfusions). See also MAJOR HISTOCOMPATIBILITY COMPLEX (MHC), OLIGOSACCHARIDES, and SERUM LIFETIME.

Seronegative Refers to negative results of a serology test. See also SEROLOGY, HUMORAL IMMUNITY, and ANTIBODY.

Serotonin An important neurochemical whose effects upon the human brain include mood elevation. Production of serotonin in the brain is increased by ingestion of the amino acid tryptophan (a chemical precursor to serotonin) and the pharmaceutical anti-depressant Prozac (trademarked product of Eli Lilly & Company). In 1997, Marianne Regard and Theodor Landis discovered that humans afflicted with hemorrhagic lesions in the brain (cause of abnormal serotonin activation/production) often became "passionate culinary afficionados." See also TRYPTOPHAN (trp), ESSENTIAL AMINO ACIDS, BLOOD-BRAIN BARRIER (BBB), and NEUROTRANSMITTER.

Serotypes A variety (sub-strain) of a microorganism that is distinguished from others in (strain) via its serological effects (within immune system of the host organism it inhabits). See also BACTERIA, STRAIN, *ESCHERICHIA COLIFORM* 0157:H7 *(E. COLI* 0157:H7*)*, SEROLOGY, and HUMAN IMMUNODEFICIENCY VIRUS TYPES 1 AND 2.

Serum Blood plasma that has had its clotting factor removed. See also FACTOR VIII and PLASMA.

Serum Half Life See SERUM LIFETIME.

Serum Immune Response See HUMORAL IMMUNITY.

Serum Lifetime The average length of time that a molecule circulates in an organism's bloodstream before it is cleared from the bloodstream. See also IMMUNE RESPONSE and ANTIGEN.

Sessile (Micro)organisms that are attached to a (support) substrate directly by their base; not attached via an intervening peduncle (i.e., stalk). Can also refer to fruit or leaves that are attached directly to the main stem or branch of a plant. See also VAGILE.

Sex Chromosomes Those chromosomes whose content is different in the two sexes of a given species. They are usually labeled X and Y (or W and Z); one sex has XX (or WW), the other sex has XY (or WZ). XX (WW) is female and XY (WZ) is male.

Sexual Conjugation An infrequent occurrence in which two adjacent bacteria stretch out portions of their (cell) membranes to touch one another, fuse; and then pass transposons, jumping genes, or plasmids to each other. See also ASEXUAL, BACTERIA, CELL, CONJUGATION, PLASMID, TRANSPOSON, and JUMPING GENES.

Shotgun Cloning Method A technique for obtaining the desired gene that involves "chopping up" the entire genetic complement of a cell using restriction enzymes, then attaching each (resultant) DNA fragment to a vector and transferring it into a bacterium, and finally screening those (engineered) bacteria to locate the bacteria that are producing the desired product (e.g., a protein). See also GENETIC ENGINEERING, GENOME, RESTRICTION ENDONUCLEASES, and VECTOR.

"Shotgun" Method [to introduce foreign (new) genes into plant cells] A technique for gene-into-cell introduction in which the gene is attached to tiny "bullets" made of tungsten or other metal. By means of a special device ("gene gun") the tiny particles are then literally "shot" through the plasma membrane into plant cells with:

(a) High-pressure gas (e.g., the GENEBOOSTER® gun developed at Hungary's Agricultural Biotechnology Center utilizes nitrogen).

(b) A rather conventional firearm (sometimes called a particle gun) which uses a .22 caliber shell minus the lead tip. The tiny particles are used in place of the lead tip. For example, the BIOLISTIC® Gene Gun invented at America's Cornell University utilizes "bullets" made of tungsten.

Some plant cells are destroyed in the process and the survivors heal (provided the "bullet" is small enough), and incorporate (some) of the new genetic material into their genetic complement, and produces whatever product (i.e., a protein) the newly introduced gene codes for. See also *AGROBACTERIUM TUMEFACIENS,* CODING SEQUENCE, GENETIC ENGINEERING, VECTOR, BIOLISTIC® GENE GUN, "EXPLOSION" METHOD [TO INTRODUCE FOREIGN (NEW) GENES INTO PLANT CELLS], and ELECTROPORATION.

Shuttle Vector A vector capable of replicating in two unrelated species. See also VECTOR and REPLICATION (OF VIRUS).

Signal Transduction The "reception" and "conversion" of a "chemical message" (e.g., hormone) by a cell. For example, G-proteins (which are embedded in the surface membrane of certain cells, but extend through to outside and inside of the membrane) accomplish signal transduction. When a hormone, drug, neurotransmitter, or other signal chemical binds to the receptor (on the exterior of the cell membrane), the receptor activates the G-protein, which causes an effector inside cell to produce a "signal" chemical inside cell, which causes the cell to react to the original external chemical signal received. See also CELL, RECEPTORS, NUCLEAR RECEPTORS, SIGNALLING, G-PROTEINS, MAST CELLS, CD95 PROTEIN, HORMONE, SUBSTANCE P, and LECITHIN.

Signalling The "communication" that occurs between and within cells of an organism, e.g., via hormones. Such signalling "tells" certain cells to grow, change, or produce specific proteins at specific times. See also RECEPTORS, PROTEIN, NUCLEAR RECEPTORS, G-PROTEINS, SIGNAL TRANSDUCTION, CELL, CD95 PROTEIN, HORMONE, PARKINSON'S DISEASE, SUBSTANCE P, LECITHIN, and NITRIC OXIDE.

Silent Mutation A mutation in a gene that causes no detectable change in the biological characteristics of that gene's product (e.g., a protein). See also EXPRESS, GENE, and PROTEIN.

Silk A natural, protein polymer with a predominance of alanine and glycine amino acids. Silk is produced by silkworms that have fed on mulberry tree leaves. The body of a silkworm can retain proteins (i.e., raw material for silk) amounting to as much as 20% of its body weight. It is thought that silk may be altered, via genetic engineering of silkworms, to produce fibers of very high strength. See also GENETIC ENGINEERING, PROTEIN ENGINEERING, and AMINO ACID.

Simple Protein A protein that yields only amino acids on hydrolysis (i.e., cleavage into fragments) and does not have other constituents such as lipids or polysaccharide attachments. See also PROTEIN, AMINO ACID, GLYCOPROTEIN, and POLYSACCHARIDES.

Simple Sequence Repeat (SSR) DNA Marker Technique A "genetic mapping" technique that utilizes the fact that microsatellite sequences "repeat" (appear repeatedly in sequence within the DNA molecule) in a manner enabling them to be used as "markers." See GENETIC MAP, SEQUENCE (OF A DNA MOLECULE), RANDOM AMPLIFIED POLYMORPHIC DNA (RAPD) TECHNIQUE, RESTRICTION FRAGMENT LENGTH POLYMORPHISM (RFLP) TECHNIQUE, DEOXYRIBONUCLEIC ACID (DNA), PHYSICAL MAP (OF GENOME), LINKAGE, LINKAGE GROUP, MARKER (GENETIC MARKER), LINKAGE MAP, TRAIT, and MICROSATELLITE DNA.

Single-Cell Protein (SCP) Protein that is derived from single-celled organisms with a high protein content. Yeast is an example. Generally used in regard to those organisms that are edible by domesticated animals, or humans.

Single-Domain Antibodies (dAbs) VH "heavy chains" (portion of antibody molecules) produced by genetically engineered *Escherichia coli* cells that act to bind antigens in a manner similar to antibodies or monoclonal antibodies (MAbs). Similar to MAbs, dAbs can be produced in large quantities, to be used as human or animal therapeutics (e.g., to combat diseases). See also ANTIBODY, MONOCLONAL ANTIBODY (MAb), ANTIGEN, and *ESCHERICHIA COLI (E. COLI)*.

Site-Directed Mutagenesis (SDM) A technique that can be used to make a protein that differs slightly in its structure from the protein that is normally produced (by an organism or cell). A single mutation (in the cell's DNA) is caused by hybridizing the region in a codon to be mutated with a short, synthetic oligonucleotide. This causes the codon to code for a *different* specific amino acid in the protein gene product. Site-directed mutagenesis holds the potential to enable man to create modified (engineered) proteins that have desirable properties not currently available in the proteins produced by existing organisms. See also MUTANT, MUTATION, POINT MUTATION, PROTEIN, GENE, INFORMATIONAL MOLECULES, HEREDITY, GENETIC CODE, GENETIC MAP, AMINO ACID, DEOXYRIBONUCLEIC ACID (DNA), CODON, OLIGONUCLEOTIDE, and PROTEIN ENGINEERING.

Sitostanol A chemical (ester) that is derived from sitosterol (a sterol that is present in pine trees, and can be extracted from those trees). It can also be extracted from fibers (e.g., the hull or seed coat) of corn/maize (*Zea mays*) or soybeans (*Glycine max L.*). When sitostanol is ingested by humans in sufficient quantities, it causes their total serum cholesterol and their low-density lipoproteins (LDLP) to be lowered by approximately 10%, via inhibition (i.e., the sitostanol is absorbed by the gastrointestinal system instead of cholesterol). See also ABSORPTION, DIGESTION (WITHIN ORGANISMS), SOYBEAN PLANT, LOW-DENSITY LIPOPROTEINS (LDLP), SERUM LIFETIME, and CHOLESTEROL.

SK See SUBSTANCE K.

Slime An extracellular (i.e., outside of the cell) material that is produced by some (micro)organisms, characterized by a slimy consistency. The slime is of varied chemical composition. However, usual components are polysaccharides (polysugars) and specific protein molecules.

Sodium Dodecyl Sulfate See SDS.

Sodium Lauryl Sulfate See SDS.

Soluble CD4 A synthetic version of the CD4 protein that may interfere with the ability of HIV (i.e., AIDS) viruses to infect the human immune system cells with the acquired immune deficiency syndrome (AIDS) virus. See also CD4 PROTEIN, ADHESION MOLECULES, SELECTINS, LECTINS, and PROTEIN.

Soluble Fiber See WATER SOLUBLE FIBER.

Somaclonal Variation The genetic variation (i.e., new traits) that results from the growing of entire new plants from plant cells or tissues (e.g., maintained in culture). Frequently encountered when plants are regenerated (grown) from plant cells that have been altered via genetic engineering. However, somaclonal variation (i.e., new genetic traits) can occur even when plants are regenerated from cells that were part of the same original plant. See also CELL CULTURE, SOMATIC VARIANTS, CLONE (AN ORGANISM), *AGROBACTERIUM TUMEFACIENS*, BIOLISTIC® GENE GUN, "EXPLOSION" METHOD [TO INTRODUCE FOREIGN (NEW) GENES INTO PLANT CELLS], and "SHOTGUN" METHOD [TO INTRODUCE FOREIGN (NEW) GENES INTO PLANT CELLS].

Somatacrin See GROWTH HORMONE–RELEASING FACTOR (GRF or GHRF).

Somatic Cells All eucaryote body cells except the gametes and the cells from which they develop. See GAMETE and OOCYTES.

Somatic Variants Regenerated plants (i.e., clones) that were derived (produced) from cells that originally came from the same plant—that are *not* genetically identical. Such plants (clones) are called "sports" or somatic variants because they vary (genetically) from the "parent" plant. Sometimes, such somatic variants are developed by man to become a new plant variety (e.g., the nectarine is an example of this). See also SOMACLONAL VARIATION, CELL CULTURE, CLONE (AN ORGANISM), and GENOTYPE.

Somatomedins A family of peptides that mediates the action of growth hormone on skeletal tissue, and stimulates bone formation. See also HUMAN GROWTH HORMONE (HGH), PEPTIDE, and BONE MORPHOGENETIC PROTEIN (BMP).

Somatostatin A 14 amino acid peptide that inhibits the release of growth hormone. See also HUMAN GROWTH HORMONE (HGH) GROWTH HORMONE–RELEASING FACTOR (GRF or GHRF), and PEPTIDE.

Somatotropin Category of hormone that is produced naturally in the bodies of all mammals, including man. See HORMONE, GROWTH HORMONE, BOVINE SOMATOTROPIN (BST), and PORCINE SOMATOTROPIN (PST).

SOS Protein See SOS RESPONSE (IN *ESCHERICHIA COLI* BACTERIA).

SOS Response (in *Escherichia coli* bacteria) The "switching on" of genetic repair machinery in this bacteria when its DNA has been damaged (e.g., by radiation). See also *ESCHERICHIA COLI (E. COLI)*.

Soybean Cyst Nematodes (SCN) Microscopic round worms living in the soil, which feed parasitically on roots of the soybean plant. The nematodes use a spear-like mouthpart, called a stylet, to puncture the plant's root cells so the nematodes can eat their cell contents. That root damage causes the soybean's growth to be stunted, and the plants turn yellow because of a reduction in nodule formation by the nitrogen-fixing bacteria (which normally colonize roots of soybean plants). SCN can combine with a fungus to cause a soybean plant disease known as "sudden death syndrome."

As part of Integrated Pest Management (IPM), farmers can utilize the parasitic *Pasteuria* bacteria to help control the soybean cyst nematodes. The *Pasteuria* bacteria must attach their spores (for reproduction) to juvenile nematodes, so that the *Pasteuria* offspring can consume the SCN when the spores later germinate. See also SOYBEAN PLANT, NITROGEN FIXATION, BACTERIA, and FUNGUS.

Soybean Meal See SOYBEAN PLANT.

Soybean Oil An edible oil that is produced within its beans (seeds) by the soybean plant (botanical name *Glycine max (L.) Merrill*). When removed from soybeans via crushing and refining processes, soybean oil is (historical average) composed of 60.8% polyunsaturated fatty acids (PUFA), 24.5% monounsaturated fatty acids, and 15.1% saturated fatty acids. However, soybean varieties have recently been created that possess as little as 7% saturated fatty acids. See also POLYUNSATURATED FATTY ACIDS (PUFA), FATTY ACID, ESSENTIAL FATTY ACIDS, LECITHIN, HYDROGENATION, SOYBEAN PLANT, and HIGH-OLEIC OIL SOYBEANS.

Soybean Plant *Glycine max (L.) Merrill.* A green, bushy legume that is the world's single largest provider of protein and edible oil for mankind's use. This summer annual plant varies in height from less than a foot (0.3 meter) to more than three feet (1.0 meter) tall. The seeds (soybeans) are borne in pods, and historically have contained 13%–26% oil and 38%–45% protein (on a moisture-free basis).

The traditional soybean, possessing (average) 20% oil content, contains an average of 3% stachyose within its meal (i.e., the solids remaining after the soybean oil is removed). See also FATTY ACID, PROTEIN, LECITHIN, NITROGEN FIXATION, SOYBEAN OIL, SOYBEAN CYST NEMATODES (SCN), RESVERATROL, PHYTOPHTHORA ROOT ROT, PEROXIDASE, ISOFLAVONES, LOW-STACHYOSE SOYBEANS, GENISTEIN, LIPOXYGENASE (LOX), SAPONINS, CANOLA, CHLOROPLAST TRANSIT PEPTIDE (CTP), and HERBICIDE-TOLERANT CROP.

SP See SUBSTANCE P.

Species A single type (taxonomic group) or organism as determined by the distinguishing characteristics used for the particular group of life forms (e.g., the horse is one species among the mammals). See also STRAIN and SYSTEMATICS.

Species Specific Refers to a compound (e.g., a protein) or a disease (e.g., a viral infection) or some other effect that only acts in/on one specific species of organism. For example, the antibiotic penicillin kills bacteria by blocking an enzyme which is critical for growth and repair of the bacterial cell wall (i.e., peptidoglycan layer), but penicillin does not harm other species (e.g., man). For instance, bovine somatotropin is a protein hormone that increases growth rate of young cattle and also increases the efficiency of mature cows in converting their feed into milk. Bovine somatotropin has no effect in humans, and (if eaten) is simply digested like any other food protein. It appears that most growth hormones are species specific. See SPECIES, HORMONE, and PENICILLIN G.

Specific Activity An enzyme unit defined as the number of moles of substrate converted to product by an enzyme preparation per unit time under specified conditions of pH, substrate concentration, temperature, etc. Specific enzyme activity units may be expressed as: moles of product produced/minute/mg of protein used (or mole of enzyme used if the preparation is pure). See also MOLE, ENZYME, and SUBSTRATE (CHEMICAL).

Spectrophotometer An instrument that measures the concentration of a compound that has been dissolved in a solvent (such as water, alcohol, etc.). The instrument shines a light through the solution, measures the fraction of the light that is absorbed by the solution, and calculates the concentration from that absorbance value. See also OPTICAL DENSITY (OD) and ABSORBANCE.

Splicing The removal of introns and joining of exons in RNA (e.g., genes). Thus, introns are spliced out, while exons are spliced together. See also EXON, INTRON, GENETIC ENGINEERING, and RIBONUCLEIC ACID (RNA).

Splicing Junctions The sequences (in RNA molecules) of nucleotides immediately surrounding the exon-intron boundaries. See also EXON, INTRON, and SPLICING.

Spontaneous Assembly See SELF-ASSEMBLY (OF A LARGE MOLECULAR STRUCTURE).

SPS Acronym for the Sanitary and Phytosanitary Standards Agreement of the World Trade Organization (WTO), a multinational trading agreement that "sets the rules" that govern international trade. Sanitary (i.e., human and animal) and phytosanitary (i.e., plant) standards are important in preventing the transfer of diseases from one nation to another via international trade. SPS standards are designed to protect animal, plant, and human life/health (within WTO member countries) from:

- entry of pests (e.g., insects, weeds, etc.)
- entry of disease-carrying organisms (e.g., European Corn Borer)
- entry of disease-causing organisms (e.g., *Aspergillus flavus*)
- toxins, contaminants, or disease-causing organisms in foods, beverages, or feedstuffs

See also INTERNATIONAL PLANT PROTECTION CONVENTION (IPPC), INTERNATIONAL OFFICE OF EPIZOOTICS (OIE), CODEX ALIMENTARIUS COM-

MISSION, MAXIMUM RESIDUE LEVEL (MRL), WORLD TRADE ORGANIZATION (WTO), EUROPEAN CORN BORER (ECB), and *ASPERGILLUS FLAVUS*.

Squalamine A potent antimicrobial agent (steroid, antibiotic) that was discovered in the tissues of the dogfish shark in 1992. It has been found to be active against a broad spectrum of bacteria, protozoa, and fungi. Squalamine was chemically synthesized by man in 1993. See also MAGAININS, STEROID, FUNGUS, BACTERIA, BACTERIOCINS, PROTOZOA, and ANTIBIOTICS.

SRB (sulfate reducing bacterium) Any organism that metabolically reduces sulfate to H_2S (hydrogen sulfide). This includes a variety of microorganisms. See also REDUCTION (IN A CHEMICAL REACTION), METABOLISM, MICROORGANISM, and FERROBACTERIA.

SSR See SIMPLE SEQUENCE REPEAT (SSR) DNA MARKER TECHNIQUE.

Stachyose A carbohydrate (oligosaccharide) that is naturally produced in soybeans. It is relatively insoluble in water, and less available for digestion by monogastric animals (e.g., swine, poultry) than the other carbohydrate components within soybeans. See also CARBOHYDRATES (SACCHARIDES), LOW-STACHYOSE SOYBEANS, and OLIGOSACCHARIDES.

"Stacked" Genes Refers to the insertion of two or more (synthetic) genes into the genome of an organism. One example of that would be a plant into which has been inserted a gene from *Bacillus thuringiensis (B.t.)* and a gene for resistance to a specific herbicide. See GENE, BIOTECHNOLOGY, GENETIC ENGINEERING, *BACILLUS THURINGIENSIS (B.t.)*, *B.t. KURSTAKI*, GENETICALLY ENGINEERED MICROBIAL PESTICIDES (GEMP), EPSP SYNTHASE, PAT GENE, and BAR GENE.

Staggered Cuts Scissions (cuts) made in duplex DNA when the two strands of DNA that make up the duplex DNA are cleaved at different points near each other by restriction endonucleases. What is produced is a single-stranded structure (in which the single strands are a number of nucleotide bases long) with a double-stranded core section. This core section is much longer than the single-stranded region. See also DEOXYRIBONUCLEIC ACID (DNA), RESTRICTION ENDONUCLEASES, and STICKY ENDS.

Stanol Ester See SITOSTANOL.

Stanol Fatty Acid Esters See SITOSTANOL and FATTY ACID.

Starch A polymer of glucose molecules (i.e., a polysaccharide) used by plants to store energy. Starch is broken down by enzymes (amylases) to yield glucose, which can be used as an energy source. The analogous polymer that is used by mammalian systems is called glycogen or, in old usage, animal starch. See also GLUCOSE (GLc), ENZYME, CORN, and AMYLOSE.

Startpoint Refers to the position on a DNA molecule corresponding to the first base incorporated into mRNA. See also DEOXYRIBONUCLEIC ACID (DNA), MESSENGER RNA (mRNA), EXON, and RIBONUCLEIC ACID (RNA).

Stearate (stearic acid) A saturated fatty acid containing eighteen carbon atoms in its molecular "backbone"; which is essentially neutral in effect on coronary heart disease in humans (i.e., doesn't appreciably increase low-density lipoproteins in the bloodstream). Because of the heart disease neutrality, stearate-containing oils (e.g., high-stearate soybean oil) are an excellent cooking oil choice; with the resistance to oxidation/breakdown of a saturated fatty acid, but no bloodstream-cholesterol increasing effect. In the mid-1990s, the American Cocoa Research Institute/Chocolate Manufacturers Association filed a petition with America's Food and Drug Administration (FDA) to differentiate stearate from the other saturated long-chain fatty acids used as food ingredients. See also FATTY ACID, LOW-DENSITY LIPOPROTEINS (LDLP), SATURATED FATTY ACIDS, and FOOD AND DRUG ADMINISTRATION (FDA).

Stearic Acid See STEARATE.

Stearoyl-ACP Desaturase An enzyme that is naturally produced in oilseed plants. It plays the central role in determining the ratio of saturated to unsaturated fatty acids (in the vegetable oils produced from such plants). Scientists may be able to eventually genetically engineer this enzyme into commercially important vegetable oil producing plants, to modify the ratio of saturated to unsaturated fatty acids. See also FATTY ACID, ENZYME, GENETIC ENGINEERING, GENETIC CODE, and LAURATE.

Stem Cell Growth Factor (SCF) A growth factor (glycoprotein hormone) that acts upon stem cells in a wide variety of ways to increase growth, proliferation, and maturity (into red blood cells or white blood cells). See also STEM CELLS, GROWTH FACTOR, HORMONE, GLYCOPROTEIN, TOTIPOTENT STEM CELLS, and COLONY STIMULATING FACTORS (CSFs).

Stem Cell One The single stem cell in the bone marrow of a fetus from which every immune system cell in the adult has been derived. The primordial stem cell is stimulated to develop into the mature immune system's differentiated, specialized cells by interleukin-7. See also STEM CELLS, TOTIPOTENT STEM CELLS, and INTERLEUKIN-7 (IL-7).

Stem Cells Bone marrow cells, some of which eventually mature into red blood cells or white blood cells. The stem cells that remain in the bone marrow maintain their own numbers by self-renewal divisions, yielding more cells to start the maturation process. This maturation process is stimulated and controlled by stem cell growth factor (SCF), granulocyte colony stimulating factor (G-CSF), and by granulocyte-macrophage colony stimulating factor (GM-CSF). See GRANULOCYTE-MACROPHAGE COLONY STIMULATING FACTOR (GM-CSF), WHITE BLOOD CELLS, STEM CELL ONE, STEM CELL GROWTH FACTOR (SCF), and TOTIPOTENT STEM CELLS.

Stereoisomers Molecules that have the same structural formula but different spatial arrangements of dissimilar groups (of atoms) bonded to a common atom (in the molecule). Many of the physical and chemical properties of stereoisomers are the same, but there are differences in the crystal structures, in the direction in which they rotate polarized light (which has been passed through a solution of the stereoisomer), and in their use in an enzyme-catalyzed (biological) reaction. See also RACEMATE, POLARIMETER, DEXTROROTARY (D) ISOMER, EPIMERS, ISOMER, LEVOROTARY (L) ISOMER, ISOMERASE, and DIASTEREOISOMERS.

Steric Hindrance This term refers to the compression that a group (chemical entity) suffers by being too close to its nonbonded neighbors. If an enzyme and a substrate try to come together in order to react, but the substrate has on it a bulky group that disallows close contact between the two (because the group bumps into the enzyme), then the reaction will not occur because of steric hindrance. Seen in another way, two chemical groups bump into each other and cannot get by each other because they are held in place by the bonds binding them to other atoms. Hindrance of movement or activity occurs because chemical groups bump into each other and cannot occupy the same space. See also REPRESSION (OF AN ENZYME), INHIBITION, and COREPRESSOR.

Sterile (environment) One that is free of any living organisms or spores. For example, a hypodermic needle that has been sterilized (e.g., by heating it) and is free of living microorganisms is said to be sterile.

Sterile (organism) One that is unable to reproduce. For example, a bull that is castrated is rendered sterile. See also TRIPLOID.

Sterilization See STERILE (ENVIRONMENT) and STERILE (ORGANISM).

Steroid A chemical compound composed of a series of four carbon rings joined together to form a (molecular) structural unit called cyclopentanoperhydrophenanthrene. Any of a group of naturally occurring, fat-soluble substances, essential to life, usually classed as lipids. Steroids of importance to the body are the sterols, which are bile acids (produced by the liver, characterized by the presence of a carboxyl group in the molecule's side chain), and the hormones of the sex glands and the adrenal cortex. In addition, the plant kingdom possesses a wide variety of steroid glycosides. See also GLYCOSIDE, LIPIDS, and HORMONE.

Sticky Ends Complementary single strands of DNA (deoxyribonucleic acid) that protrude from opposite ends of a DNA duplex or from ends of different DNA duplex molecules. They can be generated be staggered cuts in DNA. They are called "sticky" because the exposed single strands can bind (stick) to complementary single strands on another DNA molecule. A hybrid piece of DNA is hence produced (by that binding). See also STAGGERED CUTS, HYBRIDIZATION (MOLECULAR GENETICS), DUPLEX, ANNEAL, DEOXYRIBONUCLEIC ACID (DNA), BLUNT-END LIGATION, and RESTRICTION ENDONUCLEASES.

Strain A group or organisms of the same species that possesses distinctive genetic characteristics that set it apart from others within the same species, but which differences are not "severe" enough for it to be considered a different breed or variety (of that species). The basic taxonomic unit of microbiology. Can also be used to designate a population of cells derived from a single cell. See also SPECIES, CELL, and CLONE (AN ORGANISM).

Stress Proteins Discovered by Italian biologist Ferruchio Ritossa in the 1960s, these molecules are also called heat-shock proteins. Proteins made by many organisms' (plant, bacteria and mammal) cells when those cells are stressed by environmental conditions such as certain chemicals, pathogens, or heat.

When corn/maize (*Zea mays L.*) is stressed during its growing season by high nighttime temperatures, that plant switches from its normal production of (immune system defense) chitinase to production of heat-shock (i.e., stress) proteins, instead.

Stress proteins are also produced by tuberculosis and leprosy bacteria after these bacteria have invaded (i.e., infected) cells in the human body, in an attempt by those bacteria to mimic the stress proteins that (mammal) cells would normally manufacture to repair damage done to the (mammal) cells. This mimicry makes it more difficult for the immune system to recognize and attack those pathogenic bacteria (and/or repair misshaped protein molecules in the body's cells). In 1996, Richard I. Morimoto discovered that two stress proteins known as HSP 90 and HSP 70 help to ensure that certain crucial proteins in cells are folded into the configuration/conformation needed by that cell. See also ANTIGEN, IMMUNE RESPONSE, PATHOGEN, PROTEIN, PROTEIN FOLDING, CONFORMATION, CHAPERONES, PROTEIN STRUCTURE, ABSOLUTE CONFIGURATION, PRION, CHITINASE, AFLATOXIN, and LIPOXYGENASE (LOX).

Stromelysin (MMP-3) A collagenase (enzyme) that "clears a path" through living tissue, ahead of tumor cells, thereby enabling a cancer to spread within the body. See also COLLAGENASE, ENZYME, CANCER, and TUMOR.

Structural Biology See STRUCTURAL GENE.

Structural Gene A gene that codes for any RNA (ribonucleic acid) or protein product other than a regulator molecule. It determines the primary sequences (i.e., the amino acid sequences) of a polypeptide (protein). See also EXPRESS, POLYPEPTIDE (PROTEIN), AMINO ACID, and RIBONUCLEIC ACID (RNA).

Structural Genomics Study of, or discovery of where (gene) sequences are located within the genome, and what (DNA) subunits comprise those sequences. See GENE, SEQUENCE (OF A DNA MOLECULE), DEOXYRIBONUCLEIC ACID (DNA), SEQUENCING (OF DNA MOLECULES), and GENOME.

STS Sulfonylurea (Herbicide)-Tolerant Soybeans These are soybeans that have been bred (via insertion of ALS gene) to resist the (weed killing) effects of sulfonylurea-based herbicides. The ALS gene was discovered by Scott Sebastian in 1986. See also GENE, GENETIC ENGINEERING, HTC, ALS GENE, BAR GENE, PAT GENE, EPSP SYNTHASE, GLYPHOSATE OXIDASE, and HERBICIDE-TOLERANT CROP.

Stx Shiga-like toxins. See TOXIN, TOXIGENIC *E. COLI,* ENTEROHEMORRHAGIC *E. COLI*, *ESCHERICHIA COLIFORM* 0157:H7.

Substance K See TACHYKININS.

Substance P A neuropeptide (i.e., peptide produced by cells of the nervous system) which is involved in activation of the immune system, pain sensation, and (when in excess) some psychiatric disorders.

In the case of chronic, intractable pain (hypersensitivity), approximately one percent of the nerve cells in the human spine process substance P (thereby "transmitting" its pain message via signal transduction). In 1997, Patrick Mantyh showed that killing those (one percent) cells relieved chronic pain hypersensitivity without impairing sense of touch or normal (beneficial) pain sensation, in humans. See TACHYKININS, PROTEIN, POLYPEPTIDE (PROTEIN), SIGNAL TRANSDUCTION, SIGNALLING, PEPTIDE, and NEUROTRANSMITTER.

Substantial Equivalence See CANOLA and ORGANIZATION FOR ECONOMIC COOPERATION AND DEVELOPMENT (OECD).

Substantially Equivalent See SUBSTANTIAL EQUIVALENCE.

Substrate (chemical) The substance acted upon, for example, by an enzyme. For example, the enzyme amylase breaks starch down into glucose molecules; starch is the substrate (of the enzyme amylase). See ENZYME, CATALYST, and SUBSTRATE (STRUCTURAL).

Substrate (in chromatography) The (usually solid or gel) substance that attracts and noncovalently binds (interacts) with one or more of the molecules in a solution that is passed over that substrate (e.g., in a chromatography column). This preferential binding (interaction with the substrate) enables one or more of the solution's molecular ingredients to be separated from the other(s). See also CHROMATOGRAPHY.

Substrate (structural) The substance (support) to which the agent of interest (a molecule) is attached. For example, some catalyst molecules are chemically attached to nonreactive solids to preserve the catalyst from being flushed away when the chemical substrate (the molecule to be converted by the catalyst) is washed by the catalyst immobilized on the structural substrate. See SUBSTRATE (CHEMICAL), CATALYST, and HYBRIDIZATION SURFACES.

Sudden Death Syndrome A disease that sometimes afflicts soybean plants. See SOYBEAN PLANT and SOYBEAN CYST NEMATODES (SCN).

Sugar Molecules See OLIGOSACCHARIDES, POLYSACCHARIDES, MONO-

SACCHARIDES, CARBOHYDRATES (SACCHARIDES), ALDOSE, GLYCOBIOLOGY, PYRANOSE, GLUCOSE (GLc), FURANOSE, and GLYCOPROTEIN.

Sulfate Reducing Bacterium See SRB.

Superantigens Certain types of antigens that activate a large proportion of an organism's immune system T cells. These superantigens, which thus over-activate the organism's immune system, are thought to be responsible for some autoimmune diseases (in which T cells attack and destroy the organism's own, healthy tissues). See also ANTIGEN, T CELLS, and AUTOIMMUNE DISEASE.

Supercoiling Also known as superhelicity. The coiling of a closed duplex DNA (deoxyribonucleic acid molecule) in space so that it crosses over its own axis. See also DEOXYRIBONUCLEIC ACID (DNA), HELIX, DUPLEX, DOUBLE HELIX, and POSITIVE SUPERCOILING.

Supercritical Carbon Dioxide A solvent that, when combined with water and an appropriate surfactant (e.g., fluoroethers), forms a solvent system that can effectively dissolve large biological molecules without causing those molecules to lose biological activity. Carbon dioxide is a gas at normal (atmospheric) pressure and ambient temperature, but in its *supercritical state*—temperature above 31.3°C (88°F) and pressure greater than 72.9 atmospheres—carbon dioxide becomes a dense (sort of) liquid. Some coffee processors have used supercritical carbon dioxide as a solvent to remove caffeine from coffee.

In 1995, Keith Johnston added the surfactant ammonium carboxylate perfluoropolyether to a supercritical carbon dioxide system containing water; and proved that the large biological molecule *bovine serum albumin* dissolved inside the micelles that form via water droplet surrounded by fluoroether molecules. Subsequent to that, Eric Beckman proved that the protease *subtilisin Carlsberg* can be extracted from crude (impure) cell broth because that protease preferentially dissolves in a supercritical carbon dioxide/water system containing fluoroether amphiphiles as surfactants. See also BIOLOGICAL ACTIVITY, SURFACTANT, MICELLE, REVERSE MICELLE (RM), BROTH, PROTEASE, and SUPERCRITICAL FLUID.

Supercritical Fluid A material that has been heated to a temperature above its (normal atmospheric pressure) boiling point, but which is kept in a state that resembles a liquid via the application of high pressure. For example, water will remain "liquid" up to a temperature of 375°C (617°F) if it is placed under enough pressure. Ammonia will remain "liquid" up to a temperature of 133°C (271°F) if it is placed under enough pressure, despite the fact that ammonia normally becomes a gas (at std. atmospheric pressure) whenever the temperature is higher than –33.35°C (–30°F).

These supercritical fluids have unique properties (e.g., they are often better solvents than their true liquid forms). Some supercritical fluids (e.g., supercritical carbon dioxide) can be used to extract biological molecules (e.g., chlorophyll) from mixtures (e.g., ground up plant leaves). After the biological molecule has dissolved out of the mixture, the biological molecule is recovered by releasing pressure so the carbon dioxide returns to gaseous form, and drifts away. See also SUPERCRITICAL CARBON DIOXIDE.

Superoxide Dismutase (SOD) See HUMAN SUPEROXIDE DISMUTASE (hSOD).

Suppressor Gene A gene that can reverse the effect of a specific type of mutation in other genes, such as a premature termination sequence. See GENE and TRANSWITCH®.

Suppressor Mutation A mutation that totally or partially restores a function that was lost by a primary mutation. It is located at a site in the gene different from the site of the primary mutation. See also GENE.

Suppressor T Cells Those T cells (thymus-derived lymphocytes) that are triggered (after other types of T cells and other immune system cells have successfully fought off an infection) to gradually slow down and halt the body's immune response (to the now-conquered pathogen). Discovered by Tomio Tada in 1971, suppressor T cells suppress B cell activity. Failure to halt the immune response in time could lead to harm to the body by its own immune system. The B and T lymphocytes are indistinguishable in size and general morphology. Only the existence or nonexistence of certain proteins on their cell surfaces distinguishes the two classes of lymphocytes. See also CELLULAR IMMUNE RESPONSE, PATHOGEN, B LYMPHOCYTES, T CELLS, and AUTOIMMUNE DISEASE.

Surfactant Acronym for surface active agent. Amphipathic molecules (i.e., molecules that contain both a polar and nonpolar domain) which, due to their unique properties, position themselves at interfacial regions (surfaces) such as an oil/water interface. When surfactants are dissolved above a certain critical concentration in either water or nonpolar solvents they may form micelles or reverse micelles, respectively. Surfactants are commonly used to solubilize cell membrane components and other hard to solubilize molecules. See also AMPHIPATHIC MOLECULES, AMPHIPHILIC MOLECULES, MICELLE, REVERSE MICELLE (RM), SDS, and ADJUVANT (TO A HERBICIDE).

"Switch" Proteins Special protein molecules that signal a plant when environmental conditions are so dry that the plant needs to protect itself. See also TREHALOSE, PROTEIN, and SIGNALLING.

Syk Protein See MAST CELLS.

Symbiotic Refers to the mutually beneficial living together of organisms, in an intimate association or union. For example, lichen are a life form consisting of algae and a fungus growing together as a unit on a solid surface (e.g., a tree trunk or a rock). Each helps the other to survive and grow. See also ALGAE, FUNGUS, and ANTIBIOSIS.

Synthase See ACC SYNTHASE, EPSP SYNTHASE, ENZYME, CP4 EPSPS, CITRATE SYNTHASE (CSb) GENE, and GLUTAMINE SYNTHETASE.

Synthesizing (of DNA molecules) The building (i.e., polymerization manufacture) of a known sequence of nucleotides into a chain called an oligonucleotide (of which genes are made) or DNA (deoxyribonucleic acid). Invented by Har Goribind Khorana and his colleagues at the University of Wisconsin-Madison in 1968, this process enables scientists to create genes or gene fragments for use in research.

In 1973, Robert Bruce Merrifield of Rockefeller University developed a means to partially automate the oligonucleotide assembly process. This led to automated machines that can now rapidly manufacture a gene fragment, gene, or DNA probe. See also GENE MACHINE, NUCLEOTIDE, OLIGOMER, OLIGONUCLEOTIDE, SYNTHESIZING (OF PROTEINS), DEOXYRIBONUCLEIC ACID (DNA), DNA PROBE, and SYNTHESIZING (OF OLIGOSACCHARIDES).

Synthesizing (of oligosaccharides) Chemical synthesis (i.e., manufacture) of a known oligosaccharide (structure). For example, a synthesis of a defined-sequence oligosaccharide (molecular) "branch" at a specific site on a glycoprotein in order to "cover up" an antigenic site on that glycoprotein molecule (e.g., so the glycoprotein can be used as a pharmaceutical). See also OLIGOSACCHARIDES, GLYCOPROTEIN, ANTIGEN, ANTIGENIC DETERMINANT, and RESTRICTION ENDOGLYCOSIDASES.

Synthesizing (of proteins) Chemical synthesis (manufacture) of a known protein molecule. Devised by Robert Bruce Merrifield, the desired proteins are assembled by repetitive coupling of the constituent amino acids to a growing polypeptide backbone which itself is attached to a polymeric support (substrate). This procedure has been automated, so it is now possible to make proteins via automated synthesizers. For biological synthesis, see also PROTEIN, POLYPEPTIDE (PROTEIN), AMINO ACID, and SUBSTRATE (STRUCTURAL). See also SYNTHESIZING (OF DNA MOLECULES).

Synthetase See SYNTHASE.

Systematics An extension of taxonomy, it is the scientific classification of living organisms. See CLADISITICS.

Systemic Inflammatory Response Syndrome See SEPSIS.

T

T Cell Growth Factor (TCGF) Also known as Interleukin-2. See INTERLEUKIN-2 (IL-2).

T Cell Modulating Peptide (TCMP) A short protein chain that is thought to restrain certain types of T cells from attacking an (arthritis) afflicted patient's tissues (mainly cartilage). Arthritis is caused by the arthritis sufferer's own immune system attacking the body's cartilage tissues. See also CYTOTOXIC T CELLS, HELPER T CELLS (T4 CELLS), LYMPHOCYTE, SUPPRESSOR T CELLS, T CELL RECEPTORS, AUTOIMMUNE DISEASE, and TUMOR NECROSIS FACTOR (TNF).

T Cell Receptors Antibody-like transmembrane (i.e., across the cell's surface membrane) proteins located on the surface of T cells. These trigger the (cellular) immune response that is mounted by T cells when these receptors bind to antigens (foreign pieces of antigenic protein) which have been "presented" to these receptors by an MHC protein which itself is located on the surface of phagocytic (i.e., scavenging, pathogen-ingesting) B lymphocyte. Antibodies in the blood recognize native antigen macromolecules (i.e., large molecules), whereas T cell receptors recognize fragments derived from those antigen macromolecules (upon presentation at the surface of B lymphocytes following ingestion and digestion by the B lymphocytes). See also ANTIBODY, ANTIGEN, MAJOR HISTOCOMPATIBILITY COMPLEX (MHC), PROTEIN, T CELLS, CELLULAR IMMUNE RESPONSE, PHAGOCYTE, B LYMPHOCYTES, CYTOTOXIC T CELLS, HELPER T CELLS, and SUPPRESSOR T CELLS.

T Cells A class of (thymus-derived) lymphocytes which include helper T cells (also known as T helper cells or T_H cells), suppressor T cells, and cytotoxic T cells (also known as killer cells or CTL for cytotoxic T lymphocyte). These cells mediate (i.e., control/direct) the cellular response of the human immune system in very complex ways. T cells are involved in the activation of B cells. See CELLULAR IMMUNE RESPONSE, CYTOTOXIC T CELLS, HELPER T CELLS (T4 CELLS), LYMPHOCYTE, SUPPRESSOR T CELLS, T CELL RECEPTORS, and T CELL MODULATING PEPTIDE (TCMP).

T Lymphocytes See T CELLS, LYMPHOCYTE, LYMPHOKINES, and THYMUS.

t-IND Treatment Investigational New Drug Application to America's Food and Drug Administration (FDA). See "TREATMENT" IND REGULATIONS.

T4 Cells See HELPER T CELLS.

Tachykinins A class of neuropeptides (i.e., peptides produced by cells of the nervous system; neurons) which includes neurokinin A, neurokinin B, eledoisin, physalaemin, kassinin, substance P, and substance K. Some of these neuropeptides (e.g., Substance P) are picked up by mast cells, lymphocytes, and/or monocytes; and cause those three types of immune system cells to release certain lymphokines (e.g., tumor necrosis factor, interleukin-1 etc.), thus activating the immune system. See also MAST CELL, LYMPHOCYTE, MONOCYTES, TUMOR NECROSIS FACTOR (TNF), and INTERLEUKIN-1.

TAG See TRIACYLGLYCEROLS.

***Taq* DNA Polymerase** A 94 kilodalton DNA polymerase, which was originally isolated from the thermophilic bacteria *Thermus aquaticus*.

Commonly utilized to catalyze PCR reactions due to its heat resistance (needed for thermal cycles utilized in the PCR technique). See also DNA POLYMERASE, POLYMERASE, KILODALTON, DEOXYRIBONUCLEIC ACID (DNA), BACTERIA, THERMOPHILIC BACTERIA, PCR, and POLYMERASE CHAIN REACTION (PCR) TECHNIQUE.

Target The molecule (e.g., receptor) or moiety that a given drug or therapeutic regimen (e.g., gene delivery) is "aimed" at. See also GENE DELIVERY (GENE THERAPY), RECEPTORS, MOIETY, COMBINATORIAL CHEMISTRY, COMBINATORIAL BIOLOGY, SIGNALLING, SIGNAL TRANSDUCTION, G-PROTEINS, and TUMOR NECROSIS FACTOR (TNF).

TAT See TATA HOMOLOGY.

TATA Homology An adenine-thymidine-rich (gene) sequence present 20 to 30 nucleotides "upstream" of the transcription start site on most eucaryotic protein coding genes; it is required for correct expression. Recent research indicates that blocking this portion of the (gene) sequence may inhibit ability of the AIDS virus to reproduce. See also GENE, GENETIC CODE, NUCLEOTIDE, ADENINE, SEQUENCE (OF A DNA MOLECULE), TRANSCRIPTION, STARTPOINT, EUCARYOTE, CODING SEQUENCE, HOMOLOGY, PRIBNOW BOX, PROMOTER, and SEQUENCE (OF A PROTEIN MOLECULE).

Taxol Coined during the 1960s by Monroe E. Wall when it was originally isolated from the Pacific yew tree (genus *Taxus*) this word is now a trademark of the Bristol-Myers Squibb Co. Taxol now refers to the anticancer preparation sold by Bristol-Myers Squibb Company. The active compound from Pacific yew tree is now known as paclitaxel. See also PACLITAXEL and CANCER.

TCGF See T CELL GROWTH FACTOR.

Telomerase An enzyme that enables the "repair" of telomeres (thereby stabilizing their length, and preventing "shortening" of the telomeres). The telomerase enzyme is only present in cancerous cells (thereby enabling the "immortality" of cancerous cells). Human telomerase contains an RNA component and a catalytic-protein component (i.e., a member of the reverse transcriptase "family" of enzymes). See also REVERSE TRANSCRIPTASE, CANCER, NEOPLASTIC GROWTH, ZYGOTE, TELOMERES, ENZYME, ONCOGENES, HYBRIDOMA, MONOCLONAL ANTIBODIES (MAb), and AGING.

Telomeres DNA sequences, that do not code for proteins, which are located at the (end) tips of chromosomes. Telomeres consist of the sequence GGGGTT repeated many times. With the exception of certain types of cells (e.g., zygotes, cancerous cells, "immortal" hybridoma cells), portions of each telomere "break off" each time that the cell containing that chromosome divides. This "shortening" process serves to limit the lifetime (i.e., number of replications) of those (noncancerous, nonzygote, nonhybridoma, etc.) cells. See also DEOXYRIBONUCLEIC ACID (DNA), CODING SEQUENCE, PROTEIN, CHROMOSOMES, SEQUENCE (OF A DNA

MOLECULE), TELOMERASE, MITOSIS, MITOGEN, CANCER, GAMETE, AGING, RETINOIDS, HYBRIDOMA, and ZYGOTE.

Template In general terms it is a mold or pattern that can be copied or its shape reproduced. When used with reference to molecular dimensions it is a macromolecular mold or pattern for the synthesis of another macromolecule. See also DEOXYRIBONUCLEIC ACID (DNA), STRUCTURAL GENE, INFORMATIONAL MOLECULES, HEREDITY, GENE, GENETIC CODE, GENETIC MAP, BIOSENSORS, GENOSENSORS, RIBONUCLEIC ACID (RNA), CODON, EXON, and NANOTECHNOLOGY.

Termination Codon (sequence) One of three triplet sequences (U-A-G, U-A-A, or U-G-A) found in DNA molecules (genes) that cause termination of protein synthesis; they are also called nonsense codons. The sequences cause the termination of the peptide chain and its release in free form. See also CODING SEQUENCE, CODON, DEOXYRIBONUCLEIC ACID (DNA), GENETIC CODE, NONSENSE CODON, and SEQUENCING (OF DNA MOLECULES).

Terminator See TERMINATION CODON (SEQUENCE).

Tertiary Structure The three-dimensional folding of the polypeptide (i.e., protein) molecular chains that characterizes a protein molecule in its native state. See also PROTEIN STRUCTURE, PROTEIN, POLYPEPTIDE (PROTEIN), CONFORMATION, PROTEIN FOLDING, and NATIVE CONFORMATION.

Testosterone An androgen (steroid hormone) that is biochemically synthesized (made) from androstenedione, which is itself synthesized from progesterone. Testosterone is responsible for the development of male secondary sex characteristics in humans such as greater strength, larger body size, facial hair and a deeper voice, etc. See also STEROID and ESTROGEN.

Tetrahydrofolic Acid The reduced, active coenzyme form of the vitamin folic acid; involved in C_1 transfers. Tetrahydrofolate (also known as FH_4) serves as an intermediate carrier (molecule) of methyl, hydroxymethyl, or formyl groups (all containing one carbon atom) in a relatively large number of enzymatic reactions in which such one-carbon groups are transferred from one metabolite to another.

TG See TRIGLYCERIDES.

TGA The government regulatory agency charged with approving all pharmaceutical products sold within Australia. See also FOOD AND DRUG ADMINISTRATION (FDA), KOSEISHO, COMMITTEE FOR PROPRIETARY MEDICINAL PRODUCTS (CPMP), EUROPEAN MEDICINES EVALUATION AGENCY (EMEA), MEDICINES CONTROL AGENCY (MCA), COMMITTEE ON SAFETY IN MEDICINES, BUNDESGESUNDHEITSAMT (BGA), and GENE TECHNOLOGY OFFICE.

TGF See TRANSFORMING GROWTH FACTOR (ALPHA AND BETA).

Thermoduric An organism that can survive high temperatures but does not necessarily grow at such temperatures. See also THERMOPHILE, MESOPHILE, EXTREMOPHILIC BACTERIA, and PSYCHROPHILE.

Thermophile An organism whose optimum temperature for growth is close to, or exceeds, the boiling point of water (100°C, 212°F). See also EXTREMOPHILIC BACTERIA, THERMOPHILIC BACTERIA, THERMODURIC, MESOPHILE, and PSYCHROPHILE.

Thermophilic Bacteria Literally "heat loving" bacteria. They are a category of thermophiles generally found near geothermal vents beneath bodies of water. See also THERMOPHILE, THERMODURIC, EXTREMOPHILIC BACTERIA, MESOPHILE, and PSYCHROPHILE.

Thiol Group (on a molecule) See CYSTEINE (cys) and CYSTINE.

Threonine (thr) A crystalline, α-amino acid considered essential for normal growth of animals. It is biosynthesized (i.e., made) from aspartic acid and is a precursor of isoleucine in microorganisms. See also ESSENTIAL AMINO ACIDS.

Thrombin The key to thrombus (blood clot) formation. Thrombin is a proteolytic enzyme that cleaves fibrinogen into (molecular) pieces, which then spontaneously assemble themselves into fibrin, which forms a clot. See also THROMBUS, THROMBOSIS, THROMBOMODULIN, THROMBOLYTIC AGENTS, FIBRIN, and FIBRINOLYTIC AGENTS.

Thrombolytic Agents Blood-borne compounds (such as tissue plasminogen activator) that work to disintegrate (break up or lyse) blood clots. See also FIBRIN, FIBRINOLYTIC AGENTS, and TISSUE PLASMINOGEN ACTIVATOR (tPA).

Thrombomodulin A cell surface protein found on endothelial cells that plays a key role in modulating the final step in the coagulation process. After thrombin binds to thrombomodulin, thrombin loses its ability to cleave fibrinogen to form fibrin. In addition, once thrombin binds to thrombomodulin, thrombin's activation of protein C is increased 200-fold and this activated protein C then degrades factors Va and VIIIa which are both required for the production of thrombin from prothrombin. Hence, thrombomodulin modulates the activity of the enzyme thrombin causing a cessation of full-blown clotting activity. See also THROMBIN, PROTEIN, PROTEIN C, and THROMBOSIS.

Thrombosis The intravascular (i.e., inside of blood vessel) formation of a blood clot. See also THROMBIN, THROMBUS, THROMBOLYTIC AGENTS, TRIGLYCERIDES, FIBRIN, FIBRINOLYTIC AGENTS, and TISSUE PLASMINOGEN ACTIVATOR (tPA).

Thrombus The blood clot itself. The mass of blood coagulated *in situ* in the heart or other blood vessel. For example, such a clot causes a heart attack when the coagulation occurs in the vessels feeding the heart. See also THROMBIN, THROMBOSIS, THROMBOLYTIC AGENTS, FIBRIN, TRIGLYCERIDES, and FIRBRINOLYTIC AGENTS.

Thymine (thy) A pyrimidine component of nucleic acid first isolated from the thymus. See also NUCLEIC ACIDS and PYRIMIDINE.

Thymoleptics A class of drugs that primarily exerts their effect on the brain influencing "feeling" and behavior.

Thymus A gland that enables cells of the immune system of mammals to mature. In humans, it lies behind the breast bone and extends upwards as far as the thyroid gland. The thymus is the place in the body where T lymphocytes are "taught" to distinguish "foreign" (e.g., pathogen's) antigens from "self" cell antigens, to avoid immune responses in which the body's immune system attacks organs and other cells within the body (resulting in autoimmune diesease). Any T lymphocytes that remain "autoreactive" (i.e., would tend to attack "self" cells, such as organs in the body) are destroyed by the thymus via a cytotoxic mechanism.

An example of an autoimmune disease is multiple sclerosis (MS), where the body's acetylcholine receptors are attacked by the body's immune system. Since acetylcholine is crucial in the transmission of nerve impulses to the body's muscles, such destruction of acetylcholine receptors results in loss of control of the body's muscles. See also T LYMPHOCYTES, CYTOTOXIC, RECEPTORS, T CELLS, IMMUNE RESPONSE, PATHOGEN, ANTIGEN, and NEUROTRANSMITTER.

Thyroid Gland A gland that is found on both sides of the trachea ("windpipe") in humans. This gland secretes the hormone thyroxine, which increases the rate of metabolism. See also THYROID STIMULATING HORMONE (TSH).

Thyroid Stimulating Hormone (TSH) A hormone that causes the thyroid gland to secrete additional amounts of thyroxine. See also THYROID GLAND and GRAVE'S DISEASE.

Tissue Culture The growth and maintenance (by researchers) of cells from higher organisms *in vitro*, that is, in a sterile test tube environment that contains the nutrients necessary for cell growth. See CULTURE MEDIUM.

Tissue Plasminogen Activator (tPA) A glycoprotein that possesses thrombolytic (i.e., blood clot-dissolving) activity. It is used as a drug to dissolve clots and acts by first binding to fibrin (clots). It then activates (i.e., proteolytically cleaves) plasminogen (molecules) to yield plasmin, a bloodborne enzyme that itself cleaves molecular bonds in the fibrin clot. The plasmin molecules diffuse through the fibrin clot and cause the clot to dissolve rapidly. With the dissolution of the clot, blood flow to the formerly blocked blood vessel (e.g., the heart) is restored. See also THROMBUS, THROMBIN, THROMBOLYTIC AGENT, GLYCOPROTEIN, FIBRIN, and FIBRINOLYTIC AGENTS.

TME (N) Abbreviation for "true metabolizable energy (corrected for nitrogen)," a measure of the amount of energy that a given animal (e.g., chicken) can extract from a given feed ration. See METABOLISM and CHEMOMETRICS.

TMEn See TME (N).

Tobacco Budworm See *HELIOTHIS VIRESCENS*.

Tobacco Mosaic Virus (TMV) One the of smallest viruses, consisting of some 2,200 chains of identical polypeptides and a molecule of RNA. All of the genetic/heredity information of the Tobacco Mosaic Virus is contained in its RNA.

The first discovery of a self-assembling, active biological structure occurred in 1955, when Heinz Frankel-Conrat and Robley Williams showed that TMV, will reassemble into functioning, infectious virus particles (after the TMV has been dissociated into its components via immersion in concentrated acetic acid. The TMV virus infects the leaves of tobacco plants, causing disease. Tobacco plants can be genetically engineered to resist TMV infection. See also GENETIC ENGINEERING, CAPSID, VIRUS, RNA, POLYPEPTIDE (PROTEIN), GENE, INFORMATIONAL MOLECULES, HEREDITY, and SELF-ASSEMBLY (OF A LARGE MOLECULAR STRUCTURE).

Tomato Fruitworm See *HELICOVERPA ZEA*.

Topotaxis See TROPISM.

Totipotent Stem Cells Bone marrow cells that (when signalled) mature into both red blood cells and white blood cells. Receptors on the surface of totipotent stem cells "grasp" passing blood cell growth factors (e.g., Interleukin-7, Stem Cell Growth Factor, etc.), bringing them inside these stem cells and thus causing the maturation and differentiation into red and white blood cells. These receptors are called FLK-Z receptors. See also STEM CELL ONE, STEM CELLS, WHITE BLOOD CELLS, GROWTH FACTOR, RECEPTORS, "HEDGEHOG" CELL-DIFFERENTIATION PROTEINS, CELL DIFFERENTIATION, and CELL.

Toxic Substances Control Act (TSCA) A 1976 American federal law under which the U.S. Environmental Protection Agency (EPA) has sought to regulate the release of genetically engineered organisms (e.g., bacteria or plants) that produce natural insecticides. This is based on supposed analogy to synthetic chemical insecticides, which are clearly regulated under TSCA. See also OAB (OFFICE OF AGRICULTURAL BIOTECHNOLOGY), FEDERAL INSECTICIDE, FUNGICIDE AND RODENTICIDE ACT (FIFRA), GENETICALLY ENGINEERED MICROBIAL PESTICIDES (GEMP), WHEAT TAKE-ALL DISEASE, and *BACILLUS THURINGIENSIS (B.t.)*.

Toxigenic *E. coli* See ENTEROHEMORRHAGIC *E. COLI*, and *ESCHERICHIA COLIFORM* 0157:H7.

Toxin A substance (e.g., produced in some cases by disease-causing microorganisms) which is poisonous to certain other living organisms. See also ANTITOXIN, ABRIN, RICIN, COLICINS, BACTERIOCINS, *ESCHERICHIA COLIFORM* 0157:H7, ENTEROHEMORRHAGIC *E. COLI*, and *PFIESTERIA PISCICIDA*.

Tracer (radioactive isotopic method) A metabolite that is labeled by incorporation of an isotopic atom into its structure. The metabolic fate of the labeled metabolite can then be traced in intact organisms. That is, one is able to ascertain where (in what kind of structure) the metabolite ends up as well as the transformation products (intermediate molecules) which were involved in its formation. Certain atoms of a given metabolite are labeled. This is done by substituting radioactive isotopes for the atom in question. Because an atom is replaced by an isotope, the metabolite as a whole is chemically and biologically indistinguishable from its normal analog. The presence of the isotope allows the metabolite and its transformation products to be detected and measured. Without this technique, many aspects of metabolism could not have been studied. These include: the process of photosynthesis, metabolic turnover rates, and the biosynthesis of proteins and nucleic acids. See also REASSOCIATION (OF DNA), RADIOACTIVE ISOTOPE, and RADIOIMMUNOASSAY.

Trait A characteristic of an organism, which manifests itself in the phenotype (physically). Many traits are the result of the expression of a single gene, but some are polygenic (result from simultaneous expression of more than one gene). For example, the level of protein content in soybeans is controlled by five genes. See also PHENOTYPE, GENOTYPE, EXPRESS, GENE, POLYGENIC, PROTEIN, and CALLIPYGE.

***trans* Fatty Acids** One of the two isomeric forms that fatty acids can exist in. *Trans* fatty acids are naturally present in some meat and dairy products (which constitute approximately 5% of the average American diet). See FATTY ACID, ISOMER, STEREOISOMERS, and HYDROGENATION.

***trans*-Acting Protein** A *trans*-acting protein has the exceptional property of acting (having an effect) only on the molecule of DNA (deoxyribonucleic acid) from which it was expressed. See also EXPRESS and *cis*-ACTING PROTEIN.

Transactivating Protein See VIRAL TRANSACTIVATING PROTEIN.

Transaminase A large group of enzymes that catalyze the transfer of the amino group from any one of at least 12 amino acids to a keto acid to form another amino acid. Also known as aminotransferases. See also ENZYME and AMINO ACID.

Transamination The reaction of the enzymatic removal and transfer of an amino group from one specific compound to another. See also TRANSAMINASE and AMINO ACID.

Transcription The enzyme-catalyzed process whereby the genetic information contained in one strand of DNA (deoxyribonucleic acid) is used as a template to specify and produce a complementary mRNA strand. Transcription may be thought of as a rewriting of the information contained in DNA into RNA. The language, however, is the same—both are nucleic acid–based. This is in contrast to translation, in which the information is translated from one language (RNA, nucleic acid–based) into another language (protein, amino acid–based). See also TRANSLATION, MESSENGER RNA (mRNA), GENETIC CODE, DEOXYRIBONUCLEIC ACID (DNA), TRANSCRIPTION FACTORS, and TRANSCRIPTION UNIT.

Transcription Factors Proteins that interact with each other and (somewhat) with DNA (when immediately adjacent to the DNA in a cell) to either facilitate (i.e., "turn on") or inhibit (i.e., "turn off") the activity (i.e., coding for proteins) of that DNA's genes. Transcription factors hold potential to cure diseases (e.g., by blocking the deleterious effects of certain disease-causing genes). See also PROTEIN, GENETIC CODE, CODING SEQUENCE, DEOXYRIBONUCLEIC ACID (DNA), CELL, INHIBITION, GENE, P53 GENE, TRANSCRIPTION, and P53 PROTEIN.

Transcription Unit A group of genes that code for functionally related RNA molecules or protein molecules. This group of genes is expressed (transcribed) together (i.e., as a unit, thus the name). See also EXPRESS, GENE, TRANSCRIPTION, TRANSLATION, GENETIC CODE, CODING, SEQUENCE, DEOXYRIBONUCLEIC ACID (DNA), RIBONUCLEIC ACID (RNA), and RIBOSOMES.

Transduction (gene) The transfer of bacterial genes (DNA) from one bacterium to another by means of a (temperature or defective) bacterial virus (bacteriophage). There exist two kinds of transduction: specialized and general. In the case of specialized transduction, a restricted group of host genes becomes integrated into the virus genome. These "guest" genes usually replace some of the virus genes and are subsequently transferred to a second bacterium. In the case of generalized transduction, host genes become a part of the mature virus particle in place of, or in addition to the virus DNA. However, in this case the genes can come from virtually any portion of the host genome and this material does not become directly integrated into the virus genome. In the case of plants, the vector can be *Agrobacterium tumefaciens*. See also BACTERIOPHAGE, VECTOR, GENETIC CODE, *AGROBACTERIUM TUMEFACIENS*, RETROVIRAL VECTORS, GENE DELIVERY (GENE THERAPY), and TRANSFECTION.

Transduction (signal) See SIGNAL TRANSDUCTION.

Transfection A special case of transformation in which an appropriate recipient strain of bacteria is exposed to (free) DNA isolated from a transducing phage with the "take-up" of that DNA by some of the bacteria and consequent production and release of complete virus particles. The process involves the direct transfer of genetic material from donor to recipient. See also MARKER (GENETIC MARKER), TRANSFORMATION, DEOXYRIBONUCLEIC ACID (DNA), and TRANSDUCTION.

Transfer RNA (tRNA) A class of relatively small RNA (ribonucleic acid) molecules of molecular weight 23,000 to about 30,000. tRNA molecules act as carriers of specific amino acids during the process of protein synthesis. Each of the 20 amino acids found in proteins has at least one specific corresponding tRNA. The tRNA binds covalently with "its" specific amino acid and "leads" it to the ribosome for incorporation into the growing peptide chain. See also RIBONUCLEIC ACID (RNA), MOLECULAR WEIGHT, AMINO ACID, and MESSENGER RNA (mRNA).

Transferases Enzymes that catalyze the transfer of functional groups to molecules (from other molecules). See also TRANSAMINASE, ENZYME, and GLYCOSYLTRANSFERASES.

Transferrin The protein molecule that is responsible for transporting iron (molecules) to tissues throughout the body, via the circulatory system. See also PROTEIN, TRANSFERRIN RECEPTOR, HEME, and BLOOD-BRAIN-BARRIER (BBB).

Transferrin Receptor The receptor molecule (located on the surface of cells throughout the body) that is responsible for binding to transferrin molecules, then bringing those iron-rich transferrin molecules into the cell where the iron is released to be used by the cell. See also TRANSFERRIN, RECEPTORS, HEME, and BLOOD-BRAIN-BARRIER (BBB).

Transformation The process in which free DNA is transferred directly into a competent recipient cell. The direct transfer of genetic material from donor to recipient. The acquisition (e.g., by bacteria cells) of new genetic markers (new traits coded for by the new DNA) via the process of transformation. See also DEOXYRIBONUCLEIC ACID (DNA), TRANSFECTION, and MARKER (GENETIC MARKER).

Transforming Growth Factor-Alpha (TGF-alpha) An angiogenic growth factor produced by tumor cells. It is able to induce specific malignant characteristics in normal cells (such as fibroblasts), thereby "transforming" those cells. TGF-alpha appears to possess a variety of potentially useful pharmaceutical properties, such as powerful stimulation of scar tissue formation following wounding of a tissue, as indicated by preliminary research. See also TRANSFORMING GROWTH FACTOR-BETA (TGF-BETA), GROWTH FACTOR, NERVE GROWTH FACTOR (NGF), TUMOR, FIBROBLASTS, and ANGIOGENIC GROWTH FACTORS.

Transforming Growth Factor-Beta (TGF-beta) An angiogenic growth factor produced by tumor cells, it is able to induce specific malignant characteristics in normal cells (such as fibroblasts), thereby "transforming" those cells. TGF-beta stimulates blood vessel growth, even though it inhibits the division of endothelial cells. TGF-beta is a strong "attracting agent" for macrophages (i.e., TGF-beta is chemotactic), and appears to be responsible for the high concentrations of macrophages that are often found in tumors. TGF-beta has shown immunosuppressive activity (i.e., it suppresses the immune system). See TRANSFORMING GROWTH FACTOR-ALPHA (TGF-ALPHA), GROWTH FACTOR, OSTEOINDUCTIVE FACTOR (OIF), IMMUNOSUPPRESSIVE, and NERVE GROWTH FACTOR (NGF).

For example, transforming growth factor-beta works together with osteoinductive factor (OIF) to promote bone-formation by first causing connective tissue cells to grow together to form a matrix of cartilage (e.g., across a bone break) then bone cells slowly replace that cartilage. See also TUMOR, FIBROBLASTS, ANGIOGENIC GROWTH FACTORS, MITOGEN, ENDOTHELIAL CELLS, CHEMOTAXIS, and MACROPHAGE.

Transgene A "package" of genetic material (i.e., DNA) that is inserted into the genome of a cell via gene splicing techniques. May include promoter(s), leader sequence, termination codon, etc. See DEOXYRIBONUCLEIC ACID (DNA), GENE SPLICING, GENOME, LEADER SEQUENCE, PROMOTER, GENETIC CODE, TERMINATION CODON (SEQUENCE), GENETIC ENGINEERING, and CASSETTE.

Transgenic (organism) An organism whose gamete cells (sperm/egg) contain genetic material originally derived from an organism other than the parents or in addition to the parental genetic material. See also GENETIC ENGINEERING and GAMETE.

Transgressive Segregation A plant breeding (propagation) technique, in which *genetically very different* members of the *same species* are mated with each other. The offspring of that mating can be more healthy, productive (e.g., fast growing), and uniform than their parents, a phenomenon known as "hybrid vigor." See GENETICS, SPECIES, F1 HYBRIDS, and HYBRIDIZATION (PLANT GENETICS).

Transition State (in a chemical reaction) That point in the chemical reaction at which the reactants (i.e., chemical entities about to react with each other) have been "brought to the brink." It is a point in the chemical reaction process in which an "activated condition" is reached. From this point the probability of the reaction going to completion and producing a product is very high. The transition state separates (energetically) products from reactants. It is viewed as being at the top of the energy barrier separating reactants and products. The reacting species in the transition state can, because of their location at the "top" of the energy barrier, "fall" to either products or reactants. See also CATALYST, ENDERGONIC REACTION, ACTIVATION ENERGY, FREE ENERGY, CATALYTIC ANTIBODY, SEMISYNTHETIC CATALYTIC ANTIBODY, and EXERGONIC REACTION.

Transit Peptide A peptide that, when fused to a protein, acts to transport that protein between compartments within eucaryotic cells. Once inside the "destination compartment," the transit peptide is cleaved off the protein and that protein is then free (to do its designed task). See PEPTIDE, PROTEIN, EUCARYOTE, CELL, FUSION PROTEIN, GATED TRANSPORT (OF A PROTEIN), VESICULAR TRANSPORT (OF A PROTEIN), and CHLOROPLAST TRANSIT PEPTIDE (CTP).

Translation The process whereby the genetic information present in an mRNA molecule directs the order of incorporation of specific amino acids, and hence the growth of the polypeptide chain during protein synthesis. One can think of translation as the process of translating one language into another. In this particular case the nucleic acid–based language represented by mRNA is translated into the amino acid–based language of proteins. See also CODING SEQUENCE, CODON, RIBOSOMES, MESSENGER RNA (mRNA), PROTEIN, GENE, and GENETIC CODE.

Translocation Genetic mutation in which a section of a chromosome "breaks off" and moves to a new (abnormal) position in that (or a different) chromosome. See also GENE, CHROMOSOMES, GENETIC CODE, CODING SEQUENCE, TRANSPOSITION, DEOXYRIBONUCLEIC ACID (DNA), MUTATION, INTROGRESSION, JUMPING GENES, and HOT SPOTS.

Transposable Element See TRANSPOSON.

Transposase An enzyme that is required for transposition to occur. It is coded-for by the transposon known as the P element. See also TRANSPOSITION, TRANSPOSON, ENZYME, GENETIC CODE, and CODING SEQUENCE.

Transposition Movement of a gene or set of genes from one site in the genome to another without a reciprocal exchange (of DNA). See also GENE, JUMPING GENES, GENOME, TRANSPOSON, TRANSPOSASE, HOT SPOTS, and DEOXYRIBONUCLEIC ACID (DNA).

Transposon A DNA (deoxyribonucleic acid) sequence (segment of molecule) able to replicate and insert one copy (of itself) at a new location in the genome (i.e., a transposition of location). Discovered in 1950 by geneticist Barbara McClintock in corn (maize) plants (*Zea mays* L.); and in bacteria a decade later by Joshua Lederberg. Transposons can either carry genes along one organism's genome, or even into another organism's genome (e.g., via sexual conjugation, in bacteria). By such sexual conjugation, transposons can carry genes that confer new phenotypic properties (e.g., resistance to certain antibiotics, for a given bacterial cell). See also DEOXYRIBONUCLEIC ACID (DNA), REPLICATION (OF VIRUS), GENOME, TRANSPOSITION, TRANSPOSASE, SEQUENCE (OF A DNA MOLECULE), CORN, JUMPING GENES, GENE, SEXUAL CONJUGATION, PHENOTYPE, and CONJUGATION.

TRANSWITCH® A "sense" technology used to "turn off" (suppress) a gene (e.g., the one that causes tomato to ripen) that causes an unwanted effect (e.g., premature softening of tomato). TRANSWITCH® and its registered trademark are owned by DNA Plant Technology Corp. See also GENE SILENCING, SUPPRESSOR GENE, and SENSE.

"Treatment" IND Regulations Food and Drug Administration (FDA) regulations promulgated in 1987, to provide a more rapid formal pharmaceutical approval mechanism than the usual IND (Investigational New Drug) regulatory approval process. Its purpose is to enable drug developers to provide promising experimental drugs to patients suffering from immediately life-threatening diseases or certain serious conditions (e.g., acquired immune deficiency syndrome, or AIDS) before complete data on that drug's efficacy or toxicity are available. See also IND, FOOD AND DRUG ADMINISTRATION (FDA), DELANEY CLAUSE, KOSEISHO, and COMMITTEE FOR PROPRIETARY MEDICINAL PRODUCTS (CPMP).

Treatment Investigational New Drug See "TREATMENT" IND REGULATIONS.

Trehalose A disaccharide (simple sugar) that is naturally synthesized (i.e., "manufactured") by many plants and animals in response to the stresses of freezing, heating, or drying. That is because trehalose protects certain proteins (needed for life) and prevents loss of crucial volatile (i.e., easily evaporated) compounds from organisms during those stressful (e.g., dry, frozen, or hot) conditions. Trehalose also provides a source of quick energy after the stressful conditions have passed. That is why dried baker's yeast (which contains up to 20% trehalose by weight) can be stored in its dry state for many years, yet quickly leavens bread dough within minutes of being rehydrated (i.e., rewetted).

Trehalose accomplishes this protection by forming a non-hygroscopic "glass" on the surfaces of cells and large molecules. It immobilizes and stabilizes large molecules (e.g., proteins), but still allows water to diffuse out so complete drying can occur. Thus, trehalose holds potential as a food additive to keep proteins (e.g., eggs) fresh in the dried form. In 1991, the UK approved trehalose for use in food. Trehalose hydrolyzes (e.g., during digestion) into two molecules of glucose. See also DISACCHARIDES, PROTEIN, GLUCOSE, HYDROLYSIS, CONFORMATION, "SWITCH" PROTEINS, TERTIARY STRUCTURE, and PROTEIN FOLDING.

Triacylglycerols See TRIGLYCERIDES.

Trichoderma harzianum A microorganism that possesses (natural) fungicide activity. See also *BACILLUS THURINGIENSIS (B.t.)*, WHEAT TAKE-ALL DISEASE, FUNGUS, and FUNGICIDE.

Trichosanthin An enzyme extracted from a specific Chinese plant. It has been discovered to "cut apart" the ribosomes in cells that are infected with the HIV (i.e., AIDS) virus, thus stopping the virus and preventing infection of additional cells. When purified into a pharmaceutical, this enzyme is called GLQ223 or compound Q. Other potential uses of this drug include the treatment of certain cancers. See also RIBOSOMES, ACQUIRED IMMUNE DEFICIENCY SYNDROME (AIDS), ENZYME, PROTEIN, and HUMAN IMMUNODEFICIENCY VIRUS (HIV).

Triglycerides Molecules that consist of three fatty acids attached to a glycerol "backbone." More accurately called triacylglycerols, although long-term historical usage of "triglycerides" has made the latter term more common (though not totally accurate). Research during the 1990s provided evidence that high blood levels of triglycerides in humans (e.g., immediately after meals) contribute to thrombosis. See also THROMBOSIS, FATTY ACID, SATURATED FATTY ACIDS, and UNSATURATED FATTY ACID.

Triploid Refers to organisms that possess three sets of chromosomes, instead of the normal two sets of chromosomes. Conversion of a diploid (i.e., two sets of chromosomes) organism to triploid can be done by man (e.g., certain fish, "seedless," grapes, etc.). For example, fish are ordinarily diploid. By exposing fish eggs to certain specific combinations of temperature and pressure, immediately after fertilization of those eggs, scientists can cause the resultant fish to become triploid. Triploid fish are unable to reproduce. This sterility is desired by man, in order to prevent certain fish (e.g., those that have been genetically engineered) from mating with wild fish.

Such induced (triploid) sterility also prevents the (genetically engineered) fish from wasting energy on the act of reproduction, so they grow faster and larger. That transfer (of energy use from reproduction to growth) also holds true for "seedless" grapes, watermelons, etc. See also DIPLOID and CHROMOSOMES.

tRNA See TRANSFER RNA.

Tropism Orientation movement of a sessile organism in response to a stimulus. Movement of curvature due to an external stimulus that determines the direction of movement. Also known as topotaxis. See also SESSILE and CHEMOTAXIS.

Trypsin A proteolytic (protein chain–cutting) enzyme that is produced by the pancreas. Trypsin cleaves polypeptide (protein) molecular chains on the carboxyl (group) side of arginine and lysine units (residues). See also ARGININE (arg), LYSINE (lys), PROTEIN, POLYPEPTIDE (PROTEIN), and PROTEOLYTIC ENZYMES.

Tryptophan (trp) An essential amino acid, it is a precursor of the important biochemical molecules: indoleacetic acid, serotonin, and nicotinic acid. L-Tryptophan is used as a common feed additive for livestock to ensure that their diet includes an adequate amount of this essential amino acid. See also ESSENTIAL AMINO ACIDS, STEREOISOMERS, LEVOROTARY (L) ISOMER, SEROTONIN, and AMINO ACID.

TSH See THYROID STIMULATING HORMONE.

Tumor A mass of abnormal tissue that resembles normal tissues in structure, but which fulfills no useful function (to the organism) and grows at the expense of the body. Tumors may be malignant or benign. Malignant tumors (which infiltrate adjacent healthy tissues) can result from oncogenes and/or carcinogens. They eventually kill their host if unchecked. See also CANCER, ANGIOGENESIS, ONCOGENES, PROTO-ONCOGENES, and CARCINOGEN.

Tumor Necrosis Factor (TNF) Literally, tumor death factor. A cytokine (protein that helps regulate the immune system) that has shown potential to combat (kill) malignant (cancer) tumors. Tumor necrosis factor was discovered to be 10,000 times more toxic in humans than in rodents, where it had been tested for toxicity prior to human clinical tests. This example illustrates one potential pitfall of nontarget animal testing in that sometimes animal testing does not accurately reflect or foretell what will happen in humans.

Another drawback to using TNF as a drug to combat human tumors is the fact that it is one of the substances released (in the disease rheumatoid arthritis) that destroys tissue in the joints.

When released as part of the AIDS (disease), TNF causes cachexia, which is a "wasting away" of the body due to the body's inability to process nutrients received via digestion. See also CYTOKINES, LYMPHOKINES, NECROSIS, TUMOR, TUMOR-INFILTRATING LYMPHOCYTES (TIL CELLS), PROTEIN, AUTOIMMUNE DISEASE, T CELL MODULATING PEPTIDE (TCMP), and DIGESTION (WITHIN ORGANISMS).

Tumor-Associated Antigens Discovered by Thierry Boon in 1991, these are distinctive protein molecules that are produced in the surface membrane of tumor cells. These protein molecules are used by the body's cytotoxic T cells to recognize (and destroy) tumor cells, so such proteins hold promise for use in vaccines. See also MAJOR HISTOCOMPATIBILITY COMPLEX (MHC), MACROPHAGE, TUMOR, T CELL RECEPTORS, ANTIGEN, T CELLS, PROTEIN, CELL, CYTOTOXIC T CELLS, and HUMAN LEUKOCYTE ANTIGENS (HLA).

Tumor-Infiltrating Lymphocytes (TIL cells) The white blood cells of a cancer patient which have been:

(1) Taken from that patient's tumor (where those white blood cells had been attempting to combat the cancer, albeit unsuccessfully)
(2) Stimulated with doses of interleukin-2 (to make the lymphocytes more effective against the cancer)
(3) Multiplied *in vitro* (i.e., outside of the patient's body) to make them more numerous (and thus more likely to successfully combat the cancer).

When these "souped up" lymphocytes (white blood cells) are reintroduced into that same patient's body, the lymphocytes (now called TIL cells because they have been "souped up") attack the cancer tumor (malignant growth) more vigorously than before. See also TUMOR, WHITE BLOOD CELLS, LYMPHOCYTE, LYMPHOKINES, T CELLS, and CYTOTOXIC T CELLS.

Tumor-Suppressor Genes Also called anticancer genes. Genes within a cell's DNA that code for (i.e., cause to be manufactured in cell's ribosomes) proteins that hold the cell's growth in check. If these genes are damaged (e.g., by radiation, by a carcinogen, or by chance accident in normal cell division), they no longer hold cell growth in check—and the cell becomes malignant (if the cell's DNA also contains a gene called an oncogene).

Oncogenes must be present for the cell to become malignant, but oncogenes cannot cause a cell to become malignant until a tumor-suppressor gene is damaged.

As with all genes, tumor-suppressor genes are inherited in two copies (alleles, one from each parent) and either copy can code for the proteins necessary for cell growth control. However, an organism that is born with one defective copy of a tumor-suppressor gene (or in whom one copy is damaged early in life) is especially prone to cancer (malignancy). See also GENE, p53 GENE, GENETIC CODE, MEIOSIS, DEOXYRIBONUCLEIC ACID (DNA), CARCINOGEN, RIBOSOMES, ONCOGENES, CANCER, TUMOR, PROTO-ONCOGENES, and PROTEIN.

Tumor-Suppressor Proteins Proteins that are coded-for (i.e., caused to be manufactured in the cell's ribosomes) by tumor-suppressor genes (e.g., the p53 gene). Such proteins (e.g., the p53 protein) then act upon the cell's DNA in order to prevent uncontrolled cell growth and division (i.e., cancer). See also TUMOR-SUPPRESSOR GENES, GENE, p53 GENE, PROTEIN, GENETIC CODE, MEIOSIS, DEOXYRIBONUCLEIC ACID (DNA), RIBOSOMES, ONCOGENES, CANCER, TUMOR, CELL, and PROTO-ONCOGENES.

Turnover Number The number of molecules of a product produced per minute by a single-enzyme molecule when that enzyme is working at its maximum rate. That is, the number of substrate molecules converted into a product by one enzyme molecule per minute when that enzyme is "going (catalyzing) as fast as it can." See also ENZYME, TRANSFERASES, PROTEASE, PROTEIN KINASES, PROTEOLYTIC ENZYMES, and TRANSAMINASE.

Type Specimen The actual physical specimen (e.g., a stuffed lizard or a dried insect) that a scientist (who describes and names a previously unknown species) must place in a museum (or other recognized repository) in order to have the right to name that newly discovered species. This "officially deposited specimen" is required for three purposes:

(1) So that comparisons can later be made if there is ever a doubt whether another "new" species is simply a member of this same species (and thus already named)

(2) So that taxonomists (who determine and keep the official scientific names by which scientists must refer to each of the world's organisms) can name each of the newly discovered species in accordance with the complex rules of the International Codes for Nomenclature; examples of such names in this glossary are *Arabidopsis thaliana*, *Escherichia coli*, and *Agrobacterium tumefaciens*.

(3) So that patent claims for genetically engineered organisms can later be enforced.

See also SPECIES, STRAIN, CLADISTICS, CHAKRABARTY DECISION, AMERICAN TYPE CULTURE COLLECTION (ATCC), and CONSULTATIVE GROUP ON INTERNATIONAL AGRICULTURAL RESEARCH (CGIAR).

Tyrosine (tyr) A phenolic α-amino acid. It is a precursor of the hormones epinephrine, norepinephrine, thyroxine, and triiodothyronine. It is also a precursor of the molecule known as melanin (which is the pigment of a suntan). See also AMINO ACID and HORMONE.

U

Ubiquitin A small protein present in all eucaryotic cells (ubiquitous) which plays an important role in "tagging" other proteins that are destined (marked) for destruction (via proteolytic cleavage). Such proteins are then removed because they are damaged or no longer needed by the body. See also EUCARYOTE, PROTEIN, PROTEOLYTIC ENZYMES, and DENATURATION.

Ultracentrifuge A high-speed centrifuge that can attain revolving speeds up to 85,000 rpm and centrifugal fields up to 500,000 times gravity. The machine is used to sediment (i.e., cause to settle out) and hence separate macromolecules (i.e., large molecules) and macromolecular structures in a mixture/solution.

In general, a centrifuge is a machine that whirls test tubes around rapidly, like a merry-go-round, to force the heavier suspended materials (in the solutions in the test tubes) to the bottoms of those test tubes before the lighter material.

Ultrafiltration A (mixture) separation methodology that uses the ability of synthetic semipermeable membranes (possessing appropriate physical and chemical natures) to discriminate between molecules in the mixture, primarily on the basis of the molecules' size/shape. Invented and developed by Dr. Roy J. Taylor in the 1950s and 1960s, ultrafiltration is typically utilized for the separation of relatively high-molecular-weight solutes (e.g., proteins, gums, polymers and other complex organic molecules) and colloidally dispersed substances (e.g., minerals, microorganisms, etc.) from their solvents (e.g., water). See also DIALYSIS, MEMBRANE TRANSPORT, MICROORGANISM, MOLECULAR WEIGHT, PROTEIN, POLYMER, and HOLLOW FIBER SEPARATION (OF PROTEINS).

Union for Protection of New Varieties of Plants (UPOV) A group of the world's countries that have jointly agreed to mutually protect the intellectual property (of owners) that is inherent in new plant varieties developed by man. The secretariat for this union is in Geneva, Switzerland. See also PLANT VARIETY PROTECTION ACT (PVP), U.S. PATENT AND TRADEMARK OFFICE (USPTO), PLANT'S NOVEL TRAIT (PNT), PLANT BREEDER'S RIGHTS (PBR), EUROPEAN PATENT CONVENTION, EUROPEAN PATENT OFFICE (EPO), and MUTUAL RECOGNITION AGREEMENTS (MRAs).

Units (U) A measure (quantitation) of biological activity of a substance, as defined by various standardized assays (tests). See also ASSAY and BIOASSAY.

Unsaturated Fatty Acid A fatty acid containing one or more double bonds (between individual atoms of the molecule). See also FATTY ACID, MONOUNSATURATED FATS, and POLYUNSATURATED FATTY ACIDS.

UPOV See UNION FOR PROTECTION OF NEW VARIETIES OF PLANTS.

Uracil A pyrimidine base important as a component of ribonucleic acid (RNA). Its hydrogen-bonding counterpart in DNA is thymine. See also PYRIMIDINE and RIBONUCLEIC ACID (RNA).

Urokinase A thrombolytic (i.e., clot-dissolving) enzyme used as a biopharmaceutical. See also THROMBOLYTIC AGENTS, TISSUE PLASMINOGEN ACTIVATOR (tPA), and FIBRINOLYTIC AGENTS.

U.S. Patent and Trademark Office (USPTO) The Washington, D.C.–based American Government agency that is responsible for common patent protection matters for all of America's 50 states and its territorial possessions. The USPTO allows the patenting of new and unique microbes, plants, and animals; as well as the new and unique *methods* to produce such biotechnology advances. See also EUROPEAN PATENT OFFICE (EPO), CHAKRABARTY DECISION, MICROBE, GENETIC ENGINEERING, PLANT'S NOVEL TRAIT (PNT), PLANT BREEDER'S RIGHTS (PBR), BIOTECHNOLOGY, and AMERICAN TYPE CULTURE COLLECTION (ATCC).

USPTO See U.S. PATENT AND TRADEMARK OFFICE.

V

Vaccinia A non-pathogenic virus that is believed to be a (modified) form of the virus that causes cowpox. *Vaccinia* readily accepts genes (inserted into its genome via genetic engineering) from pathogenic viruses so it can be used to make vaccines that do not possess the risk inherent in attenuated-virus vaccines (i.e., that the attenuated virus "revives" and causes disease). Such genetically engineered *vaccinia* codes for (presents) the proteins of the pathogenic virus on its surface, which activates the immune system (e.g., of vaccinated animal) to produce antibodies against that pathogenic virus. See also VACCINE, PATHOGENIC, VIRUS, GENE, GENE DELIVERY, GENETIC ENGINEERING, ATTENUATED (PATHOGENS), ANTIBODY, MACROPHAGE, COMPLEMENT CASCADE, CELLULAR IMMUNE RESPONSE, and PHAGOCYTE.

Vaccine Any substance, bearing antigens on its surface, that causes activation of an animal's immune system without causing actual disease. The animals' immune system components (e.g., antibodies) are then prepared to quickly vanquish those particular pathogens when they later enter the body. See also DNA VACCINES, "NAKED" GENE, "EDIBLE VACCINES," ANTIGEN, CELLULAR IMMUNE RESPONSE, and HUMORAL IMMUNITY.

Vagile Wandering or roaming (e.g., a microorganism that is not attached to a solid support tends to "wander" through its environment as it gets pushed about by currents of air or liquid). See also SESSILE and VAGILITY.

Vagility The ability of organisms to disseminate (e.g., spread throughout a given habitat). See also VAGILE.

Vaginosis The process whereby a cell internalizes an entity (such as a virus or a protein) that has bound to the cell's outer membrane. Once that "bound entity" is inside the cell, the cell membrane fuses together again. See also NUCLEAR RECEPTORS, RECEPTORS, ENDOCYTOSIS, TRANSFERRIN, VIRUS, and BLOOD BRAIN BARRIER (BBB).

Validation See PROCESS VALIDATION (FOR PRODUCTION OF A PHARMACEUTICAL).

Valine (val) An amino acid considered essential for normal growth of animals. It is biosynthesized (made) from pyruvic acid. See also AMINO ACID and ESSENTIAL AMINO ACIDS.

Value-Added Grains See VALUE-ENHANCED GRAINS.

Value-Enhanced Grains Those grains that possess novel traits that are economically valuable (e.g., higher-than-normal protein content, higher-than-normal oil content, etc.). For example, high-oil corn possesses a kernel oil content of 5.8% of greater, versus oil content of 3.5% or less for traditional No. 2 yellow corn. High-amylose corn possesses a kernel amylose (starch) content of 50% or greater, etc. See also HIGH-OIL CORN, PROTEIN, AMYLOSE, OPAGUE-2, FLOURY-2, GENETIC ENGINEERING, TRAIT, HIGH-LYSINE CORN, HIGH-METHIONINE CORN, HIGH-PHYTASE CORN/SOYBEANS, HIGH-OLEIC OIL SOYBEANS, HIGH-STEARATE SOYBEANS, and HIGH-SUCROSE SOYBEANS.

Van der Waals Forces The relatively weak forces of attraction between molecules that contribute to *inter*molecular bonding (i.e., binding-together two or more adjacent molecules). Historically, it was thought that van der Waals forces were always weaker than the hydrogen bond forces responsible for *intra*molecular bonding. However, in 1995, Dr. Alfred French discovered that van der Waals forces are primarily responsible for holding together a mass of cellulose molecules, with hydrogen bonding playing a lesser role. See also CELLULOSE, CELLULASE, MOLECULAR WEIGHT, and WEAK INTERACTIONS.

Vector The agent used (by researchers) to carry new genes into cells. Plasmids currently are the vectors of choice, though viruses and other bacteria are increasingly being used for this purpose. See also PLASMID, RETROVIRAL VECTORS, PROTOPLASM, *AGROBACTERIUM TUMEFACIENS,* and BACULOVIRUS EXPRESSION VECTORS (BEVs).

Vertical Gene Transfer See OUTCROSSING.

Vesicle See VESICULAR TRANSPORT (OF A PROTEIN).

Vesicular Transport (of a protein) One of three means for a protein molecule to pass between compartments within eucaryotic cells. The compartment "wall" (membrane) possesses a "sensor" (receptor) that detects the presence of correct protein (e.g., after that protein has been synthesized in the cell's ribosomes), then bulges outward along with that protein molecule. The membrane bulge–containing protein then "breaks off" and carries (transports) the protein to its destination in another compartment in the cell. See also PROTEIN, EUCARYOTE, CELL, RIBOSOMES, SIGNALLING, VAGINOSIS, ENDOCYTOSIS, and GATED TRANSPORT (OF A PROTEIN).

Viral Transactivating Protein The specific protein used by a lytic virus to switch on the cascade of gene regulation by which that virus "takes over" a healthy cell and subverts its molecular processes (machinery) to produce virus components. This (transactivating) protein is key to the whole lytic cycle of the virus and therefore a potential target for therapeutic intervention. See also LYTIC INFECTION.

Virus A simple, noncellular particle (entity) that can reproduce only inside living cells (of other organisms). The simple structure of viruses is their most important characteristic. Most viruses consist only of a genetic material—either DNA (deoxyribonucleic acid) or RNA (ribonucleic acid)—and a protein coating. This (combination) material is categorized as a nucleoprotein.

Some viruses also have membranous envelopes (coatings). Viruses are "alive" in that they can reproduce themselves—although only by taking over a cell's "synthetic genetic machinery"—but they have none of the other characteristic of living organisms. Viruses cause a large variety of significant diseases in plants and animals, including humans. They present a philosophical problem to those who would speak of living and nonliving systems because in and of itself a virus is not "alive" as we know life, but rather represents "life potential" or "symbiotic life." See also VACCINIA,

NUCLEOPROTEINS, RETROVIRUSES, TOBACCO MOSAIC VIRUS (TMV), VIRAL TRANSACTIVATING PROTEIN, GENE DELIVERY, and ADENOVIRUS.

Viscosity A measure of a liquid's resistance to flow, as expressed in units called poise (P; grams per cm per sec). The degree of "thickness" or "syrupiness" of a liquid.

Vitamin An organic compound required in tiny amounts (for the growth, proper biological functioning and maintenance of health of an organism). Vitamins are commonly classified into two categories, the fat soluble and the water soluble. Vitamins A, D, E, and K are fat soluble whereas vitamin C and members of the vitamin B complex group are water soluble. In general, the vitamins play catalytic and regulatory roles in the body's metabolism. Among the water-soluble vitamins, the B vitamins apparently function as coenzymes (nonprotein parts of enzymes). Vitamin C's coenzyme role, if any, has not been established. Part of the importance of vitamin C to the body may arise from its strong antioxidant action. The functions of the fat-soluble vitamins are less well understood. Some of them, too, may contribute to enzyme activity; and others are essential to the functioning of cellular membranes (on surface of cells). Vitamin A is able to regulate the expression of certain genes in the embryos of mammals, via one of its metabolites; retinoic acid. Those embryo cells contain nuclear receptors (which bring the retinoic acid "signal" from outside into the cell's nucleus) on their cell membrane surface. The retinoic acid then (via the nuclear receptors) regulates the expression of the genes that cause embryonic cell differentiation into complex body structures, such as legs and arms, of the growing embryo. See also ENZYME, CATALYST, COENZYME, METABOLISM, METABOLITE, EXPRESS, EMBRYOLOGY, RETINOIDS GENE, PROTEIN, CELL, RECEPTORS, SIGNALLING, SIGNAL TRANSDUCTION, and NUCLEAR RECEPTORS.

Vomitoxin See *FUSARIUM* and MYCOTOXINS.

W

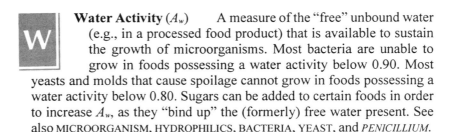

Water Activity (A_w) A measure of the "free" unbound water (e.g., in a processed food product) that is available to sustain the growth of microorganisms. Most bacteria are unable to grow in foods possessing a water activity below 0.90. Most yeasts and molds that cause spoilage cannot grow in foods possessing a water activity below 0.80. Sugars can be added to certain foods in order to increase A_w, as they "bind up" the (formerly) free water present. See also MICROORGANISM, HYDROPHILICS, BACTERIA, YEAST, and *PENICILLIUM*.

Water Soluble Fiber Food fiber (e.g., oat fiber) that dissolves in water. It apparently absorbs low-density lipoproteins (LDLP) in the intestine, before the fiber passes from the body, plus inhibits absorption of LDLP by intestinal walls due to increasing the viscosity of the intestine's contents. Those two effects thus lower the amount of "bad" cholesterol (i.e., LDLP can lead to hardening and blockage of arteries) in the body. Water soluble fiber from oat bran is a polysaccharide composed entirely of glucose (molecular) units. U.S. FDA regulations also include gums, pectins, mucilages, and certain hemicelluloses in the category of water soluble fiber. In 1997, the U.S. FDA finalized a (label) health claim that associates oat fiber with reduced blood cholesterol content and with reduced coronary heart disease.

In 1997, the U.S. FDA proposed a (label) health claim that associates soluble fiber from psyllium husks with reduced risk of coronary heart disease. See also HIGH-DENSITY LIPOPROTEINS (HDLP), LOW-DENSITY LIPOPROTEINS (LDLP), POLYSACCHARIDES, GLUCOSE (GLc), and FOOD AND DRUG ADMINISTRATION (FDA).

Weak Interactions The forces between atoms that are less strong than the forces involved in a covalent (chemical) bond (between two atoms). Weak interactions include ionic (chemical) bonds, hydrogen bonds, and van der Waals forces. See VAN DER WAALS FORCES.

Western Blot Test (e.g., for diseases such as AIDS) A test that is performed on blood (after centrifugation to remove red blood cells from the blood) to detect AIDS antibodies individually. Gel electrophoresis is used to separate the AIDS antigen proteins of killed (known) AIDS viruses. Next the protein bands (resulting from the gel electrophoresis) are exposed to the blood being tested and (AIDS) antibodies stick to specific individual antigens (bands) which are then identified (as being present in the tested blood) via dyes. See also ACQUIRED IMMUNE DEFICIENCY SYNDROME (AIDS), ANTIBODY, ANTIGEN, ELECTROPHORESIS, POLYACRYLAMIDE GEL ELECTROPHORESIS (PAGE), BASOPHILIC, and BUFFY COAT (CELLS).

Wheat Latin name *Triticum aestivum*. See WHEAT TAKE-ALL DISEASE, WHEAT SCAB, and WHEAT HEAD BLIGHT.

Wheat Head Blight See *FUSARIUM*.

Wheat Scab See *FUSARIUM*.

Wheat Take-All Disease A fungal disease that attacks wheat (*Triticum aestivum*) plant roots, and causes dry rot and premature death of the plant. Certain strains of *Brassica* plants and *Pseudomonas* bacteria act as natural antifungal agents against the wheat take-all fungus. See also FUNGUS, BACTERIA, GENETICALLY ENGINEERED MICROBIAL PESTICIDE (GEMP), and *BRASSICA*.

"Whiskers™" A trademarked method for inserting DNA into plant cells, so that those plant cells will then incorporate that new DNA and express the protein(s) coded-for by that DNA. Developed by ICI Seeds Inc. (Garst Seed Company) in 1993, Whiskers is an alternative to other methods of inserting DNA into plant cells (e.g., the Biolistic® Gene Gun, *Agrobacterium tumefaciens*, "Shotgun" Method [to introduce foreign (new) genes into plant cells], etc.), and consists of needle-like crystals of silicon carbide. The crystals are placed into a container along with the plant cells, then mixed at high speed, which causes the crystals to pierce the plant cell walls with microscopic passages. Then the new DNA is added, which causes the DNA to flow through those microscopic passages into the plant cells. The plant cells that incorporate the new DNA have thus been genetically engineered. See also BIOLISTIC® GENE GUN, *AGROBACTERIUM TUMEFACIENS*, "SHOTGUN" METHOD [TO INTRODUCE FOREIGN (NEW) GENES INTO PLANT CELLS], GENETIC ENGINEERING, GENE, BIOSEEDS, CODING SEQUENCE, PROTEIN CELL, and DEOXYRIBONUCLEIC ACID (DNA).

White Blood Cells See LEUKOCYTES.

Wide Spectrum See GRAM STAIN.

Wild Type The normal form of an organism as it is ordinarily encountered in nature. In contrast to natural mutant or laboratory mutant individuals (organisms). One example of a measurable difference between the two types is that wild strains of animals respond to the presence of EMF fields (e.g., weak magnetic fields such as those generated near power transmission cables), but laboratory strains of the same animals do not. See also STRAIN, MUTANT, PHENOTYPE, and GENOTYPE.

Wobble The ability of the third base in a tRNA (transfer RNA) anticodon to hydrogen bond with any of two or three bases at the 3′ end of a codon. This wobble (nonspecificity) allows a single tRNA species to recognize several different codons. See also TRANSFER RNA (tRNA), CODON, BASE PAIR (bp), and REDUNDANCY.

World Trade Organization (WTO) The international organization composed of the more than 100 nations that signed the General Agreement on Tariffs and Trade (GATT), which was WTO's predecessor body. WTO permits signatory countries to ban specific imports from other countries in order to protect the health of humans, animals, or plants. Such import bans are allowed based on the (GATT/WTO) Agreement on Sanitary and Phytosanitary Measures, which was approved in 1994 by GATT.

The WTO's Agreement on Sanitary and Phytosanitary (SPS) Measures requires that such import bans must be based on sound internationally-agreed science. WTO recognizes only the following three international science organizations in order to resolve SPS disputes between member nations:

(1) Codex Alimentarius Commission—for foods and food ingredients
(2) International Plant Protection Convention (IPPC)—for plants
(3) International Office of Epizootics (OIE)—for animal diseases

See SPS, CODEX ALIMENTARIUS COMMISSION, INTERNATIONAL PLANT PROTECTION CONVENTION (IPPC), and INTERNATIONAL OFFICE OF EPIZOOTICS (OIE).

WTO See WORLD TRADE ORGANIZATION (WTO).

X

X Chromosome A sex chromosome that usually occurs paired in each female cell, and single (i.e., unpaired) in each male cell in those species in which the male typically has two unlike sex chromosomes (e.g., humans). See also CHROMOSOMES.

X-ray Crystallography The use of diffraction patterns produced by X-ray scattering from crystals (of a given material's molecules) to determine the three-dimensional structure of the molecules. See also CONFIGURATION, CONFORMATION, TERTIARY STRUCTURE, and PROTEIN FOLDING.

Xanthine Oxidase An enzyme responsible for production of free radicals in the body. See HUMAN SUPEROXIDE DISMUTASE (hSOD).

Xenobiotic Compounds Those compounds (e.g., veterinary drugs, agrochemical herbicides, etc.) that are designed to be used in an ecosystem comprised of more than one species. For example, herbicides intended to kill weeds but leave commercial crops undamaged or veterinary drugs that are intended to kill parasitic worms but leave the host livestock unharmed.

Xenogeneic Organs From the Greek word *xenos*, meaning "stranger." Xenogeneic literally means "strange genes." Refers to genetically engineered (e.g., "humanized") organs that have been grown within an animal of another species. For example, several companies are working to engineer and grow—inside swine—a number of organs to be transplanted into humans that need those organs (e.g., due to loss of their own organs via disease or accident). If successful, this would free human organ transplant recipients from having to continually use immunosuppressive drugs in order to keep their body from "rejecting" the new organ. See also IMMUNOSUPPRESSIVE, GRAFT-VERSUS-HOST DISEASE (GVHD), CYCLOSPORIN, MAJOR HISTOCOMPATIBILITY COMPLEX (MHC), and GENETIC ENGINEERING.

Xenogenesis The (theoretical) production of offspring that are genetically different from, and genotypically unrelated to either of the parents of that offspring. See also GENOTYPE, TRANSGENIC (ORGANISM), HEREDITY, GENETICS, MEIOSIS, and GENETIC CODE.

Xenogenetic Organs See XENOGENEIC ORGANS.

Xenogenic Organs See XENOGENEIC ORGANS.

Xenograft See XENOTRANSPLANT.

Xenotransplant From the Greek word *xenos*, meaning "stranger." Xenotransplant is the implantation of an organ or limb from one species to another organism in a different species. When performed in animals, "rejection" of the transplant by the recipient's immune system is a common response. See also GRAFT-VERSUS-HOST DISEASE (GVHD) and XENOGENEIC ORGANS.

Y

Y Chromosome A sex chromosome that is characteristic of male zygotes (and cells) in species in which the male typically has two unlike sex chromosomes. See also CHROMOSOMES.

YAC See YEAST ARTIFICIAL CHROMOSOMES.

Yeast A fungus of the family Saccharomycetaceae that is used by man especially in the making of alcoholic liquors and as a leavening agent in bread making. Yeast cells are also commonly used in bioprocesses, because they are relatively simple to genetically engineer (via recombinant DNA) and relatively easy to propagate (via fermentation) to yield desired products (e.g., proteins). See also FERMENTATION, GENETIC ENGINEERING, YEAST ARTIFICIAL CHROMOSOMES (YAC), and RECOMBINANT DNA (rDNA).

Yeast Artificial Chromosomes (YAC) Pieces of DNA (usually human DNA) that have been cloned (made) inside living yeast cells. While most bacterial vectors cannot carry DNA pieces that are larger than 50 base pairs, YACs can typically carry DNA pieces that are as large as several hundred base pairs. See also YEAST, CHROMOSOMES, HUMAN ARTIFICIAL CHROMOSOMES (HAC), *ARABIDOPSIS THALIANA*, DEOXYRIBONUCLEIC ACID (DNA), CLONE (A MOLECULE), VECTOR, BASE PAIR (bp), and mega YAC.

Yeast Episomal Plasmid (YEP) A cloning vehicle used for introduction of constructions (i.e., genes and pieces of genetic material) into certain yeast strains at high copy number. YEP can replicate in both

Escherichia coli and certain yeast strains. See also PLASMID, CASSETTE, CLONE (AN ORGANISM), GENE, GENETIC ENGINEERING, *ESCHERICHIA COLI (E. COLI)*, and COPY NUMBER.

Z

Z-DNA A left-handed helix (molecular structure) of DNA, in contrast to A-DNA and B-DNA which are right-handed helix structures. The difference is in the direction of the double-helix twist. Z-DNA has the most base pairs per turn (in the helix), and so has the least twisted structure; it is very "skinny" and its name is taken from the zigzag path that the sugar-phosphate "backbone" follows along the helix. This is quite different from the smoothly curving path of the backbone of B-DNA. The Z-form of DNA has been found in polymers that have an alternating purine-pyrimidine sequence. The possible biological importance of Z-DNA is that it is much more stable at lower salt concentrations. See also DEOXYRIBONUCLEIC ACID (DNA), B-DNA, HELIX, DOUBLE HELIX, A-DNA, PURINE, BASE PAIR (bp), and PYRIMIDINE.

ZKBS (Central Committee on Biological Safety) The advisory body on safety in gene-splicing labs and plants for the German Government's Ministry of Health. It is the German counterpart of the American Government's Recombinant DNA Advisory Committee (RAC). The ZKBS is composed of 10 experts from the biology and ecology sectors, trade union representatives, plus representatives from the industrial sector and environmental pressure groups. The ZKBS advises the Ministry of Health and the individual German States (Länder), which regulate all recombinant DNA (i.e., gene-splicing) activities in Germany. See also GENE TECHNOLOGY OFFICE, RECOMBINANT DNA ADVISORY COMMITTEE (RAC), GENETIC ENGINEERING, RECOMBINANT DNA (rDNA), RECOMBINATION, BIOTECHNOLOGY, INDIAN DEPARTMENT OF BIOTECHNOLOGY, and COMMISSION OF BIOMOLECULAR ENGINEERING.

Zoonoses Diseases that are communicable from animals to humans.

Zoonotic See ZOONOSES.

Zygote A fertilized egg formed as a result of the union of the male (sperm) and female (egg) sex cells. See also X CHROMOSOME, Y CHROMOSOME, TELOMERES, GAMETE.

Zyme Systems Chemical reactions characterized by the presence of an inactive precursor of an enzyme. The enzyme is activated via another enzyme that normally removes an extra piece of peptide chain at a physiologically appropriate time and place. See ZYMOGENS, FIBRIN, and DIGESTION (WITHIN ORGANISMS).

Zymogens The enzymatically inactive precursors of certain proteolytic enzymes. The enzymes are inactive because they contain an extra piece of peptide chain. When this peptide is hydrolyzed (clipped away) by another proteolytic enzyme the zymogen is converted into the normal, active enzyme. The reason for the existence of zymogens may be to protect the cell, its machinery, and/or the place of manufacture within the cell from the potentially harmful or lethal effects of an active, proteolytic enzyme. In other words, the strategy is to activate the enzyme only when and especially where it is needed. See also PROTEOLYTIC ENZYMES, FIBRIN, ZYME SYSTEMS, and LIPOPROTEIN-ASSOCIATED COAGULATION (CLOT) INHIBITOR (LACI).